797,885 Books
are available to read at

www.ForgottenBooks.com

Forgotten Books' App
Available for mobile, tablet & eReader

ISBN 978-1-330-19717-2
PIBN 10050289

This book is a reproduction of an important historical work. Forgotten Books uses state-of-the-art technology to digitally reconstruct the work, preserving the original format whilst repairing imperfections present in the aged copy. In rare cases, an imperfection in the original, such as a blemish or missing page, may be replicated in our edition. We do, however, repair the vast majority of imperfections successfully; any imperfections that remain are intentionally left to preserve the state of such historical works.

Forgotten Books is a registered trademark of FB &c Ltd.
Copyright © 2017 FB &c Ltd.
FB &c Ltd, Dalton House, 60 Windsor Avenue, London, SW19 2RR.
Company number 08720141. Registered in England and Wales.

For support please visit www.forgottenbooks.com

1 MONTH OF FREE READING

at

www.ForgottenBooks.com

By purchasing this book you are eligible for one month membership to ForgottenBooks.com, giving you unlimited access to our entire collection of over 700,000 titles via our web site and mobile apps.

To claim your free month visit:
www.forgottenbooks.com/free50289

* Offer is valid for 45 days from date of purchase. Terms and conditions apply.

English
Français
Deutsche
Italiano
Español
Português

www.forgottenbooks.com

Mythology Photography **Fiction** Fishing Christianity **Art** Cooking Essays **Buddhism** Freemasonry Medicine **Biology** Music **Ancient Egypt** Evolution Carpentry Physics Dance Geology **Mathematics** Fitness Shakespeare **Folklore** Yoga Marketing **Confidence** Immortality Biographies Poetry **Psychology** Witchcraft Electronics Chemistry History **Law** Accounting **Philosophy** Anthropology Alchemy Drama Quantum Mechanics Atheism Sexual Health **Ancient History** **Entrepreneurship** Languages Sport Paleontology Needlework Islam **Metaphysics** Investment Archaeology Parenting Statistics Criminology **Motivational**

UNIVERSITY OF CALIFORNIA

FROM THE LIBRARY OF

PROFESSOR FÉLICIEN VICTOR PAGET

BY BEQUEST OF MADAME PAGET

No.

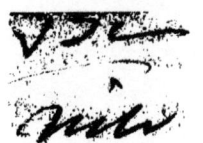

Nathan W. Moore

HARVARD

EXAMINATION PAPERS.

COLLECTED AND ARRANGED

BY

R. F. LEIGHTON, A.M.,
MASTER MELROSE HIGH SCHOOL.

SEVENTH EDITION.

BOSTON:
GINN AND HEATH.
1880.

NOTE.

In the Harvard University Catalogue, published by C. W. Sever, Cambridge, a full collection of examination papers may be found, comprising not only the papers set for Admission to College, but also nearly all the final examination papers given in the several Courses of Instruction in the College, the papers given in the Divinity, Law, and Medical Schools, those set for Admission to the Lawrence Scientific School, and those used at the Preliminary Examinations for Women. These make about 160 pages of close type each year. The price of the Catalogue is, in paper 50 cts., in cloth 75 cts.

CONTENTS.

History and Geography	3
Modern and Physical Geography.	22, 196, 216, 251, 258, 282, 308
Greek Composition	27, 197, 219, 246, 259, 283
Greek Grammar	40, 199, 219, 245, 260, 284, 311
Greek Prose	197, 220, 247, 261, 285, 312
Greek Poetry	198, 222, 263, 287, 315
Latin Composition	63, 201, 223, 238, 265, 289, 317
Latin Grammar	85, 202, 224, 237, 265, 289, 318
Latin . , ...	203–207, 225, 228, 239
French	109, 252, 279, 303, 329
German	304, 330
Arithmetic	111, 208, 229, 248, 272, 297, 323
Algebra	134, 209, 230, 249, 273, 298, 324
Advanced Algebra	157, 210, 256, 274, 299, 325
Plane Geometry . . .	162, 211, 214, 232, 250, 274, 300, 326
Solid Geometry	172, 212, 232, 255, 275, 300, 326
Analytic Geometry . . .	176, 213, 233, 254, 276, 301, 327
Logarithms and Trigonometry	178, 208, 229, 302
Physics	188, 307, 331, 332
Chemistry and Physics	306, 331
Physics and Astronomy	307, 332
Mechanics	190
Ancient History and Geography . .	195, 216, 257, 281, 308
English Composition	215, 235, 278, 302, 329
Plane Trigonometry	234, 253, 277, 328
Botany	279, 307, 332

APPENDIX.
    Requisites for Admission to Harvard College . . 397

EXAMINATION PAPERS.

HISTORY AND GEOGRAPHY.

I.

1. MENTION the principal nations that flourished before the Greeks. 2. What was the extent of Greece as compared with the territories of those nations? 3. What were the earliest governments among the Greeks? Mention the other forms of government which were afterwards adopted. 4. Mention the principal periods in Grecian history. 5. Give some account of Lycurgus; of Solon. 6. Mention the principal events in the Persian wars; mention some of the most distinguished persons engaged in them. 7. What causes led to the Peloponnesian War? Mention the principal events; the principal persons; the duration; the result of this war. 8. State what you know of the condition of Greece in the period following the Peloponnesian War. 9. State what you know of the history of Thebes. 10. State briefly what you know of the relations between Macedonia and Grecee in the time of Philip and Alexander. 11. When, and by whom, was Greece subjected to Rome? 12. Give some account of the foundation of Rome, and its first form of government. 13. What revolution put an end to the first government? and what government succeeded it? 14. Mention some of the early

Italian conquests of the Romans. 15. State some of the principal events in the Punic wars, and what was their conclusion. 16. State what you know of Catiline; at what period he lived; what political transactions he was engaged in; who were his most distinguished contemporaries; what became of him. 17. Give some account of the leaders in the Civil Wars. 18. Mention the circumstances of the death of Julius Cæsar. 19. What events followed his death? How, and by whom, were the civil conflicts composed?

II.

1. Name the following persons in proper historical order, and mention something that is recorded of each: Codrus, Flaminius, Lysander, Mardonius, Marius, Pyrrhus, Regulus, Socrates, Themistocles. 2. State briefly the origin of the First Punic War. 3. Give some account of Pericles, and what he did for Athens. 4. What famous battles were fought in Bœotia? 5. Give some account of Hannibal. 6. Who was Cleopatra? 7. Describe the battle of Pharsalia? 8. Name the first six Cæsars. 9. What is meant by the Heroic Age?

III.

1. Name the following persons in proper historical order. mentioning to what nation each belonged, and for what he was noted: Aristides, Cincinnatus, Draco, Epaminondas, Fabius, Pericles, Pompey, Solon, Sylla, Trajan. 2. Name the three persons whom you consider most noted in Grecian history; and state very briefly what each did. 3. Three in Roman history, in like manner. 4. What was the occasion of the First Persian War? 5. Describe the battle of Marathon; of Arbela. 6. Who were the kings of Rome? 7. State all you know of Jugurtha.

IV.

1. Give an account of the first invasion of Greece by the Persians. Tell when it occurred, what was the cause of it, what forces were employed, both of ships and men, what generals were engaged, and every other particular you remember. 2. Describe the Second Punic War in the same way. 3. Name "the twelve Cæsars" in the order of their reigns. 4. In what year was Julius assassinated? 5. When did Constantine become emperor?

V.

1. What were the principal countries known to the ancients? 2. Describe the situation of Palestine; of Phœnicia; of Egypt. 3. What were the principal divisions of Asia Minor? By whom was Asia Minor colonized? 4. Describe the situation of Greece; mention some of the most important mountains, rivers, plains, gulfs, seas. 5. Mention the principal divisions of Greece; give the names and situation of the most celebrated cities. 6. What countries were comprised in Greece proper? in Peloponnesus? What were the principal Greek islands? 7. Describe the situation of Italy. What countries were comprised in Italy proper? 8. Where was Rome? on what hills was it built? Mention some of the other principal cities of Italy; the principal islands. 9. Describe the Mediterranean Sea; its shape; its extent. Mention the principal ancient nations that inhabited its shores.

VI.

1. Describe Sicily. 2. What were the principal cities of Greece? 3. What countries in Africa? 4. What rivers in Cisalpine Gaul? 5. Where was Illyricum? Arca-

dia? the river Thermodon? 6. Draw, on half a page, an outline map of Greece, Macedonia, Thrace, and Asia Minor, especially of their sea-coasts; or, if you cannot draw, name the seas of the ancient world, and all the islands in each of them.

VII.

1. Describe Egypt. 2. What were the principal cities of Asia Minor, and for what was each noted? 3. What mountains in and around Thessaly? What in Peloponnesus? 4. Where was Colchis? Mount Ararat? the river Strymon? Tyre? 5. Draw an outline of the coasts of the Mediterranean and Adriatic Seas, with the principal islands; and give the ancient names of the countries and rivers in the regions now occupied by Italy, Spain and Portugal, France, and Great Britain.

VIII.

1. Where was Colchis? 2. Name the countries of Greece proper. 3. What mountains in Bœotia? 4. What was its capital? 5. What other noted places in Bœotia? 6. Draw a map about two inches in breadth representing the Peloponnesus, with the divisions and cities marked. 7. Describe Spain. 8. What was the former name of Saragossa? 9. What river between Italy proper and Cisalpine Gaul? 10. What Roman roads do you remember?

IX.

1. What mountain ranges enclose the Mississippi Valley? Describe the Mississippi River and its tributaries, giving the source and direction of each. 2. Describe Chesapeake Bay and the rivers which run into it. 3. De-

scribe the Alps. Give the name, course, and exit of each of the great rivers which rise in them. 4. State the divisions, in the order of their situation, belonging to the Peloponnesus, with the position of five principal towns. 5. Give the history and geography of the battle of Marathon. 6. Who was (or were) victorious, and over whom, at (1) Salamis; (2) Platæa; (3) Mantinea; (4) Chæronea; (5) Arbela; (6) the Caudine Forks; (7) Zama; (8) Actium? 7. Name the principal events in the life of Julius Cæsar, and such dates as you can call to mind. 8. Name the Twelve Cæsars, so called. Also the Five Good Emperors, sometimes so called. 9. Themistocles. 10. Describe the administration and policy of Pericles. 11. Give an account of the Athenian expedition to Sicily.

X.

1. Athens and Sparta; compare and contrast them. 2. The death of Socrates. 3. Give the position of the following places, and tell what has made them famous: (1) Marathon; (2) Salamis; (3) Platæa; (4) Mantinea; (5) Arbela; (6) Chæronea; (7) Pydna. 4. Who gained and who lost the battles fought at the following places: (1) Cannæ? (2) Zama? (3) Pharsalia? (4) Philippi? (5) Actium? 5. Greenwich is in longitude 0°, and in north latitude $51\frac{1}{2}°$: what are the longitude and latitude of the spot on the earth's surface opposite, or antipodal, to Greenwich? 6. Describe or bound the basin of the Mississippi River. 7. The institutions of Lycurgus and the laws of Solon. 8. The chief ties which bound together the Grecian world. 9. Epaminondas. 10. Give the chief rivers of France, with their ancient names.

XI.

1. Give the general course of (1) the Nile; (2) the Rhine; (3) the Danube; (4) the Elbe; (5) the Volga; (6) the St. Lawrence; (7) the Susquehanna; (8) the Amazon. 2. (1) What number of degrees represent the greatest possible latitude? (2) The greatest possible longitude? (3) Except at the equator, which is the greater, a degree of latitude or a degree of longitude? (4) Give, in degrees, the width of the torrid zone. (5) Which way from the north pole are London and New York? (6) What island near Africa is crossed by the Tropic of Capricorn? 3. Which of the Mediterranean islands preserve substantially their ancient names? 4. What are the modern names of (1) Lugdunum? (2) Massilia? (3) Eboracum? (4) Euboea? (5) Corcyra? (6) the Sequana? (7) the Iberus? (8) the Padus? 5. (1) Plataea; (2) Sphacteria; (3) Syracuse; (4) Ægospotami: give the geographical situation of these places, and say (in a sentence or two for each) what occurred there in the Peloponnesian War. 6. Where, and over whom, did Alexander the Great gain his greatest victories, and what were the general results of his conquests? 7. Which took place first, (1) the fall of Carthage or the captivity of Jugurtha? (2) the battle of Actium or the battle of Philippi? (3) the death of Pompey or the death of Cæsar? (4) the death of Antony or the death of Cicero? (5) the fall of Corinth or the fall of Jerusalem? 8. Describe the city of Athens. 9. Contrast the empire, government, and policy of Athens with those of Sparta, giving such instances and illustrations as may occur to you.

XII.

1. (1) What are the principal river basins of France? (2) Give the general course of the rivers. (3) Through

what waters must you pass in going from London to Canton? 2. The latitude of Boston is about 42° N.; its longitude is about 71° W. (1) What city in Europe has nearly the same latitude? (2) and what are the latitude and longitude of the point opposite, or antipodal, to Boston? 3. Point out the principal divisions, rivers, and mountains of Ancient Italy, by means of an outline map, or not, as you please. 4. Where is (1) Mount Athos? (2) Thermopylæ? (3) Artemisium? (4) Salamis? (5) Platæa? (6) Mycale? With the geography of each place, mention some event connected with the history of the place. 5. Name the important battles in the Second Punic War. 6. Give a particular account of the legislation of Lycurgus, Solon, and Cleisthenes.

XIII.

1. Which way from Athens to (1) Corinth; to (2) Marathon; to (3) Delos; to (4) Thermopylæ; to (5) the Hellespont; to (6) Crete: from Rome to (7) Carthage; to (8) Carthago Nova; to (9) Cannæ; to (10) Neapolis; to (11) Tarentum; to (12) Verona; to (13) Massilia; to (14) Lugdunum; to (15) the Baleares; to (16) Gades? 2. Three statesmen: (1) Themistocles; (2) Pericles; (3) Epaminondas. 3. The expedition of Cyrus the Younger, and the retreat of the Ten Thousand. 4. The chief events in the life of Julius Cæsar. 5. The rivers of Virginia, — describe them. 6. The principal English colonies; name and situation. 7. Give a brief account of the reforms of Cleisthenes. 8. The Persian invasions, and the principal battles in each, — a short sketch. 9. After these invasions the war languished for several years, until it was finally closed by the Peace of Cimon: what can you relate of the times of that peace? Compare the Peace of Cimon with that of Antalcidas, stating the time and circumstances of the lat-

ter. 10. Compare Athens and Sparta. What were the causes and results of the Peloponnesian War? 11. Give some account of Philip of Macedon and of Alexander, and compare the two. 12. The geographical position and configuration of Greece.

XIV.

1. Give a sketch of the life of Themistocles. 2. Describe the battle of Platæa. 3. Describe the administration of Pericles, and illustrate it by events. 4. Write an account of the Sicilian expedition. 5. Give the geographical position of Byzantium, Dyrrachium, Aquileia, Tarentum, Saguntum, Cannæ, Massilia, Eboracum. 6. What is the difference between a parallel and a meridian? How far, in degrees, is each polar circle from its pole? What is the greatest possible latitude? longitude? 7. The basin of a river is the entire area or territory watered or drained by the river and all its branches: what European states lie, wholly or in part, in the basin of the Rhine, and what States of our Union are, wholly or in part, in the basin of the Mississippi? 8. Name a fact in the history of each of the following places, and give the situation of each: Marathon, Salamis, Platæa, Mantinea, Chæronea, Arbela, Cannæ, Syracuse, Zama, Pharsalia, Philippi, Actium. 9. What most notable service was rendered to his country by Leonidas, Thrasybulus, Marius, Demosthenes, Cicero?

XV.

1. Where were Corinth, Thessalonica, Philippi, Ephesus, Sardis? 2. Four important ancient battles: two from Grecian and two from Roman history. Name the victorious and the vanquished party, and show the importance of the battles. 3. The expedition of the Younger Cyrus against

Persia, and that of Alexander. 4. The position of the Alps and the Apennines; the rivers that rise in them. 5. Any four English colonies; the chief Spanish colony; the great French dependency in Africa. To what power do the Azores belong? 6. Where are Batavia, Van Diemen's Land, New Orleans, San Francisco? What historical inferences do you draw from their names? 7. The statesmanship of Themistocles. 8. The Athenian power at the beginning and at the end of the Peloponnesian War. 9. The Athenian and the Spartan polity, character, influence.

XVI.

1. Give the latitude of the tropics and of the polar circles. What makes them good boundaries for zones? Define *arctic* and *antarctic* according to their *derivation*. 2. The longitude of St. Petersburg is 30° east from Greenwich: give the longitude of two places, one 120° east, and the other 120° west, from St. Petersburg. 3. What is meant in geography by *watershed* and *basin?* What is the relation of the Po to the Alps and Apennines, and of the Mississippi to the Rocky Mountains and Alleghanies? 4. Point out four towns in this country named after foreign towns, and give the situation of the former and of the latter. 5. Candia: its situation and ancient name. Mont Blanc: in what country is it? 6. Waterloo, Sebastopol, Gettysburg, Sadowa: where? 7. Saguntum, the Trebia, Lake Trasimenus, Cannæ, Zama: geographically and historically. 8. The Rubicon, Pharsalia, Philippi, Actium: geographically and historically. 9. The Acropolis of Athens. 10. Where is Syracuse? Give an account of the failure of the Athenian expedition to Sicily. 11. What revolutions took place in the government of Athens between 477 and 403 B. C.? By whom were they effected?

12. The character of Cimon. 13. Give the history of Platæa. 14. In what year of the Peloponnesian War was the battle of Amphipolis? What were its consequences? Who was the victor? Where was Amphipolis?

XVII.

1. Themistocles, Pericles, Thrasybulus. What, *in brief*, did these men severally do for Athens, and when? 2. Give the situation of Marathon, Thermopylæ, Salamis, Sphacteria, Syracuse, Ægos-potami, Leuctra, Arbela; and tell who won and who lost there. 3. Give the position of the cities (or some of them) to which St. Paul's Epistles were directly sent. 4. With what seas are the mountains of Switzerland connected by rivers? 5. Which of the United States lie in the basin of the Mississippi River? 6. Name and place *three* of the highest mountains in the world. Knowing the height of a mountain in feet, with what divisor will you reduce the height to miles? 7. The sculptor Pheidias (Phidias). 8. The first meeting of the Peloponnesian Confederacy at Sparta (B. C. 432) just before the great war. 9. The Roman Comitia. 10. Julius Cæsar in Spain.

XVIII.

1. Bound the *basin* of the Po, of the Mississippi, of the St. Lawrence. 2. Name the chief rivers of Ancient Gaul and Modern France. Is France larger or smaller than Transalpine Gaul? What are the two principal rivers that rise in the Alps? Where is Mont Blanc? 3. Where is the source of the Danube? of the Volga? of the Ganges? of the Amazon? 4. Describe the route of the Ten Thousand, or lay it down on a map. 5. Leonidas, Pausanias, Lysander. 6. Pharsalia, Philippi, Actium: geographically and histor-

ically. 7. Supply the two names left blank in the following passage from the Oration for the Manilian Law: "Non dicam duas urbes potentissimas, *Carthaginem* et *Numantiam* ab eodem ―――― esse deletas.; non commemorabo nuper ita vobis patribusque esse visum, ut in uno ―――― spes imperii poneretur, ut idem cum *Jugurtha*, idem cum *Cimbris*, idem cum *Teutonis* bellum administraret." Who was Jugurtha? Where was Numantia? 8. Compare Athens with Sparta. 9. Pericles: the man and his policy.

XIX.

1. From Cæsar: "Gallos ab Aquitanis *Garumna* flumen, a Belgis Matrona et *Sequana* dividit." "Aquitania a *Garumna* flumine ad *Pyrenæos* montes et eam partem *Oceani*, quæ est ad *Hispaniam*, pertinet." Translate these passages. Bound Aquitania, describing geographical positions (where names are in italics), and giving modern names. (You may, if you choose, substitute a *map* for the *description*.) 2. From Cæsar: "Undique loci natura Helvetii continentur; una ex parte flumine *Rheno*, qui agrum Helvetium a Germanis dividit; altera ex parte *monte Jura* altissimo, qui est inter Sequanos et Helvetios: tertia *lacu Lemanno* et *flumine Rhodano* qui *Provinciam* nostram ab Helvetiis dividit." Deal with this as with the preceding. 3. From Virgil:

(1) "Quin Decios Drusosque procul, sævumque securi
   Adspice Torquatum, et referentem signa *Camillum:*"
(2) "Quis te, magne Cato, tacitum, aut te, Cosse, relinquat?
   Quis *Gracchi* genus, aut *geminos*, duo fulmina belli,
   Scipiadas, cladem Libyæ?"

Translate and explain. 4. Cicero enumerates the wars in which *Pompeius* had distinguished himself; among them, bellum "*Hispaniense*," bellum "*servile*," bellum "*navale*."

-Explain. 5. The expedition of the younger Cyrus against Persia, and that of Alexander: compare them. 6. Describe Athens. 7. Name in proper order the chief events of the Peloponnesian War, giving the geographical positions.

XX.

1. Where were Corinth, Thebes, Ephesus, Tarentum, Massilia, Saguntum? Where were the Pyrenæi Montes? What sea on the east of Græcia? What large islands near Italia? What large gulf in the south of Italia? Name the chief rivers of Hispania. 2. Cicero enumerates the wars in which *Pompeius* had distinguished himself; among them bellum "*Hispaniense,*" bellum "*servile,*" bellum "*navale.*" Explain. 3. Three important battles in Grecian history; — name the victorious and the vanquished party, and show the importance of the battles. 4. The expedition of Cyrus the Younger against Persia. 5. The siege of Syracuse. 6. The Athenian power at the beginning, and at the end, of the Peloponnesian War. How long did the war last? 7. The first secession of the Plebs: date, cause, and result. 8. The important battles of the Second Punic War; the commanders and victors in each. 9. The Gracchi, and their attempts at reform.

XXI.

1. From Cæsar: "Extremum oppidum Allobrogum est proximumque Helvetiorum finibus, *Geneva.*" "A lacu *Lemanno,* qui in flumen *Rhodanum* influit, ad montem *Juram* fossam perduxit." "Flumen est *Arar,* quod per fines Æduorum et Sequanorum in *Rhodanum* influit, incredibili lenitate, ita ut oculis, in utram partem fluat, judicari non possit." Translate these passages. Describe the geographical

situation of the places, etc., italicized, and give the modern names. What important town is at the junction of what were the Rhodanus and the Arar? 2. From Cicero: " Pompeius nondum tempestivo ad navigandum mari *Siciliam* adiit, *Africam* exploravit; inde *Sardiniam* cum classe venit. Inde cum se in Italiam recepisset, *duabus Hispaniis* et *Gallia Cisalpina* præsidiis ac navibus confirmata, missis item in oram *Illyrici Maris* et in *Achaiam* omnemque Græciam navibus, Italiæ *duo maria* maximis classibus firmissimisque præsidiis adornavit: ipse autem, ut a *Brundisio* profectus est, undequinquagesimo die totam ad imperium populi Romani *Ciliciam* adjunxit." Give a translation and a geographical description. 3. Where were Argos, Sparta, Salamis, Mt. Olympus, Mt. Ida, Sardis? What gulfs are separated by the Isthmus of Corinth? Name the chief rivers of Gallia, giving both ancient and modern names. 4. Marathon, Thermopylæ, Platæa, — geographically and historically. 5. Athens in the time of Pericles. 6. The Sicilian expedition. 7. Epaminondas, and the supremacy of Thebes. 8. Themistocles and Aristides. 9. Philip of Macedon, and the battle of Chæronea.

XXII.

1. " Sit Scipio ille clarus, cujus consilio atque virtute Hannibal in Africam redire atque Italia decedere coactus est; ornetur alter eximia laude Africanus, qui duas urbes huic imperio infestissimas, Carthaginem Numantiamque, delevit; habeatur vir egregius Paulus ille, cujus currum rex potentissimus quondam et nobilissimus Perses honestavit; sit æterna gloria Marius, qui bis Italiam obsidione et metu servitutis liberavit; anteponatur omnibus Pompeius, cujus res gestæ atque virtutes iisdem quibus solis cursus regionibus ac terminis continentur." Explain this passage

from Cicero by brief notes, without writing a translation of it. 2. From what places, etc., did the Bosporani, the Cyziceni, the Cretenses, the Rhodii, mentioned by Cicero, respectively derive their names? Where were those places? Where were Brundisium, Caieta, Cilicia? 3. Give the divisions of the Peloponnesus, with their relative position, and name a place in each. Connect historically Mantinea with Leuctra in Bœotia. 4. What, and where, were the chief settlements made outside of Greece by Greeks? Describe the great Sicilian expedition. 5. Point out and describe the main causes of the growth and decline of the Athenian power. 6. What were the relations at different times between the Persian kings and the Greeks?

XXIII.

1. Give a brief account of Cæsar's campaign against the Helvetii. Fix the position of the following: Lacus Lemanus, the Rhodanus, the Allobroges, the Arar, Geneva. What part did the Allobroges play in the Catilinarian conspiracy? 2. "Inde cum se in Italiam [Pompeius] recepisset, *duabus Hispaniis* et *Gallia Cisalpina* præsidiis ac navibus confirmata, missis item in oram *Illyrici Maris* et in *Achaiam* omnemque Græciam navibus, Italiæ *duo maria* maximis classibus firmissimisque præsidiis adornavit; ipse autem, ut a *Brundisio* profectus est, undequinquagesimo die totam ad imperium populi Romani *Ciliciam* adjunxit." Fix the position of the italicized provinces, towns, etc., without translating the passage. 3. Corinth, Philippi, Antioch, Sardis, Ephesus, Smyrna,— where situated? 4. Name the chief battles in which Greeks and Persians were engaged between 500 and 300 B. C. Fix the positions, give the dates, and show the importance of the several battles. 5. Name some of the chief islands belonging to

Greeks or settled by Greeks, and point out the situation of each. 6. The conquest of Greece by the Romans. Why were the Romans more successful than the Persians had been? 7. Themistocles and Pericles. 8. Pausanias and Lysander.

XXIV.

1. "Interfectus est propter quasdam seditionum suspiciones *C. Gracchus.*" (Cicero.) Translate and explain. 2. "Etenim recordamini, Quirites, omnes *civiles dissensiones*, neque (solum) eas quas audistis, sed et has quas vosmetipsi meministis et vidistis." (Cicero.) Translate and explain. 3. "Majores vestri *cum Antiocho, cum Philippo, cum Pœnis* bella gesserunt." (Cicero.) Translate and explain. 4. Brundisium, Caieta, Ostia, Gallia Cisalpina, Samos, Cilicia, Pontus (the country). (Cicero.) Give the position of each. 5. "Classes æratas, *Actia* bella,
 Cernere erat; totumque instructo Marte videres
 Fervere *Leucaten* (auroque), effulgere fluctus.
. . . .
Regina in mediis patrio (vocat) agmina sistro,
 Necdum etiam geminos a tergo respicit *angues.*" (Virgil.) Translate and explain. 6. Olympia and the Olympic Games. The Olympiads. 7. Describe the battle of Marathon and the battle of Salamis. 8. The character, policy, and works of Pericles. 9. Name the principal events which mark the decline and fall of the Athenian power. 10. The most brilliant period and the most noted men in the history of Thebes. Mention *two* battles, give the geographical site of each, and date *one* of them. 11. Demosthenes. 12. Alexander's empire, and the kingdoms into which it broke up. Date his death.

XXV.

1. "Hæc (sc. *Italia*) genus acre virûm, Marsosque, pubemque Sabellam,
Adsuetumque malo Ligurem, Volscosque verutos
Extulit; hæc *Decios, Marios,* magnosque *Camillos,*
Scipiadas duros bello, et te, maxime Cæsar." (Virgil.)
2. "Nos, quorum majores *Antiochum* regem classe *Persenque* superarunt, omnibusque navalibus pugnis *Carthaginienses* . . . . . vicerunt, ii nullo in loco jam prædonibus pares esse poteramus." (Cicero.) (Ii may be rendered, imperfectly, by *even we.*) 3. "Ego enim sic existimo: *Maximo, Marcello, Scipioni, Mario,* et ceteris magnis imperatoribus, non solum propter virtutem, sed etiam propter fortunam, sæpius imperia mandata atque exercitus esse commissos." (Cicero.) With this passage, take the following from Virgil:—
   "Tu *Maximus* ille es,
Unus qui nobis cunctando restituis rem."
4. Describe the city of Athens. 5. Name and describe some important places and events which are associated with the rivalry between Athens and Sparta. 6. What were the causes of the fall of the Athenian power? 7. The first Darius and the last Darius: how were they connected with Grecian history? 8. Name some of the Greek islands, and give their situation, with anything memorable in their history. 9. The rise and fall of the Achæan League. 10. What Romans gained great victories over Greeks? when, and where? 11. Name the sections or provinces of the Peloponnesus, and point out places of historical importance. (Draw a map, if you choose.)

XXVI.

1. By a single map (or otherwise) illustrate the following quotations from Caesar, without writing a translation of them: Gallos ab Aquitanis Garumna flumen, a Belgis Matrona et Sequana dividit. — Sequanos a Provincia nostra Rhodanus dividit. — Extremum oppidum Allobrogum est proximumque Helvetiorum finibus Geneva. Ex eo oppido pons ad Helvetios pertinet. Give the modern names of the rivers. 2. By means of a map, or a description in words, show the situation, relatively to Rome, of the Italian districts named in the following passage from one of Cicero's orations against Catiline, without writing a translation of the passage: Video, cui Apulia sit attributa, qui habeat Etruriam, qui agrum Picenum, qui Gallicum, qui sibi has urbanas insidias caedis atque incendiorum depoposcerit. 3. Write explanatory notes on the following lines from the prophecy of Anchises in the sixth book of the Aeneid:

Ille triumphata Capitolia ad alta Corintho
Victor aget currum caesis insignis Achivis.
Eruet ille Argos Agamemnoniasque Mycenas,
Ipsumque Aeaciden, genus armipotentis Achilli,
Ultus avos Trojae templa et temerata Minervae.

4. Name (and date, as far as you can) the chief occasions which brought Greeks into contact with Persians. Give the situation of places. 5. Themistocles and Aristides. 6. The causes and results of the Peloponnesian War. 7. What objects 4. The Aryan settlement of Europe. 5. Forms of government in Greece. 6. Philip and Alexander; the results of the conquests of the latter. 7. The increase of the dominion of Rome during and in consequence of the Punic Wars. 8. The Claudian,

would an Athenian be most likely to point out to a stranger visiting Athens? Describe some of them. 8. Name some turning-points or critical periods in the history of Athens. What made them such?

Flavian, and "Good" emperors. Name them, and give some account of one emperor from each class. 9. Diocletian and Constantine.

XXVII.

1. Cicero mentions, in his account of the depredations of the pirates, Cnidus, Colophon, Samos, Misenum. Where were they? 2. In what part of Gaul was the territory of the Aedui? that of the Sequani? that of the Arverni? that of the Treviri? Describe the course of the Rhodanus, the Arar, and the Rhenus, and give the modern names of these rivers. 3. Write explanatory notes on the following lines from the third book of the Aeneid:—

> Linquimus Ortygiae portus, pelagoque volamus.
> Bacchatamque jugis Naxon viridemque Donusam,
> Olearon, niveamque Paron, sparsasque per aequor
> Cycladas, et crebris legimus freta consita terris.
> Nauticus exoritur vario certamine clamor;
> Hortantur socii, Cretam proavosque petamus.

4. Name several of the rivers or mountains in Greece which are noted in mythology or history. 5. Name some places of historical interest in the Peloponnesus, and give the position of each.

4. Compare the geographical character of Greece with that of Italy. 5. B. C. 490, 480, 431–404, 334–323: to what events in the history of Greece do these dates point? 6. The relations of

6. Represent by a map, or describe otherwise, the course of the expedition of Cyrus the Younger, and of that of Alexander of Macedon. 7. The Roman Conquest of Greece. 8. B. C. 490, 480, 431 – 404, 334 – 323 : to what events in the history of Greece do these dates point? 9. Give the situation of each of the following places, and connect an event with each: Marathon, Thermopylae, Plataea, Leuctra, Syracuse, Chaeronea, the Italian States to Rome. 7. The Roman civil wars. 8. The extent of the Roman Empire. 9. The English conquest of Britain.

MODERN AND PHYSICAL GEOGRAPHY.

I.

1. Give a physical description of Italy. 2. Define *plateau, delta, steppe, bight, lagoon, glacier*. What is the *profile* of a country? 3. How many degrees apart from each other are the two polar circles? What is the breadth (in degrees) of the torrid zone? What is the shortest distance in degrees of longitude between Madras (80° E. from Greenwich) and San Francisco (122½° W.)? 4. Describe the chief physical features of the State in which you live. 5. Describe the Andes, and one of the three great river systems of South America. 6. Name the rivers connected with the lakes of Switzerland and of Northern Italy, and give their source, course, and end. 7. The Vosges, the Jura, the Carpathians; Mt. Everest, Mt. Chimborazo, Mont Blanc, Monte Rosa, Mt. St. Elias. Give their position. 8. To what powers belong the Azores, Corsica, Malta, Heligoland, Algeria, Batavia, Manilla, Sydney, Havana? 9. Constantinople, Alexandria, Gibraltar, New York, New Orleans, Hamburg, — show the convenience or importance of the position of each.

II.

1. What is meant by the terms "latitude" and "longitude"? 2. Give the approximate longitude, reckoned from the meridian of Greenwich, of London; New York; Cape Horn; the Cape of Good Hope; Melbourne; Shanghai; the Sandwich Islands. 3. Name and give the position, with reference to the various countries, of the principal moun-

tain chains of Europe. 4. Name and give the position and direction of the principal mountain ranges of North and South America. 5. Describe the principal rivers of North America, giving an approximate statement of the position of their sources, the direction in which they run, and their lengths. 6. Describe the principal rivers of Europe, in the manner indicated in the preceding question. 7. Name the principal islands of the East Indies, and state to what political powers they belong. 8. Name the West India Islands, and state to what powers they belong. 9. Describe the position of France with reference to the adjacent countries, rivers, mountains, and seas. 10. What are the advantages of London with regard to its geographical position? of St. Louis? of Chicago? of San Francisco? 11. What are the principal exports of England? of the United States? of Russia? of Central America? 12. State approximately the population of the most important states of Europe.

III.

1. Where is Manilla? Through what waters and across what countries would you pass in travelling from Manilla to New York, (*a*) entirely by water, (*b*) partly by water and partly by land? 2. Explain what is meant by *latitude* and *longitude*. What is the longitude of the point in the northern hemisphere directly opposite Washington? (Assume longitude of Washington 77° W.) What is the latitude and longitude of the point in the southern hemisphere directly opposite Cambridge? The latitude of Cambridge is 42° 23′ N., the longitude 71° 7′ W. 3. Mention the different bodies of water surrounding the British Islands, and the rivers flowing into each. 4. Give as precisely as you can the position of the following mountains, and state,

where possible, to what range each belongs: — Washington; St. Elias; Hecla; Elburz; Pike's Peak; Dwalagiri; Chimborazo; Shasta; Orizaba. 5. Enumerate the States and Territories through or by which the Mississippi, Missouri, and Ohio Rivers flow. 6. State the principal conditions which determine the head of navigation on a river. Name the town or city at the head of navigation on two rivers in the United States, and on one in Europe. 7. State what you know about the Gulf Stream.

IV.

1. State as precisely as you can where the following rivers rise and empty, their general directions, and the countries or states which they cross or bound: — Amazon; Rhine; Connecticut; Volga; Indus; Ohio; Obi; St. Lawrence. 2. Explain the terms *latitude* and *longitude*. Given the longitude of Melbourne as 145° E. when referred to Greenwich, what would be its longitude when referred to Washington? (Washington lies 77° west of Greenwich.) 3. Describe the coast of Asia from Behring's Strait to the Strait of Malacca, mentioning the peninsulas, the seas, the mouths of important rivers, and the islands lying near the mainland. (Draw a map comprising these particulars, if you prefer.) 4. What is meant by the *snow line?* Name some countries in which the snow line is very high. 5. Mention the principal islands in the Atlantic Ocean, and state to what political power each belongs. 6. Bound France, and give the name and position of four of its chief cities. 7. Name and give the position of the principal mountain ranges of North and South America. 8. Bound Pennsylvania. What mountains cross the State? What are its principal rivers? How does it rank with the other States as to area? as to population?

V.

1. What is the breadth of the north temperate zone in degrees? in miles? 2. What is the length of the longest day at the North Pole? at the Arctic Circle? at the Equator? Account for the differences. 3. What countries of South America are on the Pacific coast? Which one has no coast line? 4. Through what waters would a vessel pass in sailing from Sevastopol to St. Petersburg? 5. In what zone does Australia principally lie? What is the chief river of Australia? What gulf on the north? What important islands and groups of islands to the north and east? 6. Give the position of the following cities as precisely as you can, naming in all cases the river or other body of water on or near which the city lies:— Cayenne, Constantinople, Detroit, Lyons, Madras, Omaha, Palermo, Para, Sitka, Zanzibar. 7. Bound Illinois. What is its capital? Give the name and position of three other important cities. What are its chief rivers, and in what direction do they flow? 8. Upon what three circumstances is the climate of any region chiefly dependent? 9. To what states or countries would you go for caoutchouc? coffee? olives? opium? pepper? rice? silk? sugar? tapioca? turpentine? 10. Write what you can about coral islands and reefs.

VI.

1. What is meant by the *relief* of a country? the *profile*? What are the principal features of relief in North America? 2. Bound the three principal river basins of South America. 3. In what direction are the Bahamas from the Bermudas? the Azores from Oporto? Honolulu from San Francisco? Pekin from Yedo? 4. Through or

near what countries, islands, important cities, and bodies of water does the Tropic of Cancer pass? 5. Name and give the situation of the English colonies in Africa. 6. Bound Italy. What is its largest city? Name its principal mountains and rivers. If there is anything peculiar about any of the rivers, mention and explain it. 7. Where are the following gulfs and bays: — Finland, Bothnia, Aden, Bengal, Lyons, Chesapeake? Name the important rivers, if any, which empty into each. 8. What are the principal productions and exports of Russia? France? Cuba? Japan? Peru? 9. How do the forms of government of the five principal nations of Europe differ? 10. Write what you can about the trade winds.

GREEK COMPOSITION.

I.

1. WHAT then? When[1] the Athenians and my [fellow] citizens[2] come,[3] let us summon[4] this man also, that we may consult[5] together.[6] 2. Cyrus said, "If you go[7] now, when[8] shall you be at home?"[9] 3. O my country![10] O that all who inhabit[11] thee would love thee as I now do! 4. Not many days after this, Chares[12] came from Athens with[13] a few[14] ships; and immediately the Lacedæmonians and Athenians fought a naval battle.[15] The Lacedæmonians were victorious,[16] under the lead[17] of Hegesandridas.[18]

1. ἐπειδάν. 2. πολίτης. 3. ἔρχομαι. 4. καλέω. 5. συμβουλεύω (mid.). 6. κοινῇ. 7. εἶμι. 8. πότε. 9. οἴκοι. 10. πατρίς. 11. οἰκέω. 12. Χάρης. 13. ἔχων. 14. ὀλίγος. 15. ναυμαχέω. 16. νικάω. 17. ἡγέομαι (gen. absol.). 18. Ἡγησανδρίδας.

II.

1. After these things, Pericles rose,[1] and thus spoke. 2. Do not obey[2] these most wicked men. 3. On the next[3] day he gave them what he promised.[4] 4. All the Greeks happened[5] to be doing this. 5. Many fear lest these things should happen[6] while Philip is king.[7] 6. If these things were true,[8] it would be still more terrible.[9]

1. ἀνίστημι. 2. πείθω. 3. ὑστεραῖος. 4. ὑπισχνέομαι. 5. τυγχάνω with the participle. 6. γίγνομαι. 7. genitive absolute. 8. ἀληθής. 9. δεινός.

III.

1. Any one might justly[1] praise[2] him, not only for[3] these things, but for what he did about[4] the same time.[5] 2. If you do[6] what I just[7] now told[8] you, you will have all things which any one could wish.[9] 3. O that[10] these things had happened[11] as we wished![9] But since[12] we were unfortunate,[13] let us do what the wisest of us shall command.[14] 4. If these men had not perished,[15] the city would have been saved[16] and we should now be free.[17]

1. δικαίως. 2. ἐπαινέω. 3. ἐπί. 4. περί. 5. χρόνος. 6. ποιέω. 7. ἄρτι. 8. φράζω. 9. βούλομαι. 10. εἴθε. 11. γίγνομαι. 12. ἐπεί. 13. ἀτυχής. 14. κελεύω. 15. ἀπόλλυμι. 16. σώζω. 17. ἐλεύθερος.

IV.

1. If I appear[1] to be wrong,[2] I will pay[3] the penalty. 2. If you should turn[4] from evils, you would quickly[5] become[6] better. 3. I fear[7] lest we have forgotten[8] the road[9] home.[10] 4. If Philip had had this opinion,[11] — that it is difficult[12] to fight[13] with the Athenians, — he would have done[14] no one of the things which he has done.

1. δοκέω. 2. ἀδικέω. 3. δίκην δοῦναι. 4. ἀποτρέπομαι. 5. ἐν τάχει. 6. γίγνομαι. 7. δείδω. 8. ἐπιλανθάνομαι. 9. ὁδός. 10. οἴκαδε. 11. γνώμη. 12. χαλεπός. 13. πολεμέω. 14. πράσσω.

V.

1. Those who were looking[1] on feared[2] lest their friends[3] should suffer[4] anything. 2. They all said[5] that the king[6] had sent[7] them, and that they wished[8] to make an alliance[9] with Cyrus. 3. If another shall come[10] in his own name,[11] him ye will receive.[12] 4. When this had hap-

pened,[13] all believed [14] that an assembly [15] would be summoned.[16]

1. θεάομαι (partic.). 2. φοβέομαι. 3. φίλος. 4. πάσχω. 5. λέγω with ὅτι. 6. βασιλεύς. 7. πέμπω. 8. βούλομαι. 9. συμμαχέω. 10. ἔρχομαι. 11. ὄνομα. 12. λαμβάνω. 13. γίγνομαι. 14. οἴομαι. 15. ἐκκλησία. 16. συγκαλέω.

VI.

1. You would be approved,[1] should you appear [2] not to do those things which you would blame [3] others for doing. 2. Swear [4] by no [5] god for the sake of [6] money, not even [7] if you are not about [8] to violate [9] your faith.[10] 3. The king [11] said [12] that the messenger [13] was not then present,[14] and that, if he had been, these things would not have occurred.[15] 4. Would that I had [16] the wings [17] of an eagle,[18] that leaving [19] the earth [20] I might be numbered [21] among [22] the stars! [23]

1. εὐδοκιμέω. 2. φαίνομαι. 3. ἐπιτιμάω. 4. ὄμνυμι. 5. μηδείς or οὐδείς ἵ. 6. ἕνεκα. 7. μηδέ. 8. μέλλειν. 9. παραβαίνειν. 10. πίστις. 11. βασιλεύς. 12. λέγειν with ὅτι. 13. ἄγγελος. 14. πάρειμι. 15. γίγνομαι. 16. ἔχειν. 17. πτερόν. 18. ἀετός. 19. λείπω. 20. γῆ. 21. ἀριθμέω. 22. ἐν. 23. ἄστρον.

VII.

1. I tried [1] to show [2] him that [3] he thought [4] he was wise, but [5] was not. 2. He says [6] that these things happened [7] while Cyrus [19] was king. [8] 3. Let no one believe [9] that I now fear [10] lest our state [11] be ruined.[12] 4. If these men were not unjust,[13] they would not have condemned [14] these generals [15] to death.[16] 5. He burned [17] the vessels,[18] that Cyrus [19] might not pass over.[20]

1. πειράομαι. 2. δείκνυμι. 3. ὅτι. 4. οἴομαι (with infin.).

5. δέ (with preceding μέν). 6. φημί (with infin.). 7. γίγνομαι. 8. participle of βασιλεύω. 9. νομίζω (with infin.). 10. φοβέομαι. 11. πόλις. 12. ἀπόλλυμι (2d aor. mid.). 13. ἀδικέω. 14. καταγιγνώσκω. 15. ' στρατηγός. 16. θάνατος. 17. κατακάω. 18. πλοῖον. 19. Κῦρος. 20. διαβαίνω.

VIII.

1. The king[1] is chosen[2] in order that those who choose[2] him may be benefited[3] by[4] him. 2. They said[5] that Cyrus[6] was dead,[7] and that Ariæus[8] would flee.[9] 3. If he had been here,[10] would he have overlooked[11] these things, or have punished[12] these impious[13] men? 4. May we desire[14] only[17] those things which we shall rejoice[15] to have acquired.[16] 5. Before[18] he came,[19] the ships[20] happened[21] to have gone[22] to Caria[23] to summon[24] assistance.[25]

1. βασιλεύς. 2. αἱρέω. 3. εὖ πράττειν. 4. διά. 5. λέγω (ὅτι). 6. Κῦρος. 7. θνήσκω. 8. Ἀριαῖος. 9. φεύγω. 10. πάρειμι. 11. περιοράω. 12. κολάζω. 13. ἀσεβής. 14. ἐπιθυμέω. 15. χαίρω. 16. κέκτημαι. 17. μόνον. 18. πρίν. 19. ἔρχομαι. 20. ναῦς. 21. τυγχάνω. 22. οἴχομαι. 23. Καρία. 24. περιαγγέλλω (participle). 25. βοηθεῖν.

IX.

1. All of them fear[1] lest they may be compelled[2] to do many[3] things which now they do not wish[4] to do. 2. O that[5] this man had had[6] strength[7] equal[8] to his mind.[9] 3. They called in[10] physicians[11] when they were sick,[12] that they might not die.[13] 4. He showed[14] that he was ready[15] to fight[16] if any one should come out.[17]

1. φοβέομαι. 2. ἀναγκάζω. 3. πολύς. 4. βούλομαι. 5. εἴθε. 6. ἔχω. 7. ῥώμη. 8. ἴσος. 9. γνώμη. 10. παρακαλέω. 11. ἰατρός. 12. νοσέω (partic.). 13. ἀποθνήσκω. 14. δηλόω (with ὅτι). 15. ἕτοιμος. 16. μάχομαι. 17. ἐξέρχομαι.

GREEK COMPOSITION.

X.

1. He said[1] that he had come[2] that he might see[3] both what was doing and what had been done. 2. I told him that, if these things had been true,[4] this would not have happened.[5] 3. Would that he were alive;[7] for he would not fear[8] these dangers[9] as you do. 4. Do you wish[6] me to come?[2] Tell[1] him not to fear[8] me, thinking[10] I shall be angry.[11]

1. λέγω. 2. ἔρχομαι. 3. ὁράω. 4. ἀληθής. 5. γίγνομαι. 6. βούλομαι. 7. ζάω. 8. φοβοῦμαι. 9. κίνδυνος. 10. οἴομαι. 11. χαλεπαίνω.

XI.

1. It is said[1] that the king[2] sent them away,[3] fearing[4] lest they should perish[5] by remaining.[6] 2. Athens,[7] although it was[8] great[9] before,[10] then became[11] greater, having been freed[12] from tyrants.[13] 3. Who of all the Greeks would not justly[14] have hated[15] us, if we had fled[16] and had left[17] our city to the barbarians?[18] 4. Call[19] no one happy[20] before[21] he is dead.[22]

1. λέγω. 2. βασιλεύς. 3. ἀποπέμπω. 4. φοβέομαι. 5. ἀπόλλυμι. 6. μένω. 7. Ἀθῆναι. 8. Participle of εἰμί. 9. μέγας. 10. πρίν. 11. γίγνομαι. 12. ἀπαλλάσσω. 13. τύραννος. 14. δικαίως. 15. μισέω. 16. φεύγω. 17. καταλείπω. 18. βάρβαρος. 19. καλέω. 20. ὄλβιος. 21. πρίν. 22. τελευτάω.

XII.

1. Wish[1] to be a friend[2] of the powerful,[3] in order that you may not suffer punishment[4] if you act unjustly.[5] 2. We fear[6] lest,[7] if we do[8] this, we shall miss[9] at once[10] what we have gained[11] and what we hope[12] to gain. 3. The messenger[13] came[14] to announce[15] that the city had

been taken,[16] but that the citizens [17] were hidden [18] near [19] the sea.[20] 4. Would [21] that he had died [22] in his youth,[23] for [24] he now would be happy.[25]

1. βούλομαι. 2. φίλος. 3. to be powerful, δύνασθαι. 4. δίκην δοῦναι. 5. ἀδικεῖν. 6. φοβοῦμαι. 7. μή. 8. πράττω. 9. ἁμαρτάνω. 10. ἅμα. 11. τυγχάνω. 12. ἐλπίζω. 13. ἄγγελος. 14. ἔρχομαι. 15. ἀγγέλλω. 16. ἁλίσκομαι. 17. πολίτης. 18. κρύπτω. 19. παρά. 20. θάλαττα. 21. εἴθε. 22. ἀποθνήσκω. 23. a young man, νεανίσκος. 24. ἐπεί. 25. εὐδαίμων.

XIII.

1. I trust [1] that these things which you have heard [2] are true.[3] 2. Who would not wish [4] to leave his country,[5] when such base [6] men are in power? [7] 3. The same men were present [8] when these things happened.[9] 4. He said [10] that, although he was [11] a god, he wished [4] to die.[12]

1. πιστεύω. 2. ἀκούω. 3. ἀληθής. 4. βούλομαι. 5. πατρίς. 6. πονηρός. 7. κρατέω (partic.). 8. πάρειμι. 9. γίγνομαι. 10. εἶπον. 11. participle. 12. ἀποθνήσκω.

XIV.

1. After these things, a battle [1] having taken place,[2] the Greeks were victorious.[3] 2. The king himself came as quickly [4] as possible [5] with the army.[6] 3. The same general [7] commanded [8] the army in both [9] the battles. 4. Many of the children [10] whom he saw feared [11] lest they should be taken.[12] 5. If these things had been true,[13] it would have been still [14] more terrible.[15]

1. μάχη. 2. γίγνομαι. 3. νικάω. 4. ταχύ. 5. ὡς. 6. στράτευμα. 7. στρατηγός. 8. ἡγέομαι. 9. ἀμφότερος. 10. παῖς. 11. φοβέομαι. 12. λαμβάνω. 13. ἀληθής. 14. ἔτι. 15. δεινός.

XV.

1. I told[1] him that you all[2] were my[3] friends.[4] 2. He acts[5] thus[6] that he may not seem[7] to wrong[8] the state.[9] 3. If he had been just,[10] this would not have happened.[11] 4. Do you think[12] they will flee[13] when[14] they see[15] us?

1. λέγω. 2. πᾶς. 3. possessive dative. 4. φίλος. 5. πράττω. 6. οὕτως. 7. δοκέω. 8. ἀδικέω. 9. πόλις. 10. δίκαιος. 11. γίγνομαι. 12. οἶμαι. 13. φεύγω. 14. ὅταν. 15. ὁράω.

XVI.

1. They came[1] in order to destroy[2] their[3] enemies.[4] 2. If you should say[5] this, he would be angry.[6] 3. The men[7] reported[8] that they had seen[9] no one.[10] 4. He declares[11] that he expects[12] to die.[13]

1. ἔρχομαι. 2. ἀπόλλυμι. 3. article. 4. ἐχθρός. 5. λέγω. 6. χαλεπαίνω. 7. ἀνήρ. 8. ἀπαγγέλλω. 9. ὁράω. 10. οὐδείς. 11. ἀποφαίνω. 12. οἶμαι. 13. θνήσκω.

XVII.

1. While[1] Alexander[2] was[1] in the country[3] of the Uxii,[4] his horse Bucephalus[5] was[6] once[7] missing.[8]

1. participle. 2. Ἀλέξανδρος. 3. χώρα. 4. Οὔξιοι. 5. Βουκεφάλας. 6. γίγνομαι. 7. omit. 8. ἀφανής.

2. Accordingly,[1] he proclaimed[2] through[3] the country that he would kill[4] all the Uxii, unless they brought[5] him back his horse.

1. οὖν. 2. προκηρύττω. 3. ἀνά. 4. ἀποκτείνω. 5. ἀπάγω.

3. And such[1] fear[2] of the king had[3] the barbarians, that[4] Bucephalus was sent[5] back directly[6] upon[7] the proclamation.[8]

1. τοσόσδε. 2. φόβος. 3. use εἰμί. 4. ὥστε. 5. ἀποπέμπω. 6. εὐθύς. 7. ἐπί. 8. κήρυγμα.

XVIII.

1. Did not Homer[1] call[2] Agamemnon[3] shepherd[4] of the people,[5] because a general[6] ought[7] to take care[8] that his soldiers[9] be both[10] safe[11] and[10] prosperous?[12]

1. Ὅμηρος. 2. προσαγορεύω. 3. Ἀγαμέμνων. 4. ποιμήν. 5. λαός. 6. στρατηγός. 7. δεῖ. 8. ἐπιμελέομαι. 9. στρατιώτης. 10. τε καί. 11. σῶς. 12. εὐδαίμων.

2. For[1] you know[2] that generals are chosen[3] to be authors[4] of prosperity[5] to those who chose them.

1. γάρ. 2. οἶδα. 3. αἱρέομαι. 4. αἴτιος. 5. εὐδαιμονία.

3. It seems[1] to me, therefore,[2] that Agamemnon would not have been applauded[3] by Homer, had he not been excellent[4] in this particular.[5]

1. δοκέω. 2. οὖν. 3. ἐπαινέω. 4. from ἀγαθός. 5. omit.

XIX.

1. As[1] Xenophon[2] was[1] sacrificing,[3] a messenger[4] arrived[5] from Mantinea,[6] announcing[7] that his son[8] Gryllus[9] was dead.[10]

1. omit. 2. Ξενοφῶν. 3. θύω. 4. ἄγγελος. 5. ἥκω. 6. Μαντίνεια. 7. λέγω. 8. υἱός. 9. Γρύλλος. 10. to die, θνήσκω.

2. Then[1] he[2] laid[3] aside the garland,[4] but[5] continued[6] to sacrifice.

1. καί. 2. ἐκεῖνος. 3. ἀποτίθεμαι. 4. στέφανος. 5. δέ preceded by μέν. 6. διατελέω.

3. But when[1] the messenger had added[2] this[3] also,[4] that he had died victorious,[5] Xenophon put[6] the garland on[6] again.[7]

1. ἐπεί. 2. προστίθημι. 3. ἐκεῖνος. 4. καί. 5. νικάω (participle). 6. ἐπιτίθεμαι. 7. πάλιν.

XX.

1. Themistocles[1] said[2] that the trophies[3] of Miltiades[4] woke[5] him from his sleep.[6]

1. Θεμιστοκλῆς. 2. λέγω. 3. τρόπαιον. 4. Μιλτιάδης. 5. ἀνίστημι. 6. ὕπνος.

2. Do not hasten[1] to be[2] rich,[2] lest thou speedily[3] become[4] poor.[5]

1. σπεύδω. 2. πλουτέω. 3. ταχύ. 4. γίγνομαι. 5. πένης.

3. If he shall slay[1] his[2] enemy,[3] he will pollute[4] his hand.[5]

1. ἀποκτείνω. 2. possess. genit. 3. ἐχθρός. 4. μιαίνω. 5. χείρ.

4. A report[1] was spread[2] abroad[2] that the allies[3] had revolted[4] from the city.[5]

1. λόγος. 2. διασπείρω. 3. σύμμαχος. 4. ἀφίστημι. 5. πόλις.

XXI.

1. It became[1] evident,[2] that[3] the Greeks strongly[4] feared[5] lest he should become a tyrant.[6] 2. The god, as it seems,[7] often[8] rejoices[9] in making[10] the small great, and[11] the great small. 3. The Thebans after this raised[12] a trophy,[13] and gave up[14] the dead[15] under truce.[16] 4. He replied,[17] that he was not marching[18] that[19] he might do wrong[20] to any, but that he might assist[21] those who were wronged.[22]

1. γίγνομαι. 2. δῆλος. 3. ὅτι. 4. ἰσχυρῶς. 5. φοβέομαι. 6. τύραννος. 7. ἔοικα. 8. πολλάκις. 9. χαίρω. 10. participle of ποιέω. 11. δέ (with preceding μέν). 12. ἵστημι. 13. τρόπαιον. 14. ἀποδίδωμαι. 15. νεκρός. 16. ὑπόσπονδυς. 17. ἀποκρίνομαι. 18. στρατεύομαι. 19. ἵνα. 20. ἀδικέω. 21. βοηθέω. 22. participle.

XXII.

1. He thought[1] that he needed[2] friends[3] for this purpose,[4] that he might have helpers.[5] 2. O that[6] I had as great[7] power[8] as[9] these kings now have! 3. They were not able[10] to prevent[11] Philip from passing through.[12] 4. They announced[13] that they should treat[14] all these as enemies.[15]

1. οἴομαι (w. infin.). 2. δέομαι. 3. φίλος. 4. ἕνεκα. 5. συνεργός. 6. εἴθε. 7. τοσοῦτος. 8. δύναμις. 9. ὅσος. 10. δύναμαι. 11. κωλύω. 12. παρέρχομαι (aor.). 13. προαγορεύω (ὅτι). 14. χράομαι (use). 15. πολέμιος.

XXIII.

1. The king said that whoever killed[1] the man should rule[2] the whole city. 2. They feared[3] that the army would bring[4] aid to the inhabitants,[5] for they perceived[6] that the citizens were not despondent.[7] 3. The eagle[8] remained until[9] evening[10] came[11] on; and, terrified[12] by the sight,[13] we came to the soothsayers[14] to make[15] communication about[16] the omen.[17] 4. He hoped[18] that he should die[19] that day,[20] that he might be released[21] from his chains.[22] 5. Take[23] this soldier, and keep[24] him until[9] I come[25] with[26] the king's army. 6. Do not inflict[27] misery[28] on me who am miserable[29] already.[30]

1. ἀποκτείνω. 2. ἄρχω. 3. φοβοῦμαι. 4. βοηθέω. 5. ἐνοικέω. 6. αἰσθάνομαι. 7. ἀθυμέω. 8. ἀετός. 9. ἕως. 10. ἑσπέρα. 11. ἐπιγίγνομαι. 12. ἐκπλήσσω. 13. ὄψις. 14. μάντις. 15. κοινόω. 16. περί. 17. θεῖον. 18. ἐλπίζω. 19. ἀποθνήσκω. 20. ἡμέρα. 21. λύω. 22. δεσμός. 23. λαμβάνω. 24. σώζω. 25. ἔρχομαι. 26. ἔχω. 27. προστίθημι. 28. νόσος. 29. νοσέω. 30. ἤδη.

XXIV.

1. They say that when animals[1] were endowed[2] with voices, the sheep[3] said to her master[4]: "You do[5] a curious[6] thing,[7] because[8] to us who provide[9] you wool[10] and lambs[11] you give nothing that we don't take[12] from[13] the earth,[14] while[15] to the dog[16] you give[17] [-some-[7]] of the food[18] you have yourself." And that the dog, who had been listening,[19] said: "But I am your preserver,[20] so that you are not carried[21] off by wolves;[22] since,[23] if I should not guard[24] you, you could not feed,[25] through-fear[26] of death."[27]

1. ζῷον. 2. φωνήεις. 3. οἶς. 4. δεσπότης. 5. ποιέω. 6. θαυμαστός. 7. omit. 8. *because you* = relat. pronoun. 9. παρέχω. 10. ἔριον. 11. ἄρνες (plural). 12. λαμβάνω. 13. ἐκ. 14. γῆ. 15. δέ. 16. κύων. 17. μεταδίδωμι. 18. σῖτος. 19. ἀκούω. 20. σῴζω. 21. ἁρπάζω. 22. λύκος. 23. ἐπεί. 24. φυλάττω. 25. νέμομαι. 26. φοβοῦμαι. 27. ἀπόλλυμι.

XXV.

1. He was brought up[1] at[2] the court[3] of the king[4]; so that,[5] while[6] a boy,[7] he used to converse[8] with the best[9] of the Persians.[10] 2. Would that he had given[11] me what he promised[12] to give him! 3. Old men[13] say that life[14] is burdensome[15] to them; but if death[16] comes[17] near,[18] nobody wants[19] to die.[20]

1. παιδεύω. 2. ἐπί. 3. θύρα (plural). 4. βασιλεύς. 5. ὥστε. 6. Participle of εἰμί. 7. παῖς. 8. διαλέγομαι. 9. ἀγαθός. 10. Πέρσης. 11. δίδωμι. 12. ὑπισχνέομαι. 13. γέρων. 14. ζάω. 15. βαρύς. 16. θάνατος. 17. ἔρχομαι. 18. πλησίον. 19. βούλομαι. 20. ἀποθνήσκω.

XXVI.

1. Seuthes asked, "Would you be willing, Episthenes, to die for this boy?" And he said, holding up his hands, "Strike, if the boy commands you to strike." **2.** He feared that the men from the mountains would not make war with the Greeks.

N. B. — The sentences below need correction: write out corrected forms for them with a right translation.

3. Εἰ οὐδεις ἐρχομαι ἱνα ἀκουοι ἐμε, οὐδεις σοφωτερος εἰσιν.

4. Ἀνιστησαν οἱ Ἑλληνες και εἶπον τον παις εἰς τω πολεως τουτῳ εἰναι.

5. And Xenophon, on arriving, said to Seuthes, that the men were friendly, and would have sent mercenaries if he had asked it.

XXVII.

1. The general with all his soldiers sailed away from the island, thinking that Cimon had come from Athens with twenty-seven ships. **2.** The gods know well what it is best for man to have: to some they give much gold, to others a beautiful body, to others neither of these gifts. **3.** (Write the following sentence in a *corrected* form, with the accents.)

Γαρ ἐδωσα ἐμαυτος αὐτους ἑνα ταλαντον δε οἱ ἀνθρωποι ἐπαυσονται μαχοντες.

XXVIII.

And immediately these soldiers came back and informed Xenophon that the enemy in great numbers had occupied the heights before Cleonymus had been able to lead his horsemen into the village. Then Xenophon said that if it seemed best to the other generals, he himself would march with his own men against the barbarians, that they might not attack the Greeks with their whole force.

XXIX.

Xenophon, the Athenian, fearing that, if he should leave Chirisophus and proceed by himself to attack the enemy, the barbarians would easily master the rest of the Grecian force before he could send them assistance, said that if the gods were willing to save them, they could save them where they were; but if not, that it was fitting that they should remain and suffer whatever the gods appointed until death came.

GREEK GRAMMAR.

I.

1. Decline θάλασσα, πλόος, and λέων *throughout*, and γύψ in the *singular*. Explain the Accent of the oblique cases of γύψ. 2. Decline the Adjective μέλας. Compare σώφρων and σοφός, and give the rules. Compare μέγας. 3. Decline οὗτος and ἐμαυτοῦ. 4. Give the rule for the Augment of verbs compounded with a preposition. Give the Imperfect Indicative Active of ἐπιγράφω and περιγράφω. 5. Give the rules for the formation of the Future Active. Inflect the Future Indicative Active of ἀγγέλλω. 6. Give the Second Aorist Indicative (Active, Passive, and Middle) of λείπω. Inflect the Present Optative Active of τιμάω. 7. Give a synopsis of the Second Aorist Active of τίθημι, and inflect the Imperative. 8. Mention any classes of verbs which are followed by the Genitive. How is the *Agent* expressed after verbals in -τέος ? 9. Κόμαι Χαρίτεσσιν ὁμοῖαι (hair resembling that of the Graces) : Explain the Dative Χαρίτεσσιν. 10. What constructions follow ἵνα and ὅπως to denote a purpose? When is each construction used ? 11. Divide the following verses into feet, and name the feet :—

Χαίρετε, κήρυκες, Διὸς ἄγγελοι ἠδὲ καὶ ἀνδρῶν,
Ἆσσον ἴτ᾽· οὔτί μοι ὔμμες ἐπαίτιοι, ἀλλ᾽ Ἀγαμέμνων.

II.

1. Decline the Nouns μοῦσα and τεῖχος. Explain the change of accent in μοῦσα where it is not the same as in the Nominative Singular. What nouns in α of the first declension have ας in the Genitive Singular ? 2. Decline the Adjective χαρίεις in the Singular. 3. Decline the Pronouns σύ and οὗτος. 4. What are the *syllabic* and *temporal* augments respectively ? What is the *ordinary* reduplication, and what is the *Attic* reduplication ?

5. Conjugate the Verbs τρίβω, ἀγγέλλω, and γράφω. Inflect the Perfect Indicative Passive of γράφω, and explain the euphonic changes which the root γραφ- undergoes in that tense. 6. How is the Future Active of a *liquid* verb formed? 7. Give a synopsis of the Present and Second Aorist Active of ἵστημι, and inflect those tenses in the Indicative. 8. What is the difference between ὁ σοφὸς ἀνήρ and ὁ ἀνὴρ σοφός? What does ἀνὴρ ὁ σοφός mean? What do αὐτὸς ὁ ἀνήρ and ὁ αὐτὸς ἀνήρ mean? 9. In the phrase ἐκ τούτων ὧν λέγει, explain the case of ὧν. 10. In dependent clauses, which tenses of the Indicative are followed by the Subjunctive, and which by the Optative? 11. Translate λανθάνει ἑαυτὸν σοφὸς ὤν, and explain the use of λανθάνω with the participle.

III.

1. Which consonants are called *labials*, which *linguals*, and which *palatals*? Explain *Syncope* and *Crasis*, and give an example of each. 2. Define an *Enclitic*. Give the Enclitics which you remember. 3. Decline the Nouns μοῦσα and πρᾶγμα. Give the rule for the *accent* of the Genitive Plural of each. 4. *Compare* ἀληθής, ἀγαθός, and κακός. 5. Decline ὅδε in the *Singular*, and ὅστις in the *Plural*. 6. Explain the *Attic Reduplication*, and give an example. How are verbs beginning with a *diphthong* augmented? Give the Imperfect and First Aorist Active of αἰτέω. 7. Inflect the First Aorist Middle Indicative of βουλεύω, and the Second Aorist Passive Optative of λείπω. 8. Give a synopsis of the Second Aorist Middle of τίθημι through all the moods, and inflect the Indicative. 9. Where are εὕροι, εὑρήσοι, and εὑρεθείη formed, and from what verb? 10. What cases follow Verbs of *accusing*, *prosecuting*, and *convicting*? If these verbs are compounded with κατά, what construction follows them? 11. Translate ἐκ τούτων ὧν λέγει, and οἴχεται ὃν εἶδες ἄνδρα. Explain the *attraction* in each case. 12. Explain the Genitive Absolute. When is the *Accusative* Absolute used?

Sophomore Questions. — 1. Translate Εἶπεν ὅτι τοῦτο ποιοίη,

Εἶπεν ὅτι τοῦτο ποιήσοι, Εἶπεν ὅτι τοῦτο ποιήσειεν, and Εἶπεν ὅτι τοῦτο ποιήσει, — and explain the difference, wherever that is not made clear by the translation. 2. Translate ἐβούλετο τοῦτο ποιῆσαι, and ἔφη τοῦτο ποιῆσαι, — and explain the force of the Aorist Infinitive in each case. 3. Explain the difference between the Optative and the Secondary Tenses of the Indicative in expressions of a *wish*. Give an example of each.

IV.

1. Explain the terms *Metathesis, Epenthesis, Syncope,* and give examples of each. 2. Give the general rules for accenting the Penult of Greek words. 3. Decline τελώνης and εὔγεως, and state what nouns are indeclinable. 4. Compare the Adjectives μέλας and ἡδύς; the Adverbs μάλα and νύκτωρ. 5. Give the numeral Adverbs as far as δεκάκις, and write in full the Plural of the Article. 6. Give the Imperfect of περιγίγνομαι and ἀνέχω, also the Perfect Indicative of τρίβω, with the rule for its formation. 7. Inflect the Present Optative Active of τιμάω, and the Imperfect of τίθημι. Give the Imperative of εἰμί, and the Present Indicative of φημί. 8. Translate the words πλήθει οἵπερ δικάζουσι, and give the rule for the use of the Relative. 9. What case follows verbs of *tasting*, and what cases follow causatives of this class? 10. Explain the use of the Infinitive after verbs of saying, thinking, etc., and give an example.

SOPHOMORE QUESTIONS. — 1. What tenses can be used to express a customary action, and what is the general rule for introducing quotations? 2. Translate πρὸ Ἕλληνος οὐδὲ εἶναι τοῦτο τὸ ὄνομα δοκεῖ, and explain the Infinitive. 3. Explain in full the use of ὤφελον in the expression of a wish, and illustrate by examples.

V.

1. What consonants are called *labials*, what *linguals*, and what *palatals?* Explain *N movable*, and give an example. 2. Accent the following Nouns, γνωμαι, γνωμων, γνωμαις (from

Nom. γνώμη); παιδος, παιδα, παιδων, παισιν (from παῖς); and the following Verbs, βουλενει, βουλενοι, βουλευσασθαι, ἐλθειν, ἐλθων. 3. Decline the Noun θάλασσα, and the Adjective χαρίεις. 4. Compare καλός, σώφρων, and ἡδύς. 5. Decline the Relative ὅς in the *Singular*, and οὗτος in the *Plural*. 6. How is the Future *Passive* formed?—give an example (from βουλεύω). How is the Future *Active* of a *liquid* Verb formed? 7. Give a *synopsis* of the Second Aorist Passive of λείπω, and *inflect* the Subjunctive of that tense. *Inflect* the Aorist Imperative Active of βουλεύω, and the Present Optative Active of δίδωμι. 8. Where are βουλευθῇ, βουλευθῆναι, and τέτριψαι made? Explain the euphonic change introduced in forming τέτριψαι. 9. What is the difference between βουλεῦσαι and βούλευσαι?—between ἔστησα and ἔστην? 10. Translate βούλεσθε εἶναι σοφοί, and βούλεσθε τούτους γενέσθαι σοφούς. Explain the case of the Adjective in each. 11. Translate μέλει σοι τούτων, and explain the cases of the Pronouns. 12. Describe the *Iambic Trimeter Acatalectic*. What is a *Trochee*, a *Pyrrhic*, an *Anapæst*, and a *Cretic*?

SOPHOMORE QUESTIONS.— 1. In what constructions can the Future Optative be used? Give an example of its correct use. 2. What constructions are regularly used in dependent clauses after Verbs signifying *to strive, to take care, to effect*, etc.? 3. What is the meaning of *each tense* of the Infinitive after a verb of *saying* or *thinking*? How are the tenses that are wanting in the Infinitive supplied? Give an example of each tense, using φημί and ποιέω.

VI.

1. What Nouns of the first declension have the ending *a* of the Nominative Singular *short*? 2. Give the contracted forms *with the Accents* of the Noun πλόω (Nom. Dual), of the Adjective ἀντίπνοος, of the Verb ἔχραεν, and state the rule or exception to which the Accent of each is to be referred. 3. Decline the Noun μνάα. The Noun Θώς. The Adjective πολύς. 4. Into what eight classes are Pronouns divided? Give examples of

IX.

1. Explain the terms *Contraction, Crasis*, and *Elision*. Give an example of *Elision*. 2. Accent the following words: θαλασσαν, θαλασσαι, θαλασσων, θαλασσαις (from θάλασσα); ἀξιῳ, ἀξιοι, and ἀξιων (from ἄξιος); βουλενοι, βουλευονται, and βουλενοιτο. Give the rules for the three verbs. 3. Decline the Noun λέων and the Participle ἱστάς. 4. *Compare* χαρίεις, ἡδύς, μέγας, and πολύς. 5. Decline the Numeral εἷς and the Pronoun οὗτος· 6. How do you form the Future *Passive* and the Future *Middle*? What is the Future *Active* of ἀγγέλλω, and by what rule is it formed? 7. *Inflect* the Imperfect Passive of βουλεύω, and the Imperfect Active of δίδωμι. Give a *synopsis* of the Present Tense of εἰμί, and *inflect* the Optative. 8. Where are the following verbs made: βουλευσάτω, βουλευθείη, βουλευθήτω, and ὦμεν, ἦμεν, ἔσται? 9. How are the *gender*, *number*, and *case* of a Relative Pronoun determined? Give an example. 10. Translate ἐξιέναι ἐξ ὧν ἔχομεν, and explain the case of the Relative. Translate σὺν ᾗ ἔχεις δυνάμει, and explain the position of δυνάμει and the case of ᾗ. 11. Explain the Genitive and the Accusative Absolute, and give an example of each.

SOPHOMORE QUESTIONS. — 1. Explain the division of tenses into *primary* and *secondary*. How is the construction of a dependent sentence affected by this principle? 2. What *time* is denoted by the Aorist Infinitive? Give examples, using ἐλθεῖν. 3. Translate εἶπεν ὅτι τοῦτο ποιοῖ, εἶπεν ὅτι τοῦτο ποιήσειεν, and εἶπεν ὅτι τοῦτο ποιήσοι, and explain the difference in meaning. 4. Describe the *Iambic Trimeter Acatalectic*. What are the four feet of *two* syllables?

X.

1. Accent the following words, and give the rules for the accentuation you adopt:— τιθεις, φαγειν, λαβου, ισταντο, προσειχον, βη. 2. Decline Ἑρμέας, νῆσος, νεώς, κέρας, Περικλέης. 3. Give the Accusative of ἰχθύς, ναῦς, βοῦς, γίγας, Σωκράτης, βασιλεύς. 4. Decline ἥσυχος, δεικνύς, ἴδρις. 5. Compare καλός, μέσος, μικρός, πρέσβυς.

6. Decline σύ, ὅς. 7. Give the *ordinals* from one to ten inclusive. Decline δύο. 8. Give the synopsis of the Second Perfect of λείπω. Inflect the Second Aorist Active and the Second Aorist Passive of the same verb. 9. Give the Perfect of ὄμνυμι. Explain the *Augment*. Give the Pluperfect of περιγράφω. 10. Inflect the Present Optative, Active, and Passive of δηλῶ. 11: What is the construction after verbs of accusing? After verbs of taking away? After verbs denoting fulness and want? 12. Translate ἐθέλω χρῆσθαι οἷς ἔχεις, and explain the construction of οἷς.

SOPHOMORE QUESTIONS. — 1. When do the Present and Aorist Infinitive retain their time? What is the rule for the *time* of Participles? 2. How is an indefinite general relative sentence expressed after primary and after secondary tenses? 3. How do you express a wish referring to the Past? To the Present? To the Future? 4. In what case do you put the *object* after *verbals?* How do you express the *agent* after the same class of words?

XI.

1. What consonants are called *liquids*? What are the *mutes*, and how are they divided into *smooth, middle*, and *rough mutes* ? 2. Explain the following euphonic changes : that of the root λεγ- in ἐλέχθην and λέλεκται, that of the root λειπ- in λέλειμμαι and ἐλείφθην, and that of πειθ- in πέπεισμαι. 3. Accent the following words: γνωμην, γνωμαι, γνωμων (from γνώμη) ; πολεως and πολεις (from πόλις); οὑτινος and οὑστινας (from ὅστις); and βουλευει, βουλευοι, and βουλευωμαι (from βουλεύω). Give the rules for the three verbs. 4. Decline the Nouns δόξα and πρᾶγμα, and the Pronouns ἐγώ and οὗτος. Give the Accusative Singular and the Accusative Plural of the Relative ὅς. 5. Inflect the Future Indicative Active and the Aorist Subjunctive Active of βουλεύω. Give a synopsis of the Indicative Passive of βουλεύω, and inflect the Aorist. 6. Give a synopsis of the Second Aorist Passive of λείπω, and inflect the Subjunctive. 7. Give a synopsis of the

Second Aorist Active of δίδωμι, and inflect the Indicative and Subjunctive. 8. In what voice, mood, and tense are the following verbs: ἔθεσαν, ἱστάναι, ἴωσι, ἰέναι, and ᾔδεσαν? 9. How does the Enclitic τὶς differ from τίς? Translate ἀκούεις τι; and τί ἀκούεις; what is the difference between ἡ αὐτὴ πόλις, — αὐτὴ ἡ πόλις, — and αὕτη ἡ πόλις? 10. Explain the distinction between Primary and Secondary Tenses. Explain the terms *protasis* and *apodosis*, and give an example of each. 11. What are the Spondee, the Trochee, the Iambus, the Dactyl, and the Anapest? What is the composition of Dactylic Hexameter (Heroic), and that of the Elegiac Pentameter?

SOPHOMORE QUESTIONS. — 1. Translate εἶπον ὅτι ἔλθοι, — εἰ ἔλθοι, ἴδοι ἂν τοῦτο, — and ὁπότε ἔλθοι, τοῦτο ἐποίει. Explain the time to which ἔλθοι refers in each case. 2. When are ἵνα, ὅπως, &c., followed by the Subjunctive, when by the Optative, and when by the past tenses of the Indicative? After what class of verbs is ὅπως with the Future Indicative most frequently used? 3. Describe the Iambic Trimeter Acatalectic, stating all the substitutions allowed, and showing how the Comic Trimeter differs from the Tragic.

XII.

1. Divide the consonants of the Greek alphabet into labials, linguals, and palatals. What change does ν undergo when it precedes a labial, lingual, or liquid? 2. Accent the following words, and give the rule for each: λεγονται, τιμησαι, οικοι, οὓς from οὖας. 3. Decline the Nouns οἰκία, νεώς, and ἄστυ, and the Pronouns ἐγώ and ὅστις. 4. Give the synopsis of the Aorist Passive of βουλεύω in all the Moods, and inflect the Participle. Inflect the Perfect Imperative Passive. 5. Give a synopsis of the Second Aorist Passive of λείπω, and inflect the Subjunctive. 6. Inflect the Present Optative Passive of τιμῶ, ἵστημι, and δίδωμι. 7. Give a synopsis of the Present of τίθημι, and inflect the Second Aorist Subjunctive Active. 8. Explain the apparent irregularity in the syntax of the following sentences: πάρειμι

ἐγὼ καὶ οὗτος· ταῦτα ἐγένετο, τὸ στράτευμα μάχονται. 9. Translate οἱ ἀγαθοὶ ἄνθρωποι, ἀγαθοὶ οἱ ἄνθρωποι, and οἱ ἀγαθοὶ τῶν ἀνθρώπων, and explain the Genitive. 10. Translate πέμπει τοὺς ἀνθρώπους and πέμπει τῶν ἀνθρώπων, βουλεύων τιμᾶται, and ὁ βουλεύων τιμᾶται.

SOPHOMORE QUESTIONS. — 1. Translate εἰ ἔπραξε καλῶς ἔσχεν and εἰ ἔπραξε καλῶς ἂν ἔσχεν. Translate εἰ πράττει καλῶς ἔχει and ἐὰν πράττῃ καλῶς ἔχει. 2. Translate φοβοῦμαι μὴ γίγνεται and φοβοῦμαι μὴ οὐ γίγνηται. If the leading verb in these cases were secondary, what would be the form of the dependent verbs? 3. Give examples of the use of the Infinitive in Indirect Quotation to express an action which is past, present, or future with respect to the leading verb.

XIII.

1. Write more correctly ἔστ᾽ ὅπως, — ἀπ᾽ οὗ, — οὐκ ὑμεῖς, and explain the principle. What is *N movable?* — give an example of its use. 2. Explain the terms *oxytone, barytone, perispomenon.* Give the general rule for the accent of *Nouns.* Accent μουσης, μουσαν, μουσων, μουσαι, μουσαις.* 3. Decline the Substantive λέων, and the Adjective ἀληθής. Compare ἄξιος, ἀγαθός, κακός, and μέγας. 4. Decline ἐγώ, and the Numeral εἷς. 5. Give a synopsis (through all the moods) of the Aorist Passive of βουλεύω, and inflect the *Optative.* Give a synopsis of the Second Aorist Middle of τίθημι, and inflect the *Imperative.* 6. Give the voice, mood, and tense of λίπωσιν, λιπῶσιν, βούλευσαι, and βουλεῦσαι. 7. Give the rule for the formation of the Perfect Passive, the Future Passive, and the Future Active. What is the Future Active of μένω and of ἀγγέλλω? 8. What are the two kinds of *Augment,* and when is each used? Give an example of each. What is the ordinary Reduplication, and what is the *Attic* Reduplication? 9. Inflect the Imperative Active of τιμάω, giving both the uncontracted and the contracted forms. Inflect the Present Indicative Active of πλέω in the same way.

SOPHOMORE QUESTIONS. — 1. Explain the *three* uses of the

* From μοῦσα.

Present Infinitive, and the *two* uses of the Aorist Infinitive; and give an example of each, using ποιεῖν and ποιῆσαι. 2. How many meanings can ἔφη ποιεῖν ἂν τοῦτο and ἔφη ποιῆσαι ἂν τοῦτο have? Explain each use of the Infinitive. 3. Show the difference between a *final* clause and an *object* clause after ὅπως, and give examples. 4. In what cases is the Subjunctive used in Protasis, and what is the corresponding construction in Relative Sentences? Give an example of each. 5. Translate εἶπεν ὅτι τοῦτο ποιοίη, — εἶπεν ὅτι τοῦτο ποιήσειεν, — εἶπεν ὅτι τοῦτο ποιήσοι, — εἶπεν ὅτι τοῦτο ποιήσειεν ἄν, and explain each tense of the Optative.

XIV.

1. Decline the Nouns γνώμη and τεῖχος. Explain the change of accent where it is not the same as in the Nominative Singular. 2. Decline the Adjective ἄξιος in the Singular. Compare ἡδύς, πολύς, and χαρίεις. 3. Decline the Pronouns ἐγώ and οὗτος. 4. Give the principal parts of ἀκούω, λαμβάνω, and ἀπογράφω. Inflect the Aorist Indicative Middle of βουλεύω, and give a synopsis of that tense through all the moods. 5. Give a synopsis of the tenses of the Indicative Active of ἵστημι (in the first person). Give a synopsis of the Second Aorist Active of δίδωμι, and inflect the Optative. 6. Inflect the Perfect Indicative Passive λέλειμμαι (from λείπω), and explain the euphonic changes which the root λειπ- undergoes in that tense. 7. What is the difference between ὁ αὐτὸς ἀνήρ and ὁ ἀνὴρ αὐτός? 8. In the phrase ἐκ τούτων ὧν λέγει, explain the case of ὧν, and give the rule. 9. What are Enclitics, and what are Proclitics? Give examples of each in connection with other words. 10. What are the feet consisting of *two* syllables? Give the quantity of each.

SOPHOMORE QUESTIONS. — 1. To what *time* does the Aorist Optative refer in each of its uses? Give an example of each, using ποιέω. 2. Give the general rule for indirect quotation after ὅτι or ὡς, and examples. 3. Translate εἰ βούλοιτο, τοῦτ' ἐποίει, and εἰ βούλοιτο, τοῦτ' ἂν ποιοίη, and explain the two uses of the Optative. 4. Which is more correct, μὴ τοῦτο ποιήσῃς or μὴ τοῦτο ποίησον? Explain the reason, and give the rule for Prohibitions.

XV.

1. Write more correctly ἔστι οἵ, οὕτως σφόδρα, πέπλεκμαι, ἐνφανής. What are τυχεῖν, δίκη, γέ, called with respect to accent? 2. Write the Genitive and Accusative Singular of the nouns χαρά, δόξα, τελώνης, θώς. Decline βασιλεύς in the Singular, and τεῖχος in the Plural number. 3. Write the Dative Singular and Plural of the Adjective χαρίεις and of the Participle διδούς in all genders. Compare the Adjectives πολύς, σεμνός, ταχύς. 4. Decline οὗτος and the interrogative τίς. 5. Inflect the Future Middle Indicative of ἀγγέλλω, and the Imperative Active of τιμάω. 6. Give a synopsis (through all the moods) of the Second Aorist Active of ἵστημι, of the Present Middle of τίθημι, and the Perfect Passive of βουλεύω. 7. State the tense, mood, voice, and Present Indicative of the following verbal forms, γένοιο, στῆσαι, ἔθετο, βούλευσον, ἐφίλει. 8. Translate τὴν αὐτὴν γνώμην (opinion) ἡμῖν ἔχουσιν, and explain the case of ἡμῖν.

SOPHOMORE QUESTIONS. — 1. Translate ἐὰν βούληται, ποιεῖ τοῦτο, and ἐὰν βούληται, ποιήσει τοῦτο, and explain the two uses of the Subjunctive. 2. Translate φασὶν ἡμᾶς ἐλθεῖν, and κελεύουσιν ἡμᾶς ἐλθεῖν, and explain the two uses of the Infinitive. 3. What is the rule for mood and tense in indirect quotations with ὅτι or ὡς after verbs of saying. Give examples. 4. How is a purpose expressed in Greek, and how a result? Give examples.

XVI.

1. Explain the form of the preposition in ἐφ' ὑμῖν. What must be the quantity of the *a* in σῶμα, and why? When is a word called barytone? Accent χαρας as Genitive Singular and as Accusative Plural from χαρά. 2. Write the Genitive and Accusative Singular of the Nouns οἰκία, μοῦσα, and the

Genitive and Vocative Singular of στρατιώτης. Decline ἰχθύς in the Singular and νεώς in the Plural number. 3. Decline the Adjectives ἀληθής, ἡδίων. Compare μικρός, μέλας, σοφός. 4. Decline σύ, οὗτος, and the Masculine Singular of ὅστις. 5. Give a synopsis of the Future Active of μένω, of the Second Aorist Active of τίθημι, and of the Present Middle of δίδωμι. 6. Inflect the Imperfect Middle of φιλέω, and the Aorist Middle Indicative of βουλεύω. 7. State the tense, mood, voice, and Present Indicative First Singular of the following verbal forms: λιπῶ, νομιῶ, ἐτύψω, ἔστω. 8. Translate τὴν αὐτὴν γνώμην (opinion) ἡμῖν ἔχουσιν, and explain the case of ἡμῖν.

SOPHOMORE QUESTIONS.— 1. When is the subject of the Infinitive Mood regularly omitted? When does the Aorist Infinitive refer to past time relatively? When does ὅτι or ὡς take the Optative Mood?—give examples. 2. By what mood and tense is a supposition referring to past time, and represented as contrary to fact, expressed? Give an example. 3. Translate ἦλθεν ἵνα ἴδῃ. What might be substituted for ἴδῃ, and which would be the more regular form?

XVII.

1. Define an *Enclitic*. Give the Enclitics which you remember. State the quantity of the *a* in the following words, and give the rules: δόξα (Nom. Sing.), δόξας (Acc.), ἐλπίδας. What must be the quantity of the *a* in σῶμα, and why? 2. Decline the Nouns πρᾶγμα, χώρα, γύψ, νεώς, in the Singular, and τριήρης, ἄστυ, in the Plural. What exceptions to the usual rules for Accent occur among the forms here required? 3. Decline the Pronouns οὗτος and ὅστις. Decline in the Singular the Participle βεβουλευκώς and the Adjective γλυκύς. *Compare* γλυκύς and ἡδύς, and decline the Comparative of the latter in the Plural. 4. Inflect the Imperfect Active and the Present Optative Middle of τιμάω, writing both uncontracted and contracted forms. Give synopses of the Aorist Active, Middle, and Passive of βουλεύω

through all the moods. 5. Where are ἀγγελῶ, ἔστω, στῶ, ἐλύσω, ληφθῶ, formed (i. e. tense, mood, voice), and from what verbs? Inflect the first and the last. 6. Translate ἡ αὐτὴ γυνή, — αὕτη ἡ γυνή, — ᾗ γυνὴ αὐτή, writing the Greek with the English.

SOPHOMORE QUESTIONS. — 1. Translate ἔφη τοῦτο ποιῆσαι, and ἔφη τοῦτο ποιήσειν. Substitute εἶπεν ὅτι for ἔφη, and make the requisite changes in ποιῆσαι and ποιήσειν. 2. Translate ὅστις ἂν ἔλθῃ, ὄψεται, — ὅστις ἦλθεν, εἶδεν ἄν. To what kind of sentences are these relative clauses analogous? Explain the two uses of the adverb ἄν illustrated above. 3. How is a Purpose expressed in Greek? How a Wish referring to Future Time? How a Prohibition in the Second Person?

XVIII.

1. Write more correctly Βάχχος, τέτριβμαι, νύκτ' ὅλην, οὕτως φησί. What is *Crasis?* Give an example. Mark the quantity of the final syllables in the following words: δόξα (Nom. Sing.), δόξα (Nom. Dual), κριτάς (Acc. Pl.), λιμένας (Acc. Pl.), λύσας (Part.). 2. Decline the Nouns Πέρσης, πόλις, τεῖχος, κέρας, in the Singular, and βασιλεύς, ἀνώγεων, in the Plural. 3. Decline the Pronouns οὗ (*ī*) and τὶς. Decline the Numeral εἷς, and in the Plural the Adjectives πᾶς and πρᾶος. What exceptions to the regular rules for Accent occur among the forms here required? 4. *Compare* σώφρων, τάλας, ἀγαθός, and decline one of the Comparatives of the last in the Plural. 5. Inflect the Present Optative Active of τιμάω, and the Imperfect Middle of δίδωμι. Give synopses of the Future Middle of κτείνω, and of the Second Aorist Active of τίθημι through all the moods. 6. Where are λίπω, λιπῶ, ἐλῶ, ἔλω, ἐκρίνω, formed (i. e. tense, mood, voice), and from what verbs? Give synopses of the first and last.

SOPHOMORE QUESTIONS. — 1. Translate ἔφη καλῶς ἂν πρᾶξαι in two ways, showing what two forms in Direct Discourse are here represented. 2. Translate λέγει ἐὰν δόξῃ αὐτῷ, — λέξει ἐὰν δόξῃ αὐτῷ, and explain the two uses of the Subjunctive. 3. Translate ὅστις εἶδεν, ἐχάρη ἄν, and explain the meaning of the relative sentence. What would be the meaning, if the verbs were changed to the Optative of the same tense? What negative particle would be proper with the first verb? Describe two classes of Relative Sentences.

XIX.

1. Give an example of Elision. In what words does the accent of the elided vowel disappear with the vowel? What is the word τέ called with respect to accent? Give the other words of the same sort. Write τινός after ἀνθρώπου with the accents properly disposed. Write τέ after σῶμα. 2. Decline πόλις, Σκύθης, οἰκία, ὀστέον in the Singular, and σῦκον, νεώς, θώς, τριήρης in the Dual and Plural. Give the Genitive, Dative, and Accusative in all genders and numbers of ἄξιος, γλυκύς, of the Present Active Participle of ἵστημι, and of the Perfect Active Participle of βουλεύω (or παύω). 3. Compare σοφός, τάλας, ἀληθής, πολύς. Decline the Comparative of μέγας. Form an adverb from ἡδύς, and compare it. Decline σύ in the Dual and Plural and οὗτος in the Singular. Give the Cardinal Numerals as far as twelve. 4. Give synopses (through all the moods) of the Aorist Middle and Aorist Passive of βουλεύω (or παύω), and inflect the Imperative. Give synopses of the Perfect Passive of πλέκω and the Present Active of δίδωμι. Inflect the Perfect Passive Indicative of πλέκω and the Imperfect Passive of τιμάω. 5. Where are μενῶ, ἐπαύσω, λιπῶ, στῶ, ἴω, and ἐδίδω formed (i. e. tense, mood, voice), and from what verbs?

SOPHOMORE QUESTIONS. — 1. What is the construction in Object Clauses after verbs of *striving*? How do such clauses differ from Pure Final Clauses? What is a General Supposition? How are General Suppositions expressed? How are Prohibitions in the Second and Third Persons expressed in Greek? Translate ἔφη δώσειν εἰ δύναιτο, and state what form the last three words would have in the Direct Discourse.

XX.

1. Give an example of Crasis. When does Iota become subscript in Crasis? Write τέ after κέρας with the accents properly disposed. Write εἰμί after Κῦρος. What is ὡς called with respect to accent? 2. Decline βασιλεύς, δεσπότης, χώρα, κέρας in the Singular, and ἀνώγεων, ἰχθύς, ἄστυ in the Dual and Plural. Give the Genitive, Dative, and Accusative, in all genders and numbers, of χαρίεις and ἀληθής, of the Second Aorist Active Participle of δίδωμι, and of the Perfect Active Participle of βουλεύω (or παύω). 3. Compare ἄξιος, μέλας, σώφρων, μέγας. Decline the Comparative of ἡδύς. Form an adverb from ταχύς, and compare it. Decline ἐγώ, ὅστις, οὗτος in the Dual and Plural. 4. Give synopses (through all the moods) of the Aorist Middle of βουλεύω and the Present Passive of τιμάω and ἵστημι. Inflect the Future Optative Active of ἀγγέλλω and the Perfect Passive Indicative of τρίβω. 5. Where are εἰδῶ, ἔστω, ἐπαύσω, ἴω, λίπω, λιπῶ, ἐτιμῶ, and ἐδίδω formed (i. e. tense, mood, voice), and from what verbs?

SOPHOMORE QUESTIONS. — 1. Translate εἰ τοῦτο ποιεῖν δύναται, ποιεῖ, and ἐὰν τοῦτο ποιεῖν δύνηται, ποιεῖ, and explain the two sorts of Conditional Sentences. How would the latter be written if ποιεῖ were changed to ἐποίει? Describe two classes of Relative Sentences. 2. What is the construction in Greek after verbs of *fearing?* after verbs of *hindering?* Give the general rule for the Indirect Quotation of compound sentences after ὅτι or ὡς. Translate χαλεπὸν εὑρεῖν and φησὶν εὑρεῖν, and explain the two distinct uses of the Infinitive.

XXI.

1. Decline κριτής, πόλις, and the Singular of νῆσος. Explain the accent of the Nominative and Genitive Singular and the Nominative Plural of νῆσος. 2. Decline the Adjective χρύσεος in the Singular of all genders, and πᾶς in the Plural. Compare σοφός, ἡδύς, and μέγας; and the Adverb σοφῶς. 3. Decline the

Pronouns ἐγώ and σύ in the Plural, — οὗ in all numbers, and οὗτος in the Plural. Explain the accent of οἵδε and οἵστισι. 4. Give a synopsis of the Future and Aorist Middle of βουλεύω (in all the moods), and inflect the Optative of each. Give a synopsis of the Second Aorist Passive of λείπω, and inflect the Subjunctive. Inflect the Perfect Passive Indicative of λείπω, and explain the euphonic changes. 5. Give a synopsis of the Second Aorist Active of δίδωμι, and of the Second Aorist Middle of τίθημι, and inflect the Indicative of each. 6. Translate ἐκ τῶν πόλεων ὧν ἐτύγχανεν ἔχων, and explain the case of ὧν and the construction of ἔχων. 7. Give rules for the *position* of the Article, (1) with a noun and an adjective, (2) with a noun and a demonstrative pronoun, (3) with αὐτός. Give an example of each, and translate it. 8. Give the *names* and mark the *quantity* of the feet of *two* syllables.

SOPHOMORE QUESTIONS. — 1. Explain the ordinary difference between the Subjunctive and the Future Indicative after ὅπως, and give an example of each. 2. Translate εἰ δύναιτο, τοῦτο ἐποίει, and εἰ δύναιτο, τοῦτ' ἂν ποιοίη, and explain the two uses of the Optative. Explain the two corresponding uses of the Subjunctive, and give examples. 3. What various constructions follow πρίν? give examples. Is τοῦτο ποιήσω πρὶν ἂν ἔλθω or τοῦτο ποιήσω πρὶν ἐλθεῖν more correct? — and why?

XXII.

I. What consonants are called *liquids?* How are *mutes* divided into *labial, palatal,* and *lingual* mutes? 2. Form the Future of τρίβω, γράφω, and πείθω, and explain the euphonic changes which are made before the ending -σω. 3. Inflect the Perfect Passive (τέτριμμαι) of τρίβω, and explain the euphonic changes made in adding the endings -μαι, -σαι, -ται, -σθον, and -σθε to the stem τριβ-. 4. Decline the Nouns πολίτης and νῆσος in the Singular, and λέων and βασιλεύς in the Dual and Plural. Give the Accusative Singular of ἔρις, ἐλπίς, πόλις, and ναῦς; and give a rule for each case. 5. Decline the Pronouns σύ and ὅστις. 6. Give a synopsis (through all the moods and participles) of the Future Passive of λύω (or βουλεύω). Give a synopsis of the Second Aorist Middle of λείπω, and inflect the Indicative and Imperative. 7. Give a synopsis of the Second Aorist Middle of τίθημι, and inflect the Optative. Give the Principal Parts of ἵστημι and δίδωμι. 8. What is the difference between the use of the Article in Attic Greek and in Homer? Give a rule for the *position* of the article with adjectives and with demonstrative pronouns, and give an example of each. 9. Which tenses of the Indicative are *primary?* and which are *secondary* (or *historical*)? How does this distinction often affect the mood of a dependent verb? 10. What is the difference between the Imperfect and Aorist Indicative? and between the Present and Aorist Subjunctive?

SOPHOMORE QUESTIONS. — 1. Explain the principle of *Indirect Quotations*, as regards both the leading and the dependent verbs. When can the Subjunctive be used in *Indirect Questions* in Greek? Give an example. 2. Explain the analogy between *relative* sentences and *conditional* sentences, and give examples. 3. What form of the Infinitive construction can follow verbs denoting *hindrance* or *prevention* (like εἴργω, *to prevent*)? Give examples.

XXIII.

1. Correct the form γεγραφμαι, and form the Second and Third Persons Singular of it. Perform the operations of Crasis and Elision on the words μήτε ὁ ἀνήρ. 2. Decline in the Singular θάλασσα, δεσπότης, θρίξ, and ἄστυ, and in the Plural τριήρης. 3. Decline the Pronouns σύ in all numbers, οὗτος and ὅστις in the Singular. Decline ἡδύς in the Plural. Compare σοφός, μέλας, κακός, and ῥᾴδιος. 4. Inflect the Aorist Subjunctive Passive of τιμάω, the Present Indicative of εἰμί, the Second Perfect οἶδα in the Indicative. 5. Write the Perfect of ἀκούω, the Future and Aorist of δίδωμι, the Second Aorist of ἐκμανθάνω and of ἀποθνήσκω, — all in the Active Voice. 6. How do ὁ ἀνὴρ ἀγαθός and

ὁ ἀγαθὸς ἀνήρ differ in meaning? How is a Wish referring to future time expressed? 7. Write out the scheme of the Dactylic Hexameter Verse.

SOPHOMORE QUESTIONS. — 1. What is the difference in meaning between ἐβασίλευον and ἐβασίλευσα? Give the different constructions in use after ἵνα in Final Clauses. How do Final and Object Clauses differ in meaning? 2. How do εἴ τινας ἴδοι, οὐδὲν ἂν εἴποι and εἴ τινας ἴδοι, οὐδὲν ἔλεγεν differ in sense and grammar? What form of Direct Discourse is represented by ἔλεγεν ὅτι γράψειεν? How do you express a Wish referring to past time? Write out the scheme of the Iambic Trimeter Acatalectic.

XXIV.

1. Explain Elision and Crasis; give examples in ἐπὶ ἑτέρῳ and καὶ αὐτός. 2. What are Enclitics? Correct the accent of οὗτος ἐστίν, τιμαὶ τέ, τιμῶν τέ, τούτου γέ. 3. Decline the Nouns νῆσος, λέων, and βασιλεύς throughout, and ἀνήρ in the Singular. 4. Decline the Adjective γλυκύς. Compare σεμνός, ὀξύς, and ἀληθής. 5. Give the synopsis of the Future Passive of λύω (or βουλεύω) through all the moods, and inflect the Indicative. Inflect the Present Indicative Passive of δηλόω (in the contract form). In what places in this verb is the form δηλοῖ found? 6. Give the synopsis of the Second Aorist Middle of τίθημι in all the moods, and inflect the Optative. Inflect the Imperfect of εἶμι (to go). 7. State briefly the distinction in the uses of the adverb ἄν.

SOPHOMORE QUESTIONS. — 1. Explain the analogy between the expression of a Wish and of a Condition. What are the constructions which may follow expressions denoting *hindrance?* What are the two uses of an Aorist Infinitive? Translate "He does this whenever he pleases," and "He did this whenever he pleased." 2. What are the differences between the Homeric use of the Article and the Attic use? 3. What varieties of verse are measured by Dipodies? Explain the substitutions in the Iambic Dipody. How does the Trochaic Dipody differ from this?

XXV.

1. Write down the *smooth mutes;* the *labial mutes.* Perform Crasis and Elision on μήτε ὁ ἀνήρ. 2. Form the Accusative Plural of τιμή, Vocative Singular of πολίτης, entire Singular uncontracted and contracted of ὀστέον, Singular in all cases of βασιλεύς, Nominative Plural and Genitive Plural of σῶμα and πόλις. 3. Decline ἀληθής in the Singular. Compare ἀγαθός and κωφός. Decline σύ in all its numbers, and ὅστις in Masculine Singular. 4. Form the Second Person Singular, Aorist Imperative Middle of βουλεύω. Form the First Person Singular of λύω in Aorist Optative Passive, of φεύγω in Second Aorist Subjunctive Active, of λείπω in 2d Aorist Indic. Middle. Form the Second Person Singular of τίθημι in Second Aorist Indicative Middle. Inflect εἶμι (*I go*) in the Present Indicative. Write out the Principal Parts of γιγνώσκω and ἐκδίδωμι. 5. How is the Article used in Homer? What is the difference between καλὸς ὁ παῖς and ὁ καλὸς παῖς? Σοφώτερός ἐστιν ἢ ἐγώ; express this by changing ἐγώ into an oblique case. What constructions are used in Final Clauses? How do you express a General Supposition in present and in past time? What is a Dactyl, an Iambus, a Trochee, an Anapæst?

SOPHOMORE QUESTIONS. — 1. Explain the use of ἵνα with the past tenses of the Indicative. Explain the Accusative in μάχην νικᾶν. Express "that man" in Greek prose. What constructions are allowed with verbals in -τέος, -τέον? What tenses and moods are used to express Prohibition? Write out the scheme of the Iambic Trimeter Acatalectic both of Tragedy and of Comedy, naming the feet employed.

XXVI.

1. State the general principle for the Accent of Verbs, with such exceptions as you remember. 2. Decline the Nouns Μοῦσα, τεῖχος, and θυγάτηρ; the Singular of φλέψ, and the Plural of

λέων. 3. Decline the Pronouns σύ and τίς, and the Numeral εἷς. 4. Compare ταχύς, αἰσχρός, and κακός, and decline one of the Comparatives of the last in the Plural. 5. Inflect the Present Optative Middle of τιμάω, and the Imperfect Middle of τίθημι. Give a synopsis of the Present Active of δηλόω, and of the Aorist Passive of ἵστημι. 6. If in the sentence ἔρχεται ἵνα τὴν πόλιν ἴδῃ the first verb should be changed to ἦλθεν, how would you construct the dependent verb? 7. Translate into Greek the following: *I saw those who were present, and I say that I have seen them.*

SOPHOMORE QUESTIONS. — 1. Translate ἔφη καλῶς ἂν πρᾶξαι in two ways, showing what two forms in Direct Discourse are represented. 2. Translate ὅστις εἶδεν, ἐχάρη ἄν, and explain the meaning of the relative sentence. What negative particle would be proper with the first verb? 3. Under what circumstances is a secondary tense of the Indicative used in a Final Clause after ἵνα?

XXVII.

1. Give the accusative, *singular* and *plural*, of Μοῦσα, οἰκία, χώρα, and τιμή. Decline νῆσος in the *singular*, and λέων in the *plural*. 2. Decline the adjective γλυκύς in the *singular*, and μέγας in the *plural*. Compare κοῦφος, ὀξύς, ἀληθής, and ἡδύς. 3. How is the Future Indicative Active of a *liquid* verb formed? Give an example, and inflect it through all the numbers and persons. 4. Give the principal parts of γράφω, λαμβάνω, and φέρω. 5. Inflect the Imperfect Active of ἵστημι, and the 2d Aorist Optative Active of τίθημι. 6. What is a *stem* in grammar? In λέλειμμαι and in ἐρρῖφθαι, point out the stem and the other parts of each word; also explain *all* the euphonic changes made in any of the parts. 7. Translate πολλοὶ τῶν βουλομένων εἶναι σοφῶν, and explain the case of σοφῶν. Translate ὁ ἐμὸς τοῦ ταλαιπώρου βίος, and explain the genitive. Explain the Accusative in πλ 'ν τύπτει. 8. Translate ἐὰν ἴ νῦν πότε ἔσται οἴκοι, and ἐν

ἐγγὺς ἔλθῃ θάνατος, οὐδεὶς βούλεται θνήσκειν, and explain the subjunctive in each case. Translate φησὶν ἐλθεῖν and βούλεται ἐλθεῖν, and explain the *tense* of ἐλθεῖν in each. 9. What is a trochee, a tribrach, an anapæst, a cretic? Explain the terms *catalectic, dipody, dimeter.* How many feet are there in a trochaic dimeter, and how many in a dactylic dimeter? 10. Why is ἦλθεν ἵνα ἴδῃ more correct than ἔρχεται ἵνα ἴδοι? How is ἴδῃ to be explained? Translate into Greek: *they took care* (ἐπιμελέομαι) *that this should be done* (γίγνομαι), and explain the construction used in the dependent clause. 11. What is the difference between χρῆν σε τοῦτο ποιεῖν and χρή σε τοῦτο ποιεῖν? Express in Greek: *O that this had happened, O that this might happen,* and *O that this were true;* and explain the verbal form used in each case. 12. What is an *anapæstic dimeter acatalectic,* — an *anapæstic tetrameter catalectic,* — an *anapæstic system?* What is an *elegiac distich?*

XXVIII.

1. Decline the nouns νῆσος in the Singular, λέων in the Dual and Plural, and βασιλεύς in all numbers. Explain the accent of νῆσος and λέων wherever it varies from that of the Nominative Singular. 2. Decline the Pronouns σύ in all numbers and ὅστις in the Plural. Explain the accents of the Genitive and Dative Plural of ὅστις. 3. In what two principal ways are adjectives compared by change of termination? Give examples of each. Compare κακός, ἀγαθός, ἀληθής, and μέγας. 4. Give the principal parts of πλέκω, λαμβάνω, δίδωμι, and ἵστημι. 5. Inflect the Aorist Optative Passive of λύω, the Aorist Imperative Passive of λύω, the Imperfect Passive of ἵστημι, and the Second Aorist Optative Middle of τίθημι. 6. Explain the euphonic changes which occur in the following forms: — λύουσι, λέλειμμαι (λειπ-), δούς (δοντ-), νύξ (νυκτ-), πέπεικα (πειθ-), τέθυκα (θυ-), ἐτέθην (θε-). 7. When any forms of the substantive pronoun of the Third Person (οὗ, οἷ, σφῶν, &c.) are used in Attic prose, what is their peculiar force?

Give an example. 8. Give examples containing the correct use of the Genitive Absolute and of the Accusative Absolute. When is the latter regularly used? 9. Translate into Greek: *If these had been good men, they would not have suffered* (πάσχω), and explain the construction used. 10. What is a trochee, a spondee, an iambus, and an anapæst? What is *cæsura* in verse, and where does this generally occur in the heroic hexameter? 11. How are object clauses with ὅπως after verbs like σκοπέω distinguished, in construction and in meaning, from final clauses? Give an example of each. When do final clauses admit the Indicative? 12. Distinguish the Infinitive in Indirect Discourse from its use in other constructions. Show, by an example, how the Imperfect is expressed in the Infinitive. What two meanings can ἔφη τοῦτο ἂν ποιῆσαι have? Explain the principle in each case. 13. What is the difference between an Iambic Dipody and an Iambic Dimeter? What substitutions for the Iambus are allowed in an Iambic Dipody? what for the Trochee in a Trochaic Dipody? Explain an Anapæstic System.

LATIN COMPOSITION.

I.

1. Demaratus, the father of King Tarquin,[1] fled[2] from Corinth to Tarquinii.

1. *Tarquinius.* 2. *Fugio, fugere.*

2. I do not think[1] that immortality[2] is to be despised[3] by a mortal.[4]

1. *Arbitror, arbitrari.* 2. *Immortalitas.* 3. *Contemno, contemnere.* 4. *Mortalis, -e.*

3. Theophrastus is[1] said to have accused[2] Nature, because[3] she had given a long life to crows,[4] and so[5] short[6] a life to men.

1. *Dico, dicere.* 2. *Accuso, accusare.* 3. *Quod.* 4. *Cornix.* 5. *Tam.* 6. *Exiguus.*

4. Ignorance[1] of future[2] evils[3] is more useful[4] than knowledge[5] [of them[6]].

1. *Ignoratio.* 2. *Futurus.* 3. *Malum.* 4. *Utilis.* 5. *Scientia.* 6. Omit.

5. Do you not know[1] what[2] sort of men you charge[3] with crime?[4]

1. *Intelligo.* 2. What sort of = *qualis.* 3. *Arguo, arguere.* 4. *Scelus.*

6. If death[1] were feared,[2] Brutus would not have fallen[3] in battle,[4] and the Decii would not have exposed[5] themselves to the weapons[6] of the enemy.

1. *Mors.* 2. *Timeo, timere.* 3. *Concido, concidere.* 4. *Prœlium.* 5. *Objicio, objicere.* 6. *Telum.*

II.

1. Pompey was the first Roman who subdued[1] the Jews.[2] By right[3] of conquest[4] he entered[5] their Temple.

1. *Devinco, -ere.* 2. *Judæus.* 3. *Jus.* 4. *Victoria.* 5. *Intro, -are.*

2. They say[1] that Timotheus, a distinguished[2] man at Athens,[3] when[4] he had dined[5] at[6] Plato's[7] and had been greatly[8] gratified[9] with the entertainment,[10] and had seen him the-next-day,[11] said:[12] "Your dinners[13] are pleasant[14] not only at-the-time,[15] but also the-day-after."[16]

1. *Fero, ferre.* 2. *Clarus.* 3. *Athenæ, -arum.* 4. *Cum.* 5. *Cœno, -are.* 6. *Apud.* 7. *Plato, -onis.* 8. *Admodum.* 9. *Delecto, -are.* 10. *Convivium.* 11. *Postridie.* 12. *Dico.* 13. *Cœna.* 14. *Jucundus.* 15. *In præsentia.* 16. *Postero die.*

3. Verres also[1] ordered[2] the silver[3] tables[4] to be carried-away[5] from[6] all the shrines.[7]

1. *Idem.* 2. *Jubeo, -ere.* 3. *Argenteus, -a, -um.* 4. *Mensa.* 5. *Aufero.* 6. *De.* 7. *Delubrum.*

III.

1. The next[1] day he calls[2] the leaders of the forces[3] together, and tells[4] them that no city is more hostile[5] to the Greeks than the royal[6] (city) of the old kings.

1. *Posterus.* 2. *Convocare* = call together. 3. *Copiæ.* 4. *Docere.* 5. *Infestus.* 6. *Regius.*

2. If we grant[1] that the gods exist,[2] and that the universe[3] is ruled[4] by their mind, I do not see why[5] I should[6] say there is no divination.[7]

1. *Concedo.* 2. *Esse.* 3. *Mundus.* 4. *Regere.* 5. *Cur.* 6. I say there is no = *nego esse.* 7. *Divinatio.*

3. There is not[1] one of you who has not often[2] heard[3] how[4] Syracuse[5] was taken by Marcellus.

1. Not one = *nemo*. 2. *Sæpe*. 3. *Audire*. 4. *Quemadmodum*. 5. *Syracusæ*.

4. Demaratus, the father of our King Tarquin, fled[1] from Corinth,[2] because[3] he could not bear[4] the tyrant[5] Cypselus, to Tarquinii, and there[6] established[7] his fortunes.[8]

1. *Fugere*. 2. *Corinthus*. 3. *Quod*. 4. *Ferre*. 5. *Tyrannus*. 6. *Ibi*. 7. *Constituere*. 8. *Fortuna*.

IV.

1. They say[1] that the death of his son was[2] announced to Anaxagoras [as he was[3]] discoursing[4] among[5] his friends[6] on[7] the nature of things, and that no[8] answer was given by him except[8] that he begot[9] him mortal. A glorious[10] speech[11] in[12] truth, and worthy[13] of being uttered[14] by so great a man.

1. *Tradere*. 2. *Nuntiare*. 3. Omit. 4. *Disserere*. 5. *Inter*. 6. *Familiaris*. 7. *De*. 8. Literally, nothing else (*nihil aliud*) was answered (*respondere*) except (*nisi*). 9. *Gignere*. 10. *Præclarus*. 11. *Vox*. 12. *Vero*. 13. *Dignus*. 14. *Emittere*.

2. How much wiser[1] Xenophon [acted[2]], who, when he was engaged-in-sacred-rights,[3] and heard that his elder[4] son had fallen[5] in battle,[6] merely[7] laid-down[8] the garland[9] from[10] his head: but[11] when he heard that he had fallen fighting[12] bravely,[13] he put[14] the garland on his head again.[15]

1. *Sapienter*. 2. Omit. 3. *Sacra peragere*. 4. *Major natu*. 5. *Cadere*. 6. *Prælium*. 7. *Tantum*. 8. *Deponere*. 9. *Corona*. 10. *E*. 11. *Vero*. 12. *Pugnare*. 13. *Fortiter*. 14. Put on = *imponere* with dative. 15. *Rursus*.

V.

1. There is need[1] of magistrates,[2] without[3] whose wisdom[4] and care[5] the state[6] cannot[7] exist.[8]

1. *Opus.* 2. *Magistratus.* 3. *Sine.* 4. *Prudentia.* 5. *Diligentia.* 6. *Civitas.* 7. With *posse.* 8. *Esse.*

2. Do you see[1] how[2] the furies[3] harass[4] the impious,[5] and never[6] suffer[7] them to-stand-still?[8]

1. *Video.* 2. *Ut.* 3. *Furia.* 4. *Agito.* 5. *Impius.* 6. With *unquam.* 7. *Patior.* 8. *Consisto.*

3. Since[1] solitude[2] and a life[3] without friends[4] is full[5] of snares[6] and fear,[7] reason[8] admonishes[9] us to contract[10] friendships.[11]

1. *Cum.* 2. *Solitudo.* 3. *Vita.* 4. *Amicus.* 5. *Plenus.* 6. *Insidiæ.* 7. *Metus.* 8. *Ratio.* 9. *Moneo.* 10. *Comparo.* 11. *Amicitia.*

4. We favor[1] thee; we wish[2] thee to enjoy[3] thy virtue.[4]

1. *Faveo.* 2. *Cupio.* 3. *Fruor.* 4. *Virtus.*

5. Lucilius used[1] to say[2] that he wished[3] those things which he wrote[4] to be read[5] neither by the very unlearned[6] nor the very learned.

1. *Soleo.* 2. *Dico.* 3. *Volo.* 4. *Scribo.* 5. *Lego.* 6. *Indoctus.*

6. The decemvirate[1] and his colleagues[2] had completely[3] changed[4] Fabius,— a man formerly[5] excellent[6] both in peace[7] and in war.[8]

1. *Decemviratus.* 2. *Collega.* 3. *Plane.* 4. *Muto.* 5. *Olim.* 6. *Egregius.* 7. With *domus.* 8. *Militia.*

VI.

1. Let us consider,[1] first,[2] whether the universe[3] is governed[4] by the foresight[5] of the gods;[6] secondly,[7] whether they provide[8] for the welfare[9] of man.[10]

1. *Video.* 2. *Primum.* 3. *Mundus.* 4. *Rego.* 5. *Providentia.* 6. *Deus.* 7. *Deinde.* 8. *Consulo.* 9. *Res.* 10. *Humanus.*

2. Neoptolemus would never[1] have been able[2] to take[3] Troy, if he had been willing[4] to listen[5] to Lycomedes, in[6] whose household he had been brought[7] up.

1. *Nunquam.* 2. *Possum.* 3. *Capere.* 4. *Volo.* 5. *Audio.* 6. *Apud.* 7. *Educo.*

3. When[1] the enemy[2] saw[3] that the damages,[4] which they had hoped[5] could[6] not be repaired[7] for a long[8] time,[9] had been so[10] repaired by the toil[11] of a few[12] days[13] that there was no opportunity[14] left[15] for a sally,[16] they were eager[17] for the original[18] terms[19] of capitulation.[20]

1. *Ubi.* 2. *Hostis.* 3. *Video.* 4. *Is.* 5. *Spero.* 6. *Possum.* 7. *Reficio.* 8. *Longus.* 9. *Spatium.* 10. *Ita.* 11. *Labor.* 12. *Paucus.* 13. *Dies.* 14. *Locus.* 15. *Relinquo.* 16. *Eruptio.* 17. *Recurro.* 18. *Idem.* 19. *Conditio.* 20. *Deditio.*

4. If he is about to come[1] to Rome without[2] violence,[2] you may[3] properly[4] remain[5] at home;[6] but[7] if he is about to give[8] up the city[9] to be plundered,[10] I fear[11] that Dolabella himself[12] can[13] not fully[14] protect[15] us.

1. *Venio.* 2. *Modeste.* 3. *Possum.* 4. *Recte.* 5. *Sum.* 6. *Domus.* 7. *Sin.* 8. *Do.* 9. *Urbs.* 10. *Diripio.* 11. *Vereor.* 12. *Ipse.* 13. *Possum.* 14. *Satis.* 15. *Prosum.*

VII.

1. When[1] I was on[2] (my) Tusculan-estate,[3] and wanted[4] to use[5] certain[6] books[7] out[8] of the library[9] of Lucullus, I went[10] to his villa,[11] to take[12] them thence[13] myself,[14] as[15] I used[16] to.

1. *Cum.* 2. *In.* 3. *Tusculanum.* 4. *Velle.* 5. *Uti.* 6. *Quidam.* 7. *Liber.* 8. *E.* 9. *Bibliotheca.* 10. *Venire.* 11. *Villa.* 12. *Promere.* 13. *Inde.* 14. *Ipse.* 15. *Ut.* 16. *Solere.*

2. You know-not,[1] madman,[2] what power[3] virtue[4] has;[5] you use[6] the name[7] only[8] of virtue, you know not how[9] powerful[10] virtue itself[11] is.

1. *Nescire.* 2. *Insanus.* 3. *Vis.* 4. *Virtus.* 5. *Habere.* 6. *Usurpare.* 7. *Nomen.* 8. *Tantum.* 9. *Quid.* 10. To be powerful, *valere.* 11. *Ipse.*

3. What can[1] you say[2] in[3] your defence[4] which they have not said?

1. *Possum.* 2. *Dicere.* 3. *In.* 4. *Defensio.*

4. You are sorry[1] for others,[2] for yourself[3] you are neither[4] sorry nor[4] ashamed.[5]

1. *Miseret.* 2. *Alius.* 3. *Tu.* 4. *Nec.* 5. *Pudet.*

5. The tyrant[1] Dionysius, expelled[2] from Syracuse,[3] taught[4] boys[5] at Corinth.[6]

1. *Tyrannus.* 2. *Expello.* 3. *Syracusæ,-arum.* 4. *Docere.* 5. *Puer.* 6. *Corinthus.*

6. This state[1] has not produced[2] any[3] men more illustrious[4] in glory[5] than Africanus, Lælius, and Furius.

1. *Civitas.* 2. *Ferre.* 3. *Ullus.* 4. *Clarus.* 5. *Gloria.*

VIII.

1. Let us so[1] live[2] as always[3] to think[4] that an account[5] must be rendered[6] by us.

1. *Ita.* 2. *Vivere.* 3. *Semper.* 4. *Arbitrari.* 5. *Ratio.* 6. *Reddere.*

2. Would-that[1] I could[2] as[3] easily[4] discover[5] the truth[6] as refute[7] the falsehood.[8]

1. *Utinam.* 2. *Posse.* 3. *Tam.* 4. *Facile.* 5. *Invenire.* 6. *Verus.* 7. *Convincere.* 8. *Falsus.*

3. He exhorted[1] his friends[2] not to be-wanting[3] to the common[4] safety.[5]

1. *Hortari.* 2. *Amicus.* 3. *Deesse.* 4. *Communis.* 5. *Salus.*

4. After[1] Pompey had learned[2] what had been done[3] at Corfinium, he set-out[4] with two legions[5] from Luceria, and in five days[6] arrived-at[7] Brundisium.

1. *Posteaquam.* 2. *Reperire.* 3. *Gerere.* 4. *Proficisci.* 5. *Legio.* 6. *Dies.* 7. *Pervenire.*

5. When[1] by the supreme-authority[2] of one man there-was[3] no-longer[4] a field[5] in public-life[6] for wisdom[7] or[8] personal-influence,[9] I surrendered[10] myself neither[11] to my sorrows,[12] by which I should have been overwhelmed[13] if-I-had-not[14] resisted[15] them, nor[11] to pleasure[16] unworthy[17] of a scholar.[18]

1. *Quum.* 2. *Dominatus.* 3. *Esse.* 4. *Non jam.* 5. *Locus.* 6. *Res publica.* 7. *Consilium.* 8. *Aut.* 9. *Auctoritas.* 10. *Dedere.* 11. *Nec.* 12. *Angor.* 13. *Conficere.* 14. *Nisi.* 15. *Resistere.* 16. *Voluptas.* 17. *Indignus.* 18. *Doctus homo.*

IX.

1. I find[1] that Plato came[2] to Tarentum in the consulship[3] of Camillus and Claudius.

1. *Reperire.* 2. *Venire.* 3. Express this by the word *consul*.

2. The plays[1] of[2] Livius are not worthy[3] of being read[4] a second[5] time.

1. *Fabula.* 2. *Livianus* = of Livius. 3. *Dignus.* 4. *Legere.* 5. *Iterum.*

3. The Sicilians[1] sometimes[2] make[3] a month[4] longer[5] by one[6] day[7] or two[8] days.

1. *Siculus.* 2. *Nonnunquam.* 3. *Facere.* 4. *Mensis.* 5. *Longus.* 6. *Unus.* 7. *Dies.* 8. *Biduum.* Write out the rule for the case of *dies*..

4. The Stoics[1] think[2] it does not[3] concern[4] men[5] to know[6] what is going to happen.[7]

1. *Stoicus.* 2. *Existimare.* 3. *Nihil.* 4. *Interesse.* 5. *Homo.* 6. *Scire.* 7. *Esse.*

5. There were [some[1]] who on this day accused[2] the king[3] of rashness,[4] the consul[5] of inefficiency.[6]

1. Omit. 2. *Accusare.* 3. *Rex.* 4. *Temeritas.* 5. *Consul.* 6. *Segnitia.*

6. I am afraid[1] that I cannot[2] grant[3] that.[4]

1. *Vereri.* 2. *Possum.* 3. *Concedere.* 4. *Ille.*

X.

1. When Nasica had come[1] to the poet[2] Ennius, and the maid[3] had told[4] him[5] Ennius was not at home,[6] Nasica knew[7] that she had said so[8] at her master's[9] command,[10] and that he was within.[11]

1. *Venire.* 2. *Poeta.* 3. *Ancilla.* 4. *Dicere.* 5. *Is* (dative). 6. *Domus.* 7. *Sentire.* 8. Omit. 9. *Dominus.* 10. *Jussu* (abl.). 11. *Intus.*

2. A few[1] days[2] after,[3] when Ennius had come to Nasica and asked[4] for him, Nasica bawls[5] out that he is not at home.

1. *Paucus.* 2. *Dies.* 3. *Post.* 4. *Quærere* (with the accusative). 5. *Exclamare.*

3. Then quoth[1] Ennius: What?[2] Do I not recognize[3] your[4] voice?[5]

1. *Inquit.* 2. *Quid.* 3. *Cognoscere.* 4. *Tuus.* 5. *Vox.*

4. Hereupon[1] Nasica: You are a shameless[2] fellow:[3] when I asked for you I believed[4] your maid (when[5] she said) that you were not at home. Do you not believe my-own-self?[6]

1. *Hic.* 2. *Impudens.* 3. *Homo.* 4. *Credere* (with dative). 5. Omit. 6. My-own-self, *ego ipse.*

XI.

1. This[1] edict[2] having been published,[3] there was[4] no[5] state[6] which[7] did not send[8] a part[9] of its[10] Senate[11] to Cordova,[12] no[5] Roman citizen[13] who[7] did not come[14] to the meeting at[15] the day.[16]

1. Literally, which, *qui.* 2. *Edictum.* 3. *Pervulgare.* 4. *Esse.* 5. *Nullus.* 6. *Civitas.* 7. Which — not or who — not, *quin.* 8. *Mittere.* 9. *Pars.* 10. Omit. 11. *Senatus.* 12. *Corduba.* 13. *Civis.* 14. *Convenire.* 15. *Ad.* 16. *Dies.*

2. Nothing[1] is more praiseworthy,[2] nothing more worthy[3] of a great[4] and illustrious[5] man,[6] than clemency.[7]

1. *Nihil.* 2. *Laudabilis.* 3. *Dignus.* 4. *Magnus.* 5. *Præclarus.* 6. *Vir.* 7. *Clementia.*

3. Don't[1] you know[2] what[3] sort of dead[4] men you are accusing[5] of the worst[6] crime?[7]

1. *Nonne.* 2. *Intelligere.* 3. What sort of, *qualis.* 4. *Mortuus.* 5. *Arguĕre.* 6. *Summus.* 7. *Scelus.*

4. For many[1] ages[2] the name[3] of the Pythagoreans[4] was[5] in such high repute, that[6] no others[7] seemed[8] learned.[9]

1. *Multus.* 2. *Sæculum.* 3. *Nomen.* 4. *Pythagoreus.* 5. To be in such high repute, *sic vigere.* 6. *Ut.* 7. *Alius.* 8. *Videri.* 9. *Doctus.*

XII.

1. In-the-mean-time[1] the Romans,[2] the Scipios[3] being sent[4] to Spain,[5] first[6] drove[7] the Carthaginians[8] from the province,[9] afterwards[10] carried[11] on serious[12] wars[13] with[14] the Spaniards[15] themselves.[16]

1. *Interea.* 2. *Romanus.* 3. *Scipio, -onis.* 4. *Mittere.* 5. *Hispania.* 6. *Primo.* 7. *Expellere.* 8. *Pœnus.* 9. *Provincia.* 10. *Postea.* 11. *Gerere.* 12. *Gravis.* 13. *Bellum.* 14. *Cum.* 15. *Hispanus.* 16. *Ipse.*

2. While[1] these[2] things were carried[3] on in Asia, all[14] Greece[4] had rushed[5] to[6] arms,[7] in the hope[8] of regaining[9] liberty,[10] following[11] the authority[12] of the Lacedemonians.[13]

1. *Dum.* 2. *Hic.* 3. *Gerere.* 4. *Græcia.* 5. *Concurrere.* 6. *Ad.* 7. *Arma.* 8. *Spes.* 9. *Recuperare.* 10. *Libertas.* 11. *Sequi* (perfect participle). 12. *Auctoritas.* 13. *Lacedæmonius.* 14. *Omnis.*

3. When[1] Regulus had come[2] to Rome,[3] he set[4] forth his instructions[5] in the Senate;[6] but[7] he said[8] it was[9] not[8] expedient[10] for the captives[11] to be restored;[12] for

that they [13] were young [14] men and good [15] leaders, [16] that he (Regulus) was enfeebled [17] by age. [18]

1. *Cum.* 2. *Venire.* 3. *Roma.* 4. *Exponere.* 5. *Mandatum.* 6. *Senatus.* 7. *Sed.* 8. To say not, *negare.* 9. *Esse.* 10. *Utilis.* 11. *Captivus.* 12. *Reddere.* 13. *Ille.* 14. *Adolescens.* 15. *Bonus.* 16. *Dux.* 17. *Confectus.* 18. *Senectus.*

XIII.

1. Phormio the [1] Peripatetic, [2] when [3] Hannibal, [4] expelled [5] from Carthage, [6] had come [7] to Ephesus, [8] is said [9] to have talked [10] some [11] hours [12] about [13] the duty [14] of a commander. [15]

1. *Ille.* 2. *Peripateticus.* 3. *Cum.* 4. *Hannibal, -balis.* 5. *Expellere.* 6. *Karthago, -aginis.* 7. *Venire.* 8. *Ephesus, -esi.* 9. *Dicere.* 10. *Loqui.* 11. *Aliquot.* 12. *Hora, -ræ.* 13. *De.* 14. *Officium.* 15. *Imperator.*

2. Then, [1] when the [2] rest who had heard [3] him were greatly [4] charmed, [5] they inquired [6] of [7] Hannibal what he [8] thought [9] of [10] that [11] philosopher. [12] Hannibal is said [13] to have answered, [14] that he had often [15] seen [16] many [17] crazy [18] old [19] men, [but [20]] nobody [21] who [22] was more [23] crazy [24] than Phormio.

1. *Tum.* 2. *Cæteri.* 3. *Audire.* 4. *Vehementer.* 5. *Delectare.* 6. *Quærere.* 7. *Ab.* 8. *Ipse.* 9. *Judicare.* 10. *De.* 11. *Ille.* 12. *Philosophus.* 13. *Ferre.* 14. *Respondere.* 15. *Sæpe.* 16. *Videre.* 17. *Multus.* 18. *Delirus, -a, -um.* 19. *Senex.* 20. Omit. 21. *Nemo.* 22. *Qui.* 23. *Magis.* 24. I am crazy (by the verb) *deliro, delirare.*

XIV.

1. If the Gauls [1] had attacked [2] the town [3] that night, [4] they would have taken [5] it easily, [6] since [7] no one supposed [8] that an enemy [9] was-at-hand. [10]

1. *Gallus.* 2. *Oppugno.* 3. *Oppidum.* 4. *Nox.* 5. *Capio.* 6. *Facile.* 7. *Quum.* 8. *Puto.* 9. *Hostis.* 10. *Adsum.*

2. For three-days,[1] however,[2] they waited[3] to see[4] what the consul would do,[5] who was himself enrolling-troops[6] at Ariminum, and had ordered[7] Nero to cross[8] the Po,[9] and hinder[10] the enemy from ravaging[11] the country.[12]

1. *Triduum.* 2. *Tamen.* 3. *Exspecto.* 4. Omit. 5. *Facio.* 6. *Delectum habeo.* 7. *Impero.* 8. *Transeo.* 9. *Padus.* 10. *Prohibeo.* 11. *Populor.* 12. *Ager.*

3. After[1] the leader[2] of the Gauls saw[3] that the Romans would-not[4] risk[5] a battle,[6] he repented[7] of his own inactivity,[8] for[9] he remembered[10] the counsels[11] of his father,[12] who had feared[13] that his son[14] would not be bold[15] enough,[16] and had warned[17] him not to lose[18] a single day.

1. *Posteaquam.* 2. *Dux.* 3. *Video.* 4. *Nolo.* 5. *Committo.* 6. *Prœlium.* 7. *Pœnitet.* 8. *Inertia.* 9. *Enim.* 10. *Memini.* 11. *Consilium.* 12. *Pater.* 13. *Metuo.* 14. *Filius.* 15. *Audax.* 16. *Satis.* 17. *Moneo.* 18. *Amitto.*

XV.

1. The next[1] day[2] I was summoned[3] by Pansa to Bononia.[4] When[5] I was on[6] the way,[7] it was announced[8] to me that he was dead.[9]

1. *Posterus.* 2. *Dies.* 3. *Arcesso.* 4. First declension. 5. *Cum.* 6. *In.* 7. *Iter.* 8. *Nuntio.* 9. *Morior.*

2. You (plural) seem[1] to me not[2] even[3] to-day[4] to know[5] what[6] a crime[7] you have dared[8] against[9] me.

1. *Videor.* 2. *Ne.* 3. *Quidem.* 4. *Hodie.* 5. *Scio.* 6. Interrogative. 7. *Facinus.* 8. *Audeo.* 9. *In.*

3. This man, if[1] he had been blessed[2] with a longer[3] life,[4] would have been much[5] more illustrious[6] than his brother,[7] in peace[8] and in war.[9]

1. *Si.* 2. *Contingo*; literally, "if a longer life had fallen to him." 3. *Longus.* 4. *Ætas.* 5. *Multus.* 6. *Clarus.* 7. *Frater.* 8. With *domus.* 9. *Militia.*

4. The consul, afraid[1] of being surrounded,[2] sent[3] cavalry[4] to take[5] possession of the hills.[6]

1. *Vereor* (perfect participle). 2. *Circumvenio.* 3. *Præmitto.* 4. *Eques.* 5. *Occupo.* By what constructions may the purpose be given? 6. *Collis.*

XVI.

1. Marcellus, with[1] a small[2] body[3] of horse,[4] fought[5] [the enemy[6]] and killed[7] the king[8] of the Gauls,[9] Viridomarus by name,[10] with his[11] own hand.[3]

1. *Cum.* 2. *Parvus.* 3. *Manus.* 4. *Eques* (plural). 5. *Dimicare.* 6. Omit. 7. *Occidere.* 8. *Rex.* 9. *Gallus.* 10. *Nomen.* 11. *Suus.*

2. In the ninth[1] year[2] after[3] the banishment[4] of the kings,[5] when[6] the son-in-law[7] of Tarquinius had collected[8] a huge[9] army[10] to[11] avenge the wrong[12] done[13] his father-in-law,[14] a new[15] office[16] was created[17] at Rome.

1. *Nonus.* 2. *Annus.* 3. *Post.* 4. *Exactus* (literally, after the kings expelled). 5. *Rex.* 6. *Cum.* 7. *Gener.* 8. *Colligere.* 9. *Ingens.* 10. *Exercitus.* 11. *Ad* with gerundive of *vindicare.* 12. *Injuria.* 13. Simply the objective genitive: literally, "wrong of his." 14. *Socer.* 15. *Novus.* 16. *Dignitas.* 17. *Creare.*

3. At[1] present I will merely[2] ask[3] this,[4] whether[5] this branch-of-literature[6] is deservedly[7] suspected[8] by[9] you.

1. *Nunc.* 2. *Tantum.* 3. *Quærere.* 4. *Illud.* 5. *Ne* (enclitic). 6. *Genus scribendi.* 7. *Merito.* 8. *Suspectus.* 9. Dative.

XVII.

1. The ninth[1] year[2] after[3] the expulsion[4] of the kings,[5] when[6] the son-in-law[7] of Tarquin[8] had[9] collected an immense[10] army,[11] a new[12] dignity[13] was[14] created at Rome, which is[15] called the dictatorship,[16] — greater[17] than the consulship.[18]

1. *Nonus.* 2. *Annus.* 3. *Post.* 4. Literally, "kings expelled": *exigo.* 5. *Rex.* 6. *Cum.* 7. *Gener.* 8. *Tarquinius.* 9. *Colligo.* 10. *Ingens.* 11. *Exercitus.* 12. *Novus.* 13. *Dignitas.* 14. *Creo.* 15. *Appello.* 16. *Dictatura.* 17. *Magnus.* 18. *Consulatus.*

2. Do you suppose[1] that men[2] who are[3] said to[4] predict-the-future can[5] tell-you[6] whether[7] the[8] moon uses[9] her[10] own light[11] or[12] that[7] of the sun?[13]

1. *Censeo.* 2. *Is:* literally, "those." 3. *Dico.* 4. Predict-the-future: *divino.* 5. *Possum.* 6. Tell-you: *respondeo.* 7. Omit. 8. *Luna.* 9. *Utor.* 10. *Suus.* 11. *Lumen.* 12. *An.* 13. *Sol.*

3. It was a glorious[1] sentiment[2] and worthy[3] of being uttered[4] by that[5] great man.[6]

1. *Præclarus.* 2. *Vox.* 3. *Dignus.* 4. *Emitto.* 5. That great: *tantus.* 6. *Vir.*

XVIII.

1. When[1] Balbus had[2] said this,[3] then[4] Cotta said, with-a-smile,[5] "You are[6] late, Balbus, in telling me what to defend;[7] for[8] while[9] you were discussing[10] I was myself pondering[11] what to say in[12] reply, and[13] not so-much[14]

for-the-purpose-of [15] refuting [16] you as of finding-out [17] the-things [18] which I did not [19] understand." [20]

1. *Cum.* 2. *Dico.* 3. Relative. 4. *Tum.* 5. *Arrideo* (present participle). 6. I am late in telling, *sero præcipio.* 7. *Defendo.* 8. *Enim.* 9. Ablative absolute. 10. *Disputo.* 11. *Mecum meditor.* 12. In reply, *contra.* 13. *Neque.* 14. So much — as, *tam — quam.* 15. *Causa.* 16. *Refello.* 17. *Requiro.* 18. With *is.* 19. *Minus.* 20. *Intelligo.*

XIX.

1. I do not care [1] how [2] rich [3] Gyges is.[4]

1. Express with *refert.* 2. *Quam.* 3. *Dives.* 4. *Esse.*

2. Who [1] more [2] illustrious in Greece [3] than [4] Themistocles? who [5] when [6] he had [7] been driven into exile [8] did [9] not do harm to his thankless [10] country,[11] but did [12] the same [13] that Coriolanus had [12] done twenty [14] years [15] before.[16]

1. *Quis.* 2. *Clarus.* 3. *Græcia.* 4. Write in two ways. 5. *Qui.* 6. *Cum.* 7. *Expellere.* 8. *Exilium.* 9. Do harm to, *Injuriam ferre* with dative. 10. *Ingratus.* 11. *Patria.* 12. *Facere.* 13. *Idem.* 14. *Viginti.* 15. *Annus.* 16. *Ante.*

3. In the first [1] of the spring [2] the consul came [3] to Ephesus, and, having [4] received the troops [5] from [6] Scipio, he held [7] a speech [8] in-presence-of [9] the soldiers,[10] in [11] which, after [12] extolling their bravery,[13] he exhorted [14] them to [15] undertake a new [16] war [17] with [18] the Gauls, who had [19] [as he said [11]] helped Antiochus with [11] auxiliaries.[20]

1. *Primus.* 2. *Ver.* 3. *Venire.* 4. *Accipere.* 5. *Copiæ.* 6. *A.* 7. *Habere.* 8. *Contio.* 9. *Apud.* 10. *Miles.* 11. Omit. 12. *Collaudare* (ablative absolute). 13. *Virtus.* 14. *Adhortari.* 15. *Suscipere* with *ad* and gerundive. 16. *Novus.* 17. *Bellum.* 18. *Cum.* 19. *Juvare.* 20. *Auxilium.*

XX.

1. The plays[1] of Livius are not worth[2] reading[3] more-than-once.[4]

1. *Fabula.* 2. *Dignus.* 3. *Legere.* 4. *Iterum.*

2. What[1] style-of-speaking[2] was[3] in vogue in those[4] times[5] can[6] best[7] be[8] learned from[9] the works[10] of Thucydides.[11]

1. *Qui.* 2. *Dicendi genus.* 3. *Vigere.* 4. *Ille.* 5. *Tempus.* 6. *Posse.* 7. *Maxime.* 8. *Intelligere.* 9. *Ex.* 10. *Scriptum.* 11. *Thucydides* (genitive *-di*).

3. When[1] I had[2] been engaged a-couple-of-years[3] in[4] law[5] cases, and my name[6] was very-well-known[7] in the forum, I went[8] away from Rome. When[1] I had[9] come to Athens,[10] I stayed[11] six months[12] with[13] Antiochus, and renewed[14] the study[15] of philosophy[16] under[17] this teacher.[18]

1. *Cum.* 2. *Versari.* 3. *Biennium.* 4. *In.* 5. *Causa.* 6. *Nomen.* 7. *Jam celebratum.* 8. *Proficisci.* 9. *Venire.* 10. *Athenæ.* 11. *Esse.* 12. *Mensis.* 13. *Cum.* 14. *Renovare.* 15. *Studium.* 16. *Philosophia.* 17. Omit. 18. *Doctor* (ablative absolute).

XXI.

1. When[1] Paullus, to whom the war[2] with[3] Perses[4] had-been-allotted,[5] had[6] gone home,[7] that[8] very[9] day[10] he noticed[11] that his little[12] daughter Tertia was low-spirited.[13]

1. *Cum.* 2. *Bellum.* 3. *Cum.* 4. *Perses* (genitive *æ*). 5. To be allotted, *obtingere* (active). 6. *Redire.* 7. *Domus.* 8. *Is.* 9. *Ipse.* 10. *Dies.* 11. *Animadvertere.* 12. Diminutive of *filia.* 13. *Tristiculus.*

2. "What[1] is the matter,"[2] said[3] he, "my Tertia?" "Why[1] are you sad?"[4] "My father,"[5] said she, "Persa is[6] dead."

LATIN COMPOSITION. 79

1. *Quid.* 2. Omit. 3. *Inquit.* 4. *Tristis.* 5. *Pater.*
6. *Perire.*

3. Then[1] the-father[2] embraced[3] the girl[4] tenderly[5] and said, "I[6] accept the omen,[7] my daughter."
Now[8] this[9] Persa was a puppy,[10] which had[11] died.

1. *Tum.* 2. The father, *ille.* 3. *Complecti.* 4. *Puella.*
5. Comparative of adverb *arte.* 6. *Accipere.* 7. *Omen.*
8. *Autem.* 9. *Is.* 10. *Catellus.* 11. *Mori.*

XXII.

1. Plato, when[1] he was[2] provoked with a slave[3] of[4] his, bade[5] him doff[6] his tunic[7] forthwith[8] and hold[9] out his shoulders[10] to the scourge,[11] intending[12] to beat him himself[13] with his own hand.[14]

1. *Cum.* 2. *Irasci* with dative. 3. *Servus.* 4. Of his = *suus.* 5. *Jubere.* 6. *Ponere.* 7. *Tunica.* 8. *Statim.*
9. Hold out = *præbere.* 10. *Scapulæ, -arum.* 11. *Verber, -is,* plural. 12. Future participle of *cædere.* 13. *Ipse.*
14. *Manus.*

2. When[1] he was-aware[2] that he was provoked, he kept[3] his hand suspended,[4] just-as[5] he had raised[6] it, and stood[7] like[8] one[9] about to strike.[10]

1. *Postquam.* 2. *Intellegere.* 3. *Detinere.* 4. *Suspendere.* 5. *Sicut.* 6. *Tollere.* 7. *Stare.* 8. *Similis.*
9. Omit. 10. *Cædere.*

3. Being-asked[1] then[2] by a friend[3] who had happened[4] in what[5] he was-about:[6] "I am exacting[7] penalty,"[8] said he, "from[9] a passionate[10] man."[11]

1. *Interrogare.* 2. *Deinde.* 3. *Amicus.* 4. Happened in = *forte intervenire.* 5. *Quis.* 6. *Agere.* 7. *Exigere.*
8. *Pœna,* plural. 9. *Ab.* 10. *Iracundus.* 11. *Homo.*

XXIII.

1. While[1] this[2] was[3] done at Veii,[4] meantime[5] the citadel[6] at Rome was in great[7] danger.[8]

1. *Dum.* 2. *Hic* (neuter plural). 3. *Agere.* 4. *Veii, Veiorum.* 5. *Interim.* 6. *Arx.* 7. *Ingens.* 8. *Periculum.*

2. For[1] the Gauls,[2] having[3] observed a human[4] track,[5] climbed-up[6] to the top[7] in a glimmering[8] night[9] in such[10] silence[11] that[12] they not[13] only escaped-the-notice-of[14] the guards,[15] but[16] did not-even[17] rouse[18] the dogs,[19] — a creature[20] on-the-alert[21] for[22] noises[23] at night.[24]

1. *Namque.* 2. *Gallus.* 3. *Notare* (ablative absolute). 4. *Humanus.* 5. *Vestigium.* 6. *Evadere.* 7. *Summus,* neuter. 8. *Sublustris.* 9. *Nox.* 10. *Tantus.* 11. *Silentium.* 12. *Ut.* 13. *Non solum.* 14. *Fallere.* 15. *Custos.* 16. *Sed.* 17. *Ne — quidem.* 18. *Excitare.* 19. *Canis.* 20. *Animal.* 21. *Sollicitus.* 22. *Ad.* 23. *Strepitus.* 24. *Nocturnus.*

XXIV.

1. Death[1] alone[2] confesses[3] how puny[4] are the bodies[5] of men.[6]

1. *Mors.* 2. *Solus.* 3. *Fateor.* 4. *Quantulus.* 5. *Corpusculum.* 6. *Homo.*

2. There[1] is nothing[2] better[3] than agriculture,[4] nothing sweeter,[5] nothing worthier[6] of a free[7] man.

1. Omit. 2. *Nihil.* 3. *Bonus.* 4. *Agricultura.* 5. *Dulcis.* 6. *Dignus.* 7. *Liber.*

3. When[1] Livius Salinator was[2] going out of the city[3] to[4] carry on war[5] against[6] Hasdrubal, Fabius advising[7] him to ascertain[8] the strength[9] of the enemy[10] first,[11] he

answered [12] that he would not let [13] a chance [14] for fighting [15] pass.[13]

1. *Cum.* 2. *Egredi* (with ablative). 3. *Urbs.* 4. *Gerere.* 5. *Bellum.* 6. *Adversus.* 7. *Moneo*, ablative absolute. 8. *Agnoscere.* 9. *Vis* (plural). 10. *Hostis.* 11. *Prius.* 12. *Respondeo.* 13. *Omitto.* 14. *Occasio.* 15. *Pugno* (genitive of gerund).

XXV.

1. During[1] these events,[2] horsemen[3] had been sent[4] to Alba, to[5] transport[6] the populace[7] to Rome. Then legions[8] were brought[9] for-the-purpose[2] of destroying[10] the city.

2. When these[5] entered[11] the gates,[12] there was not that commotion[13] such[14] as is apt[15] to belong-to[16] captured[17] cities, when, on-the-capture[18] of the citadel[19] by force,[20] the rush[21] of armed[22] men[2] through the city confounds[23] all things;

3. but a sad[24] silence[25] so enchained[26] the minds[27] of all, that, forgetting[28] what to leave,[29] what to take[30] with them, they stood[31] on the thresholds,[32] or wandered[33] through their homes.[34]

1. *Inter.* 2. Omit. 3. *Eques.* 4. *Mitto.* 5. Express by a relative clause. 6. *Traduco.* 7. *Multitudo.* 8. *Legio.* 9. *Duco.* 10. *Diruo.* 11. *Intro.* 12. *Porta.* 13. *Tumultus.* 14. *Qualis.* 15. *Soleo.* 16. Expressed by the case of "cities." 17. *Capio.* 18. Express by a passive verb. 19. *Arx.* 20. *Vis.* 21. *Cursus.* 22. *Armo.* 23. *Misceo.* 24. *Tristis.* 25. *Silentium.* 26. *Defigo.* 27. *Animus.* 28. *Obliviscor.* 29. *Relinquo.* 30. *Fero.* 31. *Sto.* 32. *Limen.* 33. *Pervagor.* 34. *Domus.*

XXVI.

1. Whenever[1] the spring[2] had-set-in,[3] Verres devoted[4] himself to journeyings,[5] in which he showed[6] himself so-very[7] energetic[8] that nobody[9] ever[10] saw[11] him sitting[12] on[13] a horse.[14]

1. *Cum.* 2. *Ver.* 3. *Cœpit esse.* 4. *Do.* 5. *Iter.* 6. *Præbeo.* 7. *Usque eo.* 8. *Impiger.* 9. *Nemo.* 10. *Unquam.* 11. *Video.* 12. *Sedeo.* 13. *In.* 14. *Equus.*

2. For he used to ride[1] in a sedan and eight, in which there was a cushion[2] stuffed[3] with rose-leaves.[4] Moreover,[5] he had[6] one[7] garland[8] on his[9] head,[10] another[11] on his[9] neck,[12] and ever-and-anon[13] he gave[14] his nose a little-net[15] of the finest[16] of thread,[17] with tiny[18] meshes,[19] full[20] of rose-leaves.

1. To ride in a sedan and eight, *Lectica octophoro ferri.* 2. *Pulvinus.* 3. *Farcio.* 4. *Rosa* (singular). 5. *Autem.* 6. *Habeo.* 7. *Unus.* 8. *Corona.* 9. Omit. 10. *Caput.* 11. *Alter.* 12. *Collum.* 13. *Identidem.* 14. *Ad nares sibi admovere.* 15. *Reticulum.* 16. *Tenuis.* 17. *Linum.* 18. *Minutus.* 19. *Macula.* 20. *Plenus.*

XXVII.

1. Nasica when[1] he had come to Ennius's[2] and the girl[3] had told him that Ennius was not at home, was aware[4] that she had said it at her master's[5] order,[6] and that he was at home. A few[7] days after when[1] Ennius had come to[2] Nasica's, Nasica cries out[8] that he is not at home. Then[9] Ennius: "What! don't I know[10] your voice?"[11] Hereupon[12] Nasica: "You are a shameless[13] fellow;[14] I believed[15] your girl, don't you believe me?"

1. *Cum.* 2. Come to Ennius's, *venire ad Ennium.* 3. *Ancilla.* 4. *Sentio.* 5. *Dominus.* 6. *Jussu* (ablative).

7. *Paucus.* 8. *Exclamo.* 9. *Tum.* 10. *Cognosco.* 11. *Vox.* 12. *Hic.* 13. *Inpudens.* 14. *Homo.* 15. *Credo.*

2. It was more[1] important[2] for the Athenians to have solid[3] roofs[4] on[5] their[6] houses[7] than the loveliest[8] ivory[9] statue[10] of Minerva. Still[11] I would rather be Phidias[12] than the best possible[13] carpenter.[14]

1. *Plus.* 2. It is important, *interest.* 3. *Firmus.* 4. *Tectum.* 5. *In.* 6. Omit. 7. *Domicilium.* 8. *Pulcher* (superlative). 9. "Of ivory," *ex* and *ebur.* 10. *Signum.* 11. *Tamen.* 12. *Phidias, Phidiæ.* 13. *Vel* with superlative of *bonus.* 14. *Faber tignarius.*

XXVIII.

1. At the same[1] time[2] King Attalus, having gone[3] from Thebes[4] to Pergamus, dies[5] in his seventy-second year,[6] after[7] reigning[8] four-and-forty years. To this man fortune[9] had given no claim[10] but[11] wealth[12] toward[13] the hope[14] of the throne.[15]

1. *Idem.* 2. *Tempus.* 3. *Proficiscor.* 4. *Thebæ, Thebarum.* 5. *Morior.* 6. *Annus.* 7. *Cum* (literally, "when he had reigned"). 8. *Regno.* 9. *Fortuna.* 10. No claim, *nihil.* 11. *Præter.* 12. *Divitiæ.* 13. *Ad.* 14. *Spes.* 15. *Regnum.*

2. By using[1] this[2] at once[3] economically[4] and[5] in princely style[5] he brought it to pass[6] that he seemed[7] not unworthy[8] of the throne. Then,[9] after the Gauls were conquered[10] in a single[11] battle,[12] he assumed[13] the name[14] of King.[15]

1. *Utor.* 2. Refers to *divitiæ.* 3. At once .. and, *simul .. simul.* 4. *Prudenter.* 5. In princely style, *magnifice.* 6. Bring it to pass, *efficio.* 7. *Videor.* 8. *Indig-*

nus. 9. *Deinde.* 10. *Vinco.* 11. *Unus.* 12. *Prœlium.* 13. *Adscisco.* 14. *Nomen.* 15. *Regius, -a, -um.*

3. He ruled[1] his subjects[2] with perfect[3] justice,[4] he showed[5] unparalleled[6] fidelity[7] to his allies,[8] he was courteous[9] to wife[10] and children,[11] — four he left[12] surviving,[13] — gentle[14] and generous[15] to friends.[16]

1. *Rego.* 2. *Suus, -a, -um.* 3. *Summa.* 4. *Justitia.* 5. *Præsto.* 6. *Unicus.* 7. *Fides.* 8. *Socius.* 9. *Comis.* 10. *Uxor.* 11. *Liberi.* 12. *Relinquo.* 13. *Superstes, -stitis.* 14. *Mitis.* 15. *Munificus.* 16. *Amicus.*

LATIN GRAMMAR.

I.

1. DECLINE *carcer, deus, arcus, dies,* giving the gender of each, with the rule for it, and marking the quantities of penultimate and final syllables in all the cases.

2. Give the gender of *via, gladius, Tiberis,* with the rule for each. Give Ablative singular of *sedile, turris;* Genitive plural of *vir, pater, hostis, equa.*

3. Decline *alter, alacer, iste.* Compare *gracilis, inferus, ingens, malus;* compare *prope,* and the adverbs formed from *acer, altus.*

4. Give principal parts of *pono, sedeo, domo, vincio.* Give Future Active Participle and Future Passive Participle of *pono,* and Pluperfect Active Second Person Plural of *sedeo,* marking the quantities of all the syllables of both verbs. Inflect the Present Indicative of *eo;* of *nolo.*

5. Name some classes of verbs followed by the Genitive, by the Dative, by the Ablative, by two Accusatives. Give some of the rules for the Subjunctive after Relative Pronouns; for its use after Particles. How is *not* expressed with the Imperative? How is a Wish expressed?

II.

1. Before what vowels have *g* and *c* a soft sound? What is the gender of *Januarius?* of *Corinthus?* of *Aquilo?* Give the rule for each. What is an Epicene Noun? What words are naturally neuter? What is Declension? What are some of the general rules for Declension? Which apply to all nouns?

2. Decline *dea, Penelope, vir, vis, barbiton, sedile*. When does the Nominative plural of the third declension end in *-ia?* Give the three general rules for gender in nouns of the third declension. What is the gender of *tellus, legio, arundo, amnis?* Give the rule for each. Decline *domus, bos, Vergilius.* What are the Heterogeneous Nouns, and Heteroclites? Give some examples of each. Give the rule for the derivation of Patronymics; of Diminutives. What do the terminations *-ium, -arium, -ile,* in nouns denote?

3. Decline *alius, quisquam, tu.* Give the rules for comparing adverbs. What is a Gerund? a Gerundive? a Supine? a Participle? Give the synopsis of *possum* in the Third Person Singular throughout the verb. Inflect the Imperatives, Active and Passive, of *moneo, amo, capio, audio.* What are Irregular Verbs? Give the list of them. What compounds of *facio* have *fio* in the passive? How do you form Frequentative Verbs? how Inceptives? how Intensives? How are adverbs formed from adjectives?

4. What does *ultimus* mean? What do *hic* and *ille* mean when used together? Explain all the uses of *suus* which you know. When is the Nominative of the Third Person wanting? Translate in two ways, "A woman of remarkable beauty" (*femina, maximus, pulchritudo*). Explain the Genitive,— *pridie ejus diei.* How do you translate the name of a town to which motion proceeds? How from which? How the name of a town where an event occurs? How in each case if the name of the place is not the name of a town?

5. Tell all the ways in which a voluntary agent can be translated. Translate, "We pity (*miseret*) them." When is the Passive Voice followed by the Accusative of the

thing? What is Synecdoche? How do you translate expressions denoting time how long, and time at which something happens? What cases follow *potior, fido, doceo, peto, juvat, voco?* What two different Ablative constructions may follow a comparative? In what senses does *ut* take the Subjunctive? in what the Indicative? What construction follows *priusquam, quin, cum?* Mention four cases of a Subjunctive after *qui.*

6. Tell all the ways you know of translating a clause denoting a purpose into Latin. When is the Infinitive used without a subject? State the use of the Genitive, Dative, Accusative, and Ablative of Gerunds. What is the general order of words in a Latin sentence?

7. Give the rules for Increment in nouns and verbs. When do two consonants lengthen the preceding vowel? Give the general rules for the quantity of final syllables.

III.

1. Decline *honos.* What is its gender? Why? Is this gender natural or grammatical, and what is the difference between these two classes of genders? Mention some classes of nouns which are masculine from their signification. Some which are feminine. How do neuter nouns of the third declension end? Decline any one you think of. How do you distinguish the declensions of nouns?

2. Decline *duo.* Decline *levior.* Of what degree of comparison is it? Give the other degrees of comparison of the same word. Compare *magnus.* Mention other adjectives which are irregular in their comparison.

3. Decline *ipse.* Give all genders of the Nominative singular of *quis.* Of the Interrogative *qui.* Decline *siquis.*

4. What is an Irregular Verb? Give a synopsis of the

verb *esse*. Write out the Present tense of this verb in all modes and persons. How do you distinguish the conjugations? What are the principal stems of verbs? Give the terminations of the First Person Indicative of a verb of the third conjugation in all the tenses. What is a Frequentative Verb?

5. What is the Increment of a verb? What is the quantity of verbal increments? What is the general rule for the quantity of the increments of nouns?

IV.

1. Give the three general rules for the gender of nouns of the third declension. Gender and rule for *Boreas, manus, res, virtus*.

2. Decline the following nouns, marking the quantities of the penultimate and final syllables in all the cases: *imago, domus, poema, respublica, juvenis*.

3. Decline *tu, uterque, aliquis, brevior*. Compare *clemens, par, diu*. What does the termination *-ile* in nouns denote? *-lentus* in adjectives?

4. Give principal parts of *juvo, resisto, spondeo, haurio*, marking the quantities of all the syllables in all the forms. Inflect the Perfect Active Indicative of *resisto*, Imperative Active and Passive of *haurio*. Give the synopsis of *fio* in the present stem.

5. Give several cases of nouns which follow the verb *sum*, and the rules for them. What cases follow *utor, recordor, parco, pœnitet*? What cases do the prepositions *super, præ, inter*, govern respectively? How is a Purpose expressed?

V.

1. Decline the following nouns, marking the quantity of

the penultimate and final syllables through all the cases: *ala, genius, pars, conclave, acus, acies.* Give the gender of each noun, with rule for it.

2. Decline the adjectives *acer* and *facilis;* the pronouns *quidam* and *uter.* Compare *acer, facilis, felix, malus.* Form adverbs from *pulcher* and *prudens*, and compare them. How do you express in Latin *five, fifth,* and *five times?*

3. Give the principal parts of the following verbs; marking the quantity of all the syllables: *sto, torqueo, cado, cœdo, cedo, ordior.* Inflect the Perfect Subjunctive Active of *cedo*, and the Present Imperative Passive of *ordior*, marking the quantities throughout.

4. What case or cases follow *pudet, fungor, præsum, doceo?* Translate into Latin: 1. He asked (*rogo*) him whether (*num*) Caius had come (*venio*). 2. He said (*dico*) that Caius would come. 3. He orders (*impero*) Caius to come. 4. He was hindered (*impedio*) by Caius from (*quominus*) coming. 5. He was waiting (*opperior*) until (*dum*) Caius should come. 6. No one (*nemo*) waited who was able (*possum*) to come. 7. If he had waited, I should not have come. 8. Would that (*utinam*) Caius would come. 9. Do not come, Caius.

VI.

1. Decline *virtus, domus, puer, calcar*, giving the gender of each with the rule for it, and marking the quantity of the penultimate and final syllables in all the cases.

2. Give the gender of *juvenis, canon, ratio, flos*, with the rule for each. Give the Ablative singular of *Anchises, aper, tribus;* Genitive plural of *nubes, respublica, mater.*

3. Decline *piger, gravior, ambo, quisque.* Compare *frugi, humilis,* and the adverbs formed from *acer, durus.*

4. Give the principal parts of *veto, lacesso, pendo, pendeo, sepelio, mentior*. Mark the quantity of all the syllables of the verbal forms *adjuvare* (from *adjuvo*), *tetenderitis* (from *tendo*), and give all the voices, moods, tenses, numbers, and persons in which they may be found. Inflect the Future Perfect Indicative Passive of *moneo*, marking the quantity of all the syllables.

5. Translate into Latin in as many ways as you are able: 1. He sent (*mitto*) men to seek (*peto*) an oracle (*oraculum*). 2. He heard (*audio*) that Caius had fled (*fugio*). 3. He feared (*timeo*) that Caius had fled. 4. He was angry (*irascor*) that Caius had fled. What cases follow the prepositions *præ, sub, inter*, respectively? Give the rules which you remember for the Dative after verbs.

VII.

1. Decline *filia, vesper, navis, nemus, domus*; mark the quantity of the penultimate and final syllables through all the cases; give the gender of each noun, with the rule. Give the rules for the formation of the Genitive plural of the third declension. How are the Diminutives formed from nouns?

2. Decline *crudelis, unus, duo, idem, aliquis*; compare *crudelis, facilis, superus, vetus*. What are the meanings respectively of the terminations *-osus* (*e.g. vinosus*), *-ilis* (*e.g. mobilis*), *-ax* (*e.g. fallax*)? Give the Latin for *a hundred, two hundred*, and so on to *nine hundred* inclusive. Mark the quantity of the penultimate and final syllables in all the Latin words given in this section.

3. Give the principal parts of *juvo, veho, sentio, censeo, cædo, audeo*. Inflect the Present Subjunctive Passive of *juvo*; the Perfect Subjunctive Active of *veho*; the Future

Indicative Passive of *cædo;* the Imperfect Subjunctive Active of *volo.* Mark the quantity of the penultimate and final syllables.

4. What case or cases respectively follow the verbs *vendo, dono, pœnitet, rogo, solvo, condemno?* Translate into Latin: He orders (*impero*) Caius to be present (*adsum*). He feared (*metuo*) that Caius was not present. He sent (*mitto*) Caius to be present. He was angry (*irascor*) because (*quod*) Caius was present. He is happy (*beatus*) provided (*dummodo*) Caius is present. He did not know (*nescio*) on what day (*dies*) Caius was present. Where may the cæsural pause occur in the dactylic hexameter?

VIII.

1. Decline the following nouns, giving the gender of each with the rule, and marking the quantity of the penultimate and final syllables in all the cases: *vir, Boreas, imago, murmur, fides, rus, portus.*

2. What is denoted by the terminations -*mentum* (e. g. *documentum* from *doceo*), -*or* (e. g. *fautor* from *faveo*), -*idus* (e. g. *calidus* from *caleo*)? Compare *dexter, frugi, sacer, juvenis, merito,* and the adverbs from *alacer* and *æger.*

3. Decline *integer, alius, dispar, plus,* marking the quantity as in section one. Decline *iste, meus, quidam.*

4. Mark the quantity of all the syllables of the verbal forms in this section (4). Give the principal parts of the following verbs: *ambio, sto, maneo, arcesso.* Give a synopsis of *fio* in the present stem. Inflect the Future Perfect Indicative, Active and Passive, of *cædo.* In what places can *capere* be found? In what places *venimus,* and how distinguished by difference of quantity?

5. By what cases may *sum* be followed? Translate: He

knew (*scio*) that Caius was coming (*venio*). He begged (*oro*) Caius to come. He feared (*timeo*) that Caius would not come. He sent (*mitto*) men to hinder (*obsto*) Caius from coming. Explain the use of *ille, is, hic, iste, ipse, sui*. Give some of the rules for the case of a noun referring to the same person or thing as a preceding noun.

IX.

1. Decline *Annius, radix, fons, flos, exemplar, manus.* Give the gender of each, with the rule. 'Give the rules for the formation of the Ablative singular of the third declension.' What are the meanings of the endings *-mentum* (e. g. *impedimenta*), *-bulum* (e. g. *pabulum*), *-tor* (e. g. *doctor*), *-etum* (e. g. *rosetum*) ?

2. Decline *alacer, supplex, iste, qualis, unusquisque.* Compare *æger, suavis, dives.* What is the Latin for *four, forty, four hundred, fourth, fortieth, four hundredth*? What is the significance of the ending *-ax* (e. g. *ferax*)? *-cundus* (e. g. *verecundus*) ?

3. Give the principal parts of *cupio, cubo, tego, foveo, vincio, veho*. Inflect the Future Indicative of *redeo* and *morior;* and the Present Subjunctive of *suspicor* and *malo*.

4. Give all the rules for the construction of names of towns. What classes of verbs in Latin are constructed with the Genitive case? What classes with the Ablative? What is the difference of meaning between the Imperfect and Pluperfect tenses of the Subjunctive in Conditional Sentences?[1] How are clauses in English introduced by *that* to be translated into Latin? What is the difference between *ne* and *ut non*? Write down the following words in four columns, and mark the quantity of every syllable: *fieri, arbores, habere, desinit, fiebat, venerunt, eveho, laborat,*

improbus, dederint, perbrevis, diei, victrices, congredi, nomen, dedecori, cupidine, auditur, non, abstulerunt, peritus, requireres, dirutus, maritimus.

X.

1. Decline *locus, sol, vis, mare, motus.* Give the gender of each with the rule. What classes of nouns of the third declension form their Genitive plural in *-ium.*

2. Decline *uter.* Give the Ablative singular and Genitive plural of *celeber, crudelis, supplex.* Compare *carus, humilis, parvus.* Form adverbs from *æger* and *crudelis*, and compare them. Give, in Latin, the multiples of ten from twenty to one hundred inclusive. Decline *aliquis.*

3. Give the principal parts of *verto, veto, gaudeo, vincio, vinco.* Inflect the singular of the Present Subjunctive Active of *verto* and *veto*; of the Future Indicative Passive of *vincio*; and of the Imperfect Subjunctive of *eo.*

4. What Latin prepositions are followed by the Ablative case? By what case are *in* and *sub* followed? With what case or cases are the following verbs respectively constructed: *impero, pudet, doceo, obliviscor, ignosco?* What do *utinam adsit* and *utinam adesset* respectively mean? Give the rules for the Subjunctive mood in the following sentences: 1. Nemo est qui te non metuat. 2. Fortis est qui te non metuat. 3. Dicit adesse hominem qui te non metuat. Write out the following words, and mark the quantity of all the syllables: *transituros, sustulit, oceanus, congredi, virorum, reducit, tradiderint, mare, Cæsare, ruina, humilis, victrices, acceperas, hostilis, ratus, nemini, tenebris, reliquæ, nomina, requiris, graviora, distrahit, antiquus, mentitur.*

XI.

1. Decline *triumvir, crinis, dies, cubile, imago, domus.*

Give the gender of each noun, with the rule. Give the rules for the formation of the Genitive plural of the third declension. Give the meaning of the terminations *-ile* (*e. g. caprile*), *-ium* (*e. g. collegium*).

2. Decline *totus, dulcis, plus, quisquam*. Compare *capax, nequam, pauper*. Give the Latin for ten and multiples of ten as far as one hundred. Form adjectives from *Roma, Athenæ, civis*.

3. Give the principal parts of *depromo, jaceo, verto, ordior, jacio, spondeo*. Inflect the Perfect Subjunctive Passive of *audeo;* the Imperfect Subjunctive of *fio;* the Present Subjunctive Passive of *domo;* the Imperative of *ordior*.

4. Give the rules for the cases that follow the verbs *potior, pudet, doceo, egeo, ignosco*. Mention the various constructions by which a Purpose may be expressed in Latin. Give the rules for the use of the Subjunctive in Relative Clauses. When is a Dactylic Hexameter called Spondaic?

XII.

1. Decline *Lucius, puppis, manus, bos, September*, giving the gender of each noun with the rule, and marking the quantities of the final syllables throughout the declension of the first three. Give the rules for the genders of the following nouns: *os, sermo, lapis, dies, exemplar*.

2. Decline *acer, par, fortis, idem*. Compare *similis, pulcher, parvus*. What are the meanings of the terminations *-lentus* (*e. g. opulentus*), *-ax* (*e. g. minax*), *-ilis* (*e. g. humilis*)? Give the Latin for *eleven, nineteen, seventy-six; seven, fourteenth, twenty-fifth*.

3. Give the principal parts of *lædo, sero, seco, cædo,*

gaudeo, cado. Inflect the Present Subjunctive Active of *lœdo*; the Future Passive of *sero*; the Imperative Passive of *cœdo*.

4. What case or cases follow the following verbs respectively: *condemno, celo, pœnitet, pareo, interest?* Write in Latin "at Cannæ"; "to Cannæ"; "from Cannæ"; "at Rome." Translate: 1. Si Cæsar adest, lætor. 2. Si adsit, læter. 3. Si adesset, lætarer. 4. Si adfuisset, lætatus essem. 5. Si adfuerit, lætabor.

XIII.

1. Write down the following words and mark the quantity of the penult, giving the rules of prosody: *tempora, responderunt, dederint, discedo, iniquus, oceanus, remanet, egi, impedit, manus, brevis, cervices, protulit, nolite, vectigal*.

2. Meaning of termination *-etum* in *rosetum?* Of *-olus* in *filiolus?* Of *-ax* in *loquax?* Of *-mentum* in *tegmentum?*

3. Write the Perfects and Supines of *diligo, reperio, maneo, perfundo, indulgeo, cedo, cœdo, cado, moveo, cognosco*.

4. Compare *acer, bene, magnus, similis, gravis*.

5. Give the Present Subjunctive and Future Indicative Third Person Singular of *sum, cerno, eo, malo, caveo, venio*.

6. Decline *sedile, fructus, homo, vir, ingenium, melior*.

7. Decline *aliquis, alter, ipse*.

8. What is the Latin for *five?* For *fifth?* For *five times?* For *fifty? fiftieth? fifty times?* Write in Latin: One man in every ten.

XIV.

1. Give the gender of each of the following nouns, and the rule for it: *pax, pactio, manus, munus, salus, ager, pes*.

2. Decline the following nouns, marking the quantity of

the penultimate and final syllables in each form: *filius, iter, domus, dies.* Give the rules for the formation of the Ablative singular and Genitive plural of the third declension.

3. Decline *solus, fortis, idem, quidam.* Compare *ingens, similis, sacer.* Give the meanings of the following endings of nouns and adjectives: *-ula (cornicula), -ium (ministerium), -etum (saxetum), -icius (patricius).*

4. Give the principal parts of the verbs *fundo, veto, verto, voveo, sancio, cœdo.* Give the Third Person Singular of the Present Subjunctive Active, and of the Future Indicative Passive of *veto, verto,* and *sancio.* Inflect the Imperfect Subjunctive Passive of *facio,* and the Future Indicative Active of *transeo.*

5. By what cases respectively are these words followed: *occurro, condemno, sub, fruor, noceo?*

XV.

1. Decline *poema, domus, turris, Baiæ,* marking the quantity of all penultimate and final syllables. Give the gender of each and the rule. Write the Vocative singular and the Dative and Accusative plural of *dea, genius, locus.*

2. State the significance of the terminations in *vehiculum, orator, virtus, docilis.* Give the word from which each is derived, and the rule for the quantity of the penult. Translate *istic, istuc, istinc.* What kind of a verb is *cito?* Account for the quantity of its penultimate vowel. Give the principal parts of *tono, potior, vivo, fido, vincio,* and *faveo.*

3. What case or cases follow *similis, fungor, recordor, in, inter, interest?* What classes of verbs are followed by both Genitive and Accusative? What two constructions may follow *circumdo?*

4. Give the rules for the Subjunctive after *ut, utinam, cum, dum,* and *quod* (because). What is the meaning of *quominus,* and after what expressions is it used? When may an Infinitive with its Subject Accusative stand independent in a sentence? When may the Subject of an Infinitive be in the Nominative?

XVI.

1. Decline *deus, alius, tu, siquis,* and *audax,* marking the quantity of penultimate and final syllables. Compare *audax, multus,* and *nequam.* Compare adverbs formed from *audax, bonus, miser,* and *honorificus.* Give the rules for the gender of *formido, caput, pax, fas,* and *Tiberis.*

2. Inflect the Future Indicative and Present Subjunctive of *teneo, gero, sto,* and *fio,* marking the quantity of all the syllables. Give the Infinitives of *tollo* and *scribo.* Give all the Participles of *haurio* and *orior.* Give the principal parts of *uro, vendo, paro, pario, pareo, memini,* and *nanciscor.*

3. What case or cases follow *fido, jubeo, memini, præsum, existimo, pœnitet, contra, clam,* and the interjection *O?* By what two cases may price or value be expressed, and when is one used and when the other? What case follows the comparative when *quam* is omitted? When is it necessary that *quam* be expressed? Give five important rules for the Ablative without a preposition after verbs.

4. When is *ut* omitted before the Subjunctive? Give the rules for the Subjunctive in Relative Clauses. Translate into Latin, "The plan of setting the city on fire," using first the Gerund and then the Gerundive. Plan, *consilium.* To set on fire, *inflammare.*

XVII.

1. Decline together *frater meus*. Also decline in the singular, with the proper gender of the adjective annexed, *nox (unus), fides (Punicus), mare (uterque), Orion (nimbosus)*, marking the quantity of penultimate and final syllables. Decline in the plural, marking the quantities in the same way, *ensis (pugnax), portus (tutus), finis (extremus), mos (vetus)*. Give the rule for the gender of each of the above nouns. What is an Epicene Noun? Give the significance of the terminations *-ax* in (*pugnax*), *-osus* in (*nimbosus*), also of *-urio* in (*esurio*), and *-sco* in (*rubesco*). Form an abstract noun from *solus*. Compare *pugnax, extremus, vetus*. Compare adverbs formed from *carus, malus, similis*.

2. Give the principal parts of *reperio, ordior, cupio, circumdo, aufero, tango, arcesso*, marking the quantity of the penult. Inflect (marking the quantity of the penult) the Future Active Singular of *maneo* and *venio*; and the plural of the Present Subjunctive Passive of *facio* and *peto*. Give all the Infinitives and Participles of *purgo, pergo, morior*; and inflect the Imperative Active of *dico*.

3. What case or cases follow *ob, occurro, moneo, gaudeo, irascor, sub, pudet, pro, præditus*? Give the rules for verbs which govern two Accusatives. In what ways may the agent be expressed? State in what ways the construction of names of towns differs from that of other names of places. Give all the rules for the Subjunctive, denoting either purpose or result; after *quasi* and *priusquam*; in the Indirect Discourse.

XVIII.

1. Decline together in the singular *Marcus Tullius Cicero senex*. In the same way decline (both in singular and

plural) with the adjective annexed in the proper gender, *dies (fastus), flumen (aureus)*; in the plural: *arma (victrix), dea (immortalis).* Mark the quantity of all the vowels in the above nouns and adjectives. State the significance of the terminations *-men* in (*flumen*), *-eus* in (*aureus*), *trix* in (*victrix.*) What classes of words of the third declension form the Ablative in *-i* only?

2. Give the principal parts of *adjuvo, nolo, venio, paciscor, sperno, foveo, mordeo, scindo*, marking the quantity of the penultimate vowel. Give the synopsis of *mordeo* and *paciscor;* give all the Infinitives and Participles; and inflect the Imperatives.

3. Give all the rules you remember for verbs that govern the Dative. State the case or cases by which the price, the source, time when, and place where (including names of towns), are expressed, and give the rules. Give the rule for the Subjunctive in the following sentences: Quid enim, Catilina, est quod te jam in hac urbe delectare possit? Nunc ego mea video quid intersit. Supplicatio decreta est his verbis quod urbem incendiis liberassem. C. Sulpicium misi qui ex ædibus Cethegi, si quid telorum esset, efferret. O fortunate adolescens qui Homerum præconem inveneris.

XIX.

1. Decline in the singular: *facies, idem, ovile, sidus, filius.* Decline in the plural: *portus, dea, navis.* Write the gender over the nouns (rules not required), and mark the quantity of all penultimate and final syllables. 1. Give the significance of the terminations *-ile* in *ovile;* *-men* in *gestamen.* 2. Form an abstract noun from *felix;* from *æger.* 3. Form a noun denoting the masculine agent from *adjuvo*, and a frequentative verb from *cieo*, and account for

the quantity of their penultimate vowels. 4. Compare *humilis, juvenis*, and adverbs formed from *felix* and *æger*.

2. 1. Give the principal parts of *cado, cædo, tono, reperio, curro, pasco, paciscor*, marking the quantity of the penult. 2. Give all the Infinitives and Participles of *abeo, ulciscor;* the Present Indicative of *fio;* the Future Indicative Active and the Present Subjunctive Passive of *munio*, with the quantity of all the penults.

3. 1. What case or cases follow *super, tenus, recordor, fruor, similis?* 2. Give the principal parts of *parco* and *confido*, and the case that follows each. 3. Give the rules for the two cases after *pudet, do, doceo, moneo*. 4. Give the Latin for "at home," "at Carthage," "from Carthage," "from Italy," "to Athens." Tu discessu ceterorum nostra tamen, qui remansissemus cæde te contentum esse dicebas. 5. Give the rules for *discessu* and *cæde*. What is the antecedent of *qui?*

4. 1. Give the rules for the Subjunctive after *dum, cum, quominus*. 2. Would *ne* or *ut non* follow *restat* and *moneo*, respectively? Why? Statuisti quo quemque proficisci placeret, ... dixisti paululum tibi esse etiam nunc moræ, quod ego viverem. Reperti sunt duo equites Romani qui te ista cura liberarent. Idoneus est qui impetret quem legatum velit. Exclusi eos quos tu ad me salutatum miseras. 3. Explain the Subjunctives in the above sentences; the tense of *impetret*. 4. Give the rule for *salutatum*.

XX.

1. Decline *soror, vir, vis, vulnus, animal*. Give the gender of each of these nouns, with the rule. Mark the quantity of all the penultimate and final syllables you write in this section. Give the Genitive plural of *gens* and *hostis*, with the rules.

2. Decline *sacer, acer, alius*. Compare *similis, superus, parvus, juvenis*. Form and compare adverbs from *acer, altus*. Decline *idem, tu*, and *aliquis*. Give the Latin numerals for *sixty, seventy, eighty, six hundred, seven hundred, eight hundred*.

3. Give the principal parts of *vinco, vincio, spondeo, domo, lacesso, cædo, audeo*. All the Participles and Infinitives of *adipiscor* and *fero*. The Second Person Singular of the Future Indicative and of the Imperfect Subjunctive of *audeo, audio, fugio, eo, possum, volo*. Mark all penultimate and final syllables you write in this section.

4. How is price or value expressed in Latin? time in which? place where? What case or cases follow the verbs *miseret, obliviscor, ignosco, fungor, rogo*, respectively?

5. What is a Spondee? an Iambus? What is an Heroic Hexameter?

XXI.

1. Decline *mare, pignus, cor, fructus*. Give the gender of these nouns, with the rules. Mark the quantity of any increments that occur in their declension.

2. Compare *humilis, niger, malus*. Give the synopsis of *morior* and *gaudeo*. Give the Second Person of the Future Indicative, and of the Present, Imperfect, and Perfect Subjunctive of *spero, fero, volo*, in the Active Voice. The same of *facio* and *audio* in the Passive. Give the principal parts of *fateor, tono, peto, vincio, colo, tango*.

3. Compare *diu*. Form and compare an adverb from *brevis*. What are the meanings of the terminations of *copiosus, civilis, audacia, victrix*? What cases follow *infero, pœnitet, parco, careo, fruor, tenax, fretus, in, ante, super*?

4. How is the place to which, the price, the agent of a Passive verb expressed in Latin?

5. How is a condition contrary to the fact expressed in Latin? State one case in which a Relative Clause requires the Subjunctive. One case where the Subjunctive is used in Principal Clauses. What is a Gerundive? Give an example.

XXII.

1. Decline *Penelope, mons, cubile,* and give the gender with the rules. Mark the quantity of penults and final syllables of the above words. Decline *uterque.* Decline *acer,* and compare it. Form an adverb from it, and compare it.

2. Compare *senex* and *munificus.* Give the derivation of *filiolus, documentum, quercetum, audax, capesso,* and the meaning of the terminations. Give all the Participles and Infinitives of *vereor* and *cœdo,* and mark the quantity of the penults. Inflect the Imperative of *fero, ordior, nolo, fateor.* Give the Present and Imperfect Subjunctive First Person Singular of *adjuvo, eo, soleo,* and *fugio,* marking the quantity of the penults. Give the principal parts of *pario, pareo, paro, reddo, redeo, surgo,* and of the compound of *ab* and *fero.*

3. What case or cases follow *refert, irascor, circumdo?* How do the constructions of names of towns differ from those of other words? How is the degree of difference expressed in Latin? How the agent by the participle in *-dus?* What construction is used after verbs of Saying? Verbs of Fearing? How may a Purpose be expressed? How does a Gerund resemble a noun? How does it resemble a verb? How does the Gerundive differ from it?

XXIII.

1. Decline *filius, pectus, manus, animal.* Give the gen-

ders and mark the quantity of all penultimate and final syllables. Give the gender and the Ablative singular and Genitive plural of *imago, mons, vis, turris, sedile.* Decline *capax, æger,* and the comparative of *miser.* Compare *facilis, acer,* and an adverb formed from *piger.* Decline *uterque.*

2. Give the First Person of the Future Indicative, and all tenses of the Subjunctive of *possum, pario, sono, vereor, eo, soleo.* Mark quantities of penults. Give the Infinitives and Participles, Active and Passive, of *spondeo, morior, paro, quæro, queror, adipiscor.*

3. Explain the force of the derivative terminations in *longitudo, tenax, vehiculum, Priamides, clamito, vinolentus, filiolus.*

4. What is the construction in Latin of the place in which (including names of towns)? the price or value? the degree or measure of difference between objects compared? the agent in the Passive Voice? What case or cases follow *credo, pudet, fungor, refert, aptus, avidus, dignus, in, pro, propter, doceo, condemno, circumdo?*

5. How is a future condition with its conclusion expressed? How a condition contrary to fact? How an object clause after a verb of Fearing; of Commanding; of Saying? Translate *cave eas* and explain the peculiarity. When can you use the Gerundive for the Gerund? Give an example of each. Give an example of the use of the Supine.

XXIV.

1. Decline the following words, and give their genders respectively: *onus, collis, salus, gradus.* Decline *felix, quidam, senex.* Compare *parvus, beneficus.* Form and compare an adverb from *acer.*

2. Give a synopsis of *mordeo, scio,* in the Active Voice,

and of *hortor, orior, polliceor, nolo*. Give the principal parts of *paro, pario, pareo, ulciscor, pango, tollo*.

3. What are the meanings of the derivative terminations in *acritudo, clamito, vinculum, parvulus*?

4. What case or cases follow *moneo, prosum, rogo, in, præter*? What is the force of *num* in a question? of *ne*? Explain the mood and tense of *mansisset* in, "Mansissetque utinam fortuna." Explain the mood of *esset* and the case of *fronde* in "Nos delubra miseri, quibus ultimus esset ille dies velamus fronde." Explain the mood of *polliceantur* in "Ad eum legati veniunt, qui polliceantur obsides dare." With what other constructions could the same idea be expressed? What is the use of the supine in *-um*? in *-u*? Explain construction of *usui* and *fore* in "Magno sibi usui fore arbitrabatur." Describe the feet of two syllables. Mark the quantity of the penults and last syllables in the above extracts.

XXV.

Translate the following extract: —

Imitatus est homo Romanus veterem illum Socratem, qui cum omnium sapientissimus esset sanctissimeque vixisset, ita in judicio capitis pro se ipse dixit, ut non supplex aut reus sed magister aut dominus videretur esse judicum; quin etiam cum ei scriptam orationem disertissimus orator Lysias attulisset, quam si ei videretur edisceret ut ea pro se in judicio uteretur, non invitus legit et commode scriptam esse dixit.

Decline *veterem, magister, judicum, ei*.

Compare *invitus, sanctissime*.

Give the principal parts of *imitatus, vixisset, attulisset, edisceret, uteretur*.

Give all the Participles and Infinitives of *scriptam esse*.

What are the derivations of *Romanus, orationem, orator*, and the meaning of the derivative terminations in each?

What is the construction (i. e. where are they made and why) of *omnium, se, videretur* (in each of the two cases), *edisceret, ea, scriptam esse, ei* (first one), *attulisset?*

What are the principal rules for the change from Direct Discourse to Indirect?

XXVI.

Translate:—

Sin autem quis requirit, quæ causa nos impulerit, ut hæc tam sero literis mandaremus, nihil est, quod expedire tam facile possimus. Nam, cum otio langueremus, et is esset reipublicæ status, ut eam unius consilio atque cura gubernari necesse esset; primum, ipsius reipublicæ causa, philosophiam nostris hominibus explicandam putavi, magni existimans interesse ad decus et ad laudem civitatis, res tam graves tamque præclaras Latinis etiam literis contineri. Eoque me minus instituti mei pœnitet, quod facile sentio, quam multorum non modo discendi, sed etiam scribendi, studia commoverim. Complures enim, Græcis institutionibus eruditi, ea, quæ didicerant, cum civibus suis communicare non poterant, quod illa, quæ a Græcis accepissent, Latine dici posse diffiderent. Quo in genere tantum profecisse videmur, ut a Græcis ne verborum quidem copia vinceremur.

Decline *civibus, decus, status, quis, graves.*

Compare *minus, graves.*

Give the principal parts, Active and Passive (if any), of *requirit, impulerit, sentio, diffiderent, eruditi, vinceremur.*

Give the synopsis of *didicerant, commoverim, pœnitet.*

Explain construction (where made and why) of *possimus, esset, hominibus, magni, me, instituti, scribendi, commoverim, diffiderent.*

XXVII.

Translate (omit any words you do not remember, but give their construction):—

P. Scipionem, Marce fili, eum, qui primus Africanus appellatus est, dicere solitum scripsit Cato, qui fuit eius fere aequalis, numquam se minus otiosum esse quam cum otiosus, nec minus solum quam cum solus esset: magnifica vero vox et magno viro ac sapiente digna; quae declarat illum et in otio de negotiis cogitare et in solitudine secum loqui solitum, ut neque cessaret umquam et interdum conloquio alterius non egeret; ita duae res, quae languorem adferunt ceteris, illum acuebant, otium et solitudo. Vellem nobis hoc idem vere dicere liceret.

1. (*a*) Give the principal parts of the verbs from which come the forms *solitum, scripsit, loqui, egeret, acuebant, liceret, vellem*. (*b*) Give the Present, Imperfect, and Perfect Subjunctive, and all the participles of the same verbs. (*c*) Mark the quantity of each penult in the forms you have given.

2. Decline *aequalis, solus, viro, idem*.

3. Compare *primus, minus, vere*.

4. Account for mood and tense of *dicere, solitum, esset, cessaret, vellem, liceret*.

5. Account for case of *eius, vox, viro, otio, conloquio, nobis, ceteris*.

6. Explain derivation of *aequalis, otiosus, magnifica, negotiis, solitudine, cessaret, conloquio, acuebant*.

7. Mark the feet and quantities and explain the metre of the following lines:—

    Cetera labuntur celeri caelestia motu
    Cum caeloque simul noctesque diesque feruntur.

XXVIII.

Translate:—

M. Atilius Regulus, cum consul iterum in Africa ex insidiis captus esset duce Xanthippo Lacedæmonio, imperatore autem patre Hannibalis Hamilcare, iuratus missus est ad senatum, ut, nisi redditi essent Pœnis captivi nobiles quidam, rediret ipse Karthaginem. Is cum Romam venisset, utilitatis speciem videbat, sed eam, ut res declarat, falsam judicavit; quæ erat talis: manere in patria, esse domui suæ cum uxore, cum liberis, quam calamitatem accepisset in bello, communem fortunæ bellicæ judicantem tenere consularis dignitatis gradum. Quis hæc negat esse utilia? quem censes? magnitudo animi et fortitudo negat. Num locupletiores quæris auctores?

Decline together *Atilius Regulus; captivi nobiles quidam; speciem falsam.*

Give principal parts of the verbs from which come *redditi essent, rediret, manere, quæris.*

Give the Present and Perfect Subjunctive (1st Person) and all the participles of the above verbs, and inflect the Future Indicative.

Mark the quantity of the penults and last syllables of all the Latin words you have written.

Explain the derivation and force of derivative ending of the words *auctores, nobiles, utilitatis, consularis, utilia, falsam, bellicæ.*

What is the stem and what the root of *magnitudo?* Analyze the word by derivation as far as you can. Do you know any other words in Latin or other languages from the same root?

Explain construction of *Pœnis, duce, Romam, domui, fortunæ.*

Explain mood of *redditi essent, rediret, manere, accepisset.*

Mark the quantities and divide into feet the following lines. What verse and metre are they?

Quodcumque attigerit, siqua est studiosa sinistri
Ad vitium mores instruet inde suos.

FRENCH.

I.

I. CHARLES XII. *éprouva* ce que la prospérité a de plus grand et ce que l'adversité a de plus cruel, sans avoir été amolli par l'une ni ébranlé par l'autre. Presque toutes ses actions, jusqu'à celles de sa vie privée, ont été bien au delà du vraisemblable. C'est *peut*-être le seul de tous les hommes, et jusqu'ici le seul de tous les rois, qui *ait* vécu sans faiblesse; il a porté toutes les vertus des héros à un excès où elles sont aussi dangereuses que les vices opposés.

II. Il a été le premier qui ait eu l'ambition d'être conquérant sans avoir l'envie d'agrandir ses États; il voulait gagner des empires pour les donner. Sa passion pour la gloire, pour la guerre, et pour la vengeance, l'empêcha d'être bon politique, qualité sans laquelle on n'a jamais vu de conquérant. Avant la bataille et après la victoire, il n'avait que de la modestie; après la défaite, que de la fermeté; dur pour les autres comme pour lui-même, comptant pour rien la peine et la vie de ses sujets aussi bien que la sienne: homme unique plutôt que grand homme, admirable plutôt qu'à imiter. Sa vie doit apprendre aux rois combien un gouvernement pacifique et heureux est au-dessus de tant de gloire.

III. Charles XII. était d'une taille avantageuse et noble; il avait un beau front, de grands yeux bleus remplis de douceur, un nez bien formé, mais le bas du visage désagréable, trop souvent défiguré par un rire fréquent qui ne partait que des lèvres; presque point de barbe ni de cheveux : il parlait très-peu, et ne répondait souvent que par ce rire dont il avait pris l'habitude. On observait à sa table un silence profond. Il avait conservé dans l'inflexibilité de son caractère cette timidité qu'on nomme mauvaise honte ; il eût été embarrassé dans une conversation, parce que, s'étant donné tout entier aux travaux et à la guerre, il n'avait jamais connu la société. — VOLTAIRE.

1. Translate II. and III. of the above.

2. State mood and tense of italicized verbs in I., and give them in full.

3. Give the principal tenses of *devoir, connaître, apprendre, vivre*. (Thus, Infin., *être;* Pres. Part., *étant;* Past Part., *été;* Pres. Ind., *je suis;* Pret., *je fus*.)

4. Using mostly the words of I., translate into French: (*a*) Charles has lived in adversity. (*b*) This man is dangerous. (*c*) All heroes have not lived in the greatest prosperity. (*d*) Have you re-read (*relu*) what you have written (*écrit*) ?

ARITHMETIC.

I.

1. Reduce $\frac{3}{8}$, $\frac{5}{16}$, $\frac{7}{24}$, and $\frac{3}{48}$ to their Least Common Denominator.

2. Divide $1\frac{61}{6}$ by 42. Divide $\frac{7}{8}$ of $\frac{16}{19}$ by $\frac{4}{11}$ of $\frac{33}{28}$.

3. Reduce $\dfrac{18\frac{7}{4}}{\frac{3}{8}\ of\ \frac{3}{4}\ of\ \frac{1}{2}}$ to its simplest form.

4. Reduce $\frac{1}{22}$ of a gallon to the fraction of a gill.

5. Add $\dfrac{3\frac{1}{4}}{5\frac{3}{4}}$, $\frac{7}{8}$, and $\frac{9}{10}$ of $\frac{6}{7}$.

6. How long must $133 be on interest (simple) at 7 per cent to gain $32,585 ?

7. What is the compound interest on $1,000 for 3 years at 7 per cent (interest payable annually) ?

8. What is the cube of $\frac{7}{3}$? of .006 ?

9. Divide 46.08 by 1,000. Divide 1.096641 by 15.21.

10. What is the square root of 104.8576 ?

11. What is a Circulating Decimal ? Give an example of a Circulating Decimal.

12. What are Duodecimals ?

II.

1. Find the Greatest Common Divisor of 48 and 130.

2. Reduce $\frac{1}{2}$, $\frac{5}{6}$, $\frac{7}{12}$, and $1\frac{1}{8}$ to their Least Common Denominator.

3. What part of $1\frac{0}{15}$ is $\frac{1}{3}$?

4. Subtract $15\frac{1}{4}$ from $18\frac{2}{3}$.

5. Divide $1\frac{1}{2}$ by $1\frac{1}{8}$. Multiply the same.

6. Divide $\frac{1}{5}$ of $\frac{6}{7}$ of $2\frac{1}{2}$ by $\frac{1\frac{9}{7}}{3\frac{1}{3}}$.

7. Write $1\frac{1}{32}$ and $2\frac{1}{16}$ in a decimal form. Give the division in decimals of the first by the second.

8. Divide .09 by .0016. Multiply them.

9. Divide 876.196 by 2.12. If the decimal point were moved, in the first, two places to the left, and, in the second, one place to the right, how many times greater or less would the quotient be?

10. Find the square root of 49.2804.

11. What is the fourth power of 2? of 0.2? of .02?

12. If a man travels 64 rods in .05 of an hour, how many minutes will it take him to go a mile?

13. Find the simple interest on $1,000 for 1 yr. 2 mos. and 12 ds.

14. How many feet, board measure, in a plank 12 ft. 4 in. long, 2 ft. 3 in. wide, and 4 in. thick? (Multiplication of Duodecimals.)

III.

1. What is the Least Common Multiple of 20, 24, and 36?

2. Add $\frac{5}{6}$, $\frac{4}{9}$, $2\frac{3}{15}$, and $3\frac{3}{20}$.

3. Multiply 48 by $\frac{5}{16}$. Divide $\frac{87}{186}$ by $\frac{5}{16}$.

4. Reduce $\frac{\frac{1}{2} \text{ of } \frac{4}{7} \text{ of } 7\frac{3}{8}}{19\frac{6}{25}}$ to its simplest form.

5. Reduce $\frac{1}{6}$ of a bushel to the fraction of a pint.

6. Reduce 5 yds. 2 ft. 6 in. to the decimal of a rod, long measure.

7. Multiply 34.27 by 60,000. Divide 10634.16 by .4506.

ARITHMETIC. 113

8. At what rate per cent must $370 be put on interest to gain $55.50 in three years?

9. What is the amount of $25 for 3 yrs. 5 mos. at compound interest?

10. What is the third power of 30? of .03?

11. What is the square root of 104.8576?

12. What are the contents of a granite block that is 8 ft. 9 in. long, 3 ft. 2 in. wide, and 2 ft. 5 in. thick? (Multiplication of Duodecimals.)

IV.

1. What is the Greatest Common Divisor of 1181 and 2741?

2. Reduce $\frac{5}{6}$, $\frac{3}{13}$, and $\frac{7}{17}$ to a Common Denominator.

3. Divide $\frac{7}{8}$ of $\frac{16}{19}$ by $\frac{4}{11}$ of $\frac{33}{29}$.

4. Add $\frac{3\frac{1}{4}}{5\frac{2}{3}}$, $\frac{7}{8}$, and $\frac{9}{10}$ of $\frac{6}{7}$.

5. Reduce $\frac{53}{97}$ of a gallon to quarts, pints, etc.

6. Multiply 4 lbs. 8 oz. 16 dwt. 20 gr. by 72.

7. Find the interest on $76.72 from April 18, 1852, to January 26, 1855, at 6 per cent.

8. What principal at 6 per cent will amount to $360,585 in 16 months?

9. Multiply .427 by 345.

10. Divide 87.69 by 47, also by .47.

11. What is the square root of 747.4756?

12. Give an example of a Continued Fraction.

V.

1. Name all the Prime numbers in the series of numbers between 1 and 30 inclusive; resolve all the Composite

numbers into their Prime Factors; and name all the perfect squares, cubes, and other powers in the same series.

2. From $\frac{3}{4}$ of $\frac{4}{5}$ take $\frac{1}{2}$ of $\frac{2}{3}$.

3. Divide $\frac{3\frac{1}{4}}{6\frac{1}{2}} \times 72\frac{1}{2}$ by $\frac{2}{3}$ of $\frac{3}{8}$ of $9\frac{3}{8}$.

4. Reduce 9 rds. 1 ft. and 6 in. to the fraction of a furlong.

5. Multiply 8.764 by 40.015.

6. What is the square of 11 ? of .11 ?

7. Divide 769.428 by 200 ; by .00002.

8. Transform the Infinite Decimal .216 into its equivalent Vulgar Fraction.

9. What quantity of boards will be required to lay a floor 14 ft. 8' 3" in length and 13 ft. 6' 9" in width ? (Multiplication of Duodecimals.)

10. Find the square root of 4.190209.

11. Find the interest on $76.72 from April 18, 1852, to January 26, 1855, at 6 per cent.

12. If $50 gain $5.60 in 3 yrs. 6 mos., at simple interest, what is the rate per cent ?

13. Give an example of a Continued Fraction.

VI.

1. What are the Prime Factors of 360 ?

2. What part of a mile is one inch ?

3. Reduce $\frac{9\frac{7}{9}}{3\frac{5}{7}}$ to a Simple Fraction.

4. Add $\frac{5}{9}$ of a pound, $\frac{3}{8}$ of a shilling, and $\frac{5}{7}$ of a penny together.

5. What is the product of $\frac{3}{8}$ of $\frac{7}{11}$ of 15, and $\frac{14}{15}$ of $11\frac{5}{8}$?

6. Divide 100 by $4\frac{7}{8}$.

7. What is the square of 10.01 ?

8. Divide .1 by .0001. Divide 10 by .1.

9. Reduce $\frac{3}{32}$ to a decimal. Reduce $\frac{3}{140}$ to a Circulating Decimal.

10. What is the interest on $1461.75 for 4 yrs. 9 mos. at 8 per cent?

11. The interest on $437.21 for 9 yrs. 9 mos. is $127.884: what is the rate of interest?

12. Find the square root of 4.426816.

VII.

1. What is the Least Common Multiple of 21, 36, 50, and 64?

2. Add together $\frac{2}{6}$, $\frac{16}{21}$, and $\frac{4}{18}$, and from their sum subtract $\frac{6}{15}$.

3. Multiply $\frac{2\frac{5}{7}}{4\frac{1}{8}}$ by $\frac{8}{13}$ of $2\frac{1}{4}$.

4. Reduce $\frac{8}{11}$ of a furlong to inches.

5. Multiply 200.043 by 2.021.

6. Divide 9.00081 by 900; 4004004 by .002; .000624 by 324.

7. What are the contents of a granite block 12 ft. 2′ 3″ long, 6 ft. 8′ 9″ wide, and 4 ft. 9′ 2″ thick?

8. What is the amount of $5216.75 from January 21, 1860, to July 3, 1863, at 8 per cent, compound interest?

9. Find the cube of 10.1; of 1.01.

10. Find the square root of 49.87604.

11. Define a Circulating Decimal and give an example. What is a Continued Fraction?

12. What is the difference between an Arithmetical and a Geometrical Progression?

VIII.

1. What is a Prime Number? Find the Prime Factors of 4800.

2. What Prime Factors compose the Greatest Common Divisor and the Least Common Multiple of several numbers? Find the Greatest Common Divisor and the Least Common Multiple of 84, 126, and 140.

3. From $\frac{8}{7}$ of $\frac{7}{15}$ subtract $\frac{3}{40}$ of $1\frac{1}{9}$.

4. Divide $\frac{2}{15}$ of $\frac{28}{3}$ of $3\frac{1}{8}$ by $\dfrac{24\frac{1}{2}}{1\frac{8}{55} \times 1\frac{1}{2}}$.

5. Give the rule for pointing off in the multiplication of decimals, and explain the reason.

6. Multiply 0.0400268 by 0.260075.

7. Divide 0.011825369 by 5.884. What is the quotient of 118253690 by the same divisor?

8. Reduce $\frac{3}{220}$ to a Circulating Decimal. Verify the result by reducing it back to a Vulgar Fraction.

9. Reduce 0.845 of a mile to furlongs, rods, feet, and inches.

10. The interest on $127.50 from June 26, 1798, to May 8, 1802, was $36.975: calculate the rate of interest.

11. Find the square root of 7.333264.

12. Find the cube root of 96702.579.

13. If 6 men can build 20 feet of a stone-wall in 10 days, how many men can build 360 feet of the same wall in 90 days?

IX.

1. Reduce 10917 to the product of its Prime Factors.

2. Find the Greatest Common Divisor of 720, 336, and 1736; Least Common Denominator of $\frac{22}{18}$, $\frac{7}{82}$, $\frac{9}{24}$.

3. From $36\frac{9}{10}$ take $\frac{4}{3}$.

4. Multiply $\frac{2}{9}$ of $1\frac{3}{9}$ of $4\frac{1}{5}$ by $\frac{36\frac{3}{4}}{2\frac{8}{75} \div 1\frac{3}{4}}$. What part of $\frac{6\frac{2}{9}}{3\frac{5}{11}}$ yards is $\frac{7}{8}$ of an inch?

5. Give the rule for pointing off in multiplication of decimals, and explain its reason.

6. Reduce 0.0007648267 to a Vulgar Fraction.

7. The product of three numbers = 70.04597; two of them equal 3.91 and 3.0005 respectively. Find the third.

8. Reduce the Infinite Decimal $0.81\dot{2}4\dot{7}$ to a Vulgar Fraction.

9. Find the amount of $1000 for 2 yrs. 2 mos. 12 ds., compound interest, at 6 per cent, payable annually.

10. Find the square root of 39.037504.

11. Find the cube root of 0.000000148877.

12. Find the third power of 3; of 0.3; of 0.003.

13. If a family of 9 persons spends $305 in 4 months, how many dollars will maintain it 8 months, if 5 persons more were added to the family? Multiply 10 ft. 3' 2" by 6 ft. 7' 8".

X.

1. What is a Prime Number? When are two numbers prime to each other? What Prime Factors compose the Greatest Common Divisor and the Least Common Multiple of several numbers? Find the Greatest Common Divisor and Least Common Multiple of 156, 234, and 260.

2. From $\frac{8}{25}$ of $1\frac{2}{3}$ subtract $\frac{10}{27}$ of $\frac{9}{40}$; reduce the answer to its lowest terms, and reduce it to a decimal.

3. Divide $1\frac{4}{5}$ of $\frac{9}{56}$ of $1\frac{2}{5}$ by $\frac{\frac{3}{2}}{\frac{2\frac{9}{7}+\frac{4}{4}\cdot\frac{5}{9}}}$.

4. Reduce $\frac{17640}{29400}$ to its lowest terms. Reduce $\frac{7}{10}, \frac{11}{12}, \frac{4}{15}, \frac{6}{25}$, and $\frac{1}{60}$ to their Least Common Denominator, add them, and reduce the sum to its simplest form.

5. Multiply 6.4 by 1.5. Multiply 0.64 by 0.15. Divide 701.5 by 2.806. Divide 0.7015 by 280.6. Reduce the last answer to its lowest terms as a Vulgar Fraction.

6. The number 209.069673692836 is composed of three factors, of which two are 20083.6 and 0.260075. Find the third factor.

7. State the rule for pointing off in the multiplication of decimals, and give its reason.

8. Reduce the Infinite Decimal $0.01\dot{3}\dot{6}$ to its lowest terms as a Vulgar Fraction, and verify the result by reducing back to a decimal.

9. Calculate the date at which a sum of $450, which was put at simple interest at 8 per cent, December 30, 1797, amounted to $642.30.

10. Reduce 6 fur. 30 r. 6 ft. $7\frac{1}{5}$ in. to the decimal of a mile.

11. Divide 5 cwt. 12 lbs. 4 oz. by 7. Multiply 2 ft. 3′ 7″ by 9 ft. 5′ 11″. Reduce £17 9 s. 3 d. to Federal money, taking 4 s. 6 d. = $1.

12. Find the proportion in which sugars worth 5 cents and 8 cents a pound must be taken to form a mixture worth $6\frac{3}{4}$ cents a pound.

13. How many digits compose the 3d power of a number containing two digits? What is the reason of your answer? What is the third power of 3? of 0.3? of 0.03? of 30?

14. Find the cube root of 39512.447416.

15. Find the square root of 13 to five places of decimals.

ARITHMETIC.

16. If 3 men can build a wall 60 feet long, 8 feet high, and 3 feet thick, in 64 days of 9 hours, how many days of 8 hours will 20 men require to build a wall 400 feet long, 9 feet high, and 5 feet thick?

XI.

1. Which of the numbers 5, 9, 13, 18, 21, 25, are Prime Numbers? and which of them are prime to the number 10?

2. Find the Greatest Common Divisor and the Least Common Multiple of 630, 840, and 2772.

3. From $\frac{3}{4}$ of $\frac{28}{45}$ subtract $\frac{3}{20}$ of $2\frac{7}{9}$; reduce the answer to its lowest terms; and reduce it to a decimal.

4. Divide $1\frac{2}{25}$ of $\frac{3}{70} \times 13\frac{3}{4}$ by $\frac{\frac{11}{20}}{2\frac{1}{2} - \frac{4}{5}}$.

5. Multiply 76000 by 1.05. Multiply 0.076 by 0.0105. Divide 2926.5 by 0.3902. Divide 29.265 by 390.2. Reduce the last answer to its lowest terms as a Vulgar Fraction.

6. Reduce to their lowest terms as Vulgar Fractions the Infinite or Circulating Decimals $0.\dot{2}\dot{7}$, $0.01\dot{2}\dot{7}$, $0.00\dot{2}\dot{7}$, $0.\dot{0}02\dot{7}$.

7. Calculate the date at which a sum of $234, which was put at simple interest at 9 per cent, October 25, 1798, amounted to $351.

8. Reduce 6 fur. 30 r. 6 ft. $7\frac{1}{2}$ in. to the decimal of a mile.

9. Find the cube root of 9358 to two places of decimals.

10. If 6 men can build a wall 80 feet long, 10 feet high, and 9 feet thick, in 100 days of 9 hours, how many days of 10 hours will be required by 15 men to build a wall 200 feet long, 9 feet high, and 5 feet thick?

XII.

1. Find the Greatest Common Divisor and Least Common Multiple of 144 and 780.

2. Reduce $\frac{1}{2}$, $\frac{5}{6}$, $\frac{7}{15}$, and $\frac{11}{18}$ to their Least Common Denominator.

3. What part of $\frac{4}{5}$ is $\frac{2}{3}$?

4. Subtract $15\frac{1}{4}$ from $18\frac{2}{3}$.

5. Divide $1\frac{1}{2}$ by $1\frac{1}{8}$. Multiply the two together.

6. Divide $\frac{1}{5}$ of $\frac{6}{7}$ of $2\frac{1}{2}$ by $\frac{1\frac{9}{7}}{3\frac{1}{2}}$.

7. Write $1\frac{1}{32}$ and $2\frac{1}{16}$ in a decimal form. Give the division in decimals of the first by the second.

8. Divide .09 by .0016. Multiply them.

9. Divide 876.196 by 2.12. If the decimal point were moved, in the first, two places to the left, and, in the second, one place to the right, how many times greater or less would the quotient be?

10. Find the cube root of 51 to three places of decimals.

11. Reduce to their lowest terms as Vulgar Fractions the Infinite or Circulating Decimals, $0.\dot{2}34\dot{3}$, $0.00\dot{2}34\dot{3}$, $0.01\dot{2}34\dot{3}$, $0.00\dot{2}34\dot{3}$.

12. If a man travel 64 rods in .05 of an hour, how many minutes will it take him to go a mile?

13. Find the simple interest on $1000 for 6 yrs. 4 mos. and 15 ds. at 8 per cent.

14. How many feet, board measure, in a plank 12 ft. 4 in. long, 2 ft. 3 in. wide, and 4 in. thick?

XIII.

1. Reduce $\frac{23820}{39700}$ to its lowest terms.

2. Reduce $\frac{7}{10}$, $\frac{11}{12}$, $\frac{4}{15}$, $\frac{6}{25}$, and $\frac{1}{60}$ to their Least Common Denominator; add them, and reduce the result to a decimal form.

ARITHMETIC. 121

3. Divide $\frac{9}{10}$ of $\frac{7}{75}$ of $8\frac{1}{3}$, by $\frac{\frac{2}{3} \text{ of } 2\frac{7}{24}}{18\frac{1}{3}}$. Simplify, and reduce to lowest terms by cancelling.

4. Multiply 37900000 by 2.005. Multiply 0.0379 by 0.2005. Write the numbers 37900000 and 0.0379 in words.

5. Divide 1909.14 by 0.02708. Divide 190.914 by 27080.

6. Reduce to their lowest terms as Vulgar Fractions the Infinite or Circulating Decimals, $0.00\dot{8}\dot{1}$, $0.008\dot{1}$, $0.1\dot{0}8\dot{1}$, $0.\dot{1}0\dot{8}$.

7. Find the simple interest on $1000 for 5 yrs. 4 mos. and 15 ds. at 20 per cent. To how much will $1000 amount in 4 years, at compound interest, at 20 per cent?

8. Reduce 5 fur. 33 r. 9 ft. $10\frac{4}{5}$ in. to the decimal of a mile. Reduce £17 8 s. 9 d. to Federal money, taking 4 s. 6 d. = $1.

9. Multiply 2 ft. 3' 7" by 9 ft. 5' 11".

10. Find the cube root of 77869 to three places of decimals. Find the square root of 0.5 to five places of decimals.

XIV.

1. Reduce $\frac{16200}{24840}$ to its lowest term. What is a Prime number? When are two numbers said to be prime to each other?

2. Find the value of $\frac{7}{9} - \frac{1}{6} + 4\frac{3}{4} + \frac{29}{18} + \frac{7}{12}$; and reduce the result to its lowest terms, and also to a decimal form.

3. From $3\frac{1}{2}$ subtract $\left(\frac{7}{15} \text{ of } \frac{4\frac{1}{6}}{\frac{7}{8}} \text{ of } 1\frac{4}{5} \right) \div \frac{4\frac{7}{12}}{1\frac{7}{11}}$. Simplify by cancelling.

4. Multiply 2.708 by 0.007005. What is the product of 2.708 by 70050000 ? Write the numbers 0.007005 and 70050000 in words.

5. Divide 283891.3 by 0.07084. What is the quotient of 2.838913 divided by 708.4 ?

6. From 1 sq. rd. 5 sq. ft. subtract 7 sq. yd. 139 sq. in. Divide £ 32 16 s. 3 d. by 7.

7. Reduce 44920.9025 hours to years (of 365 days), days, hours, minutes, and seconds.

8. Find the cube root of 0.61 to five places of decimals. Find the square root of 79000 to three places of decimals.

9. Reduce to their lowest terms as Vulgar Fractions the Infinite or Circulating Decimals $0.000\dot{5}\dot{4}$ and $0.2005\dot{4}$. Add $0.0\dot{3}$ to $0.4\dot{6}\dot{2}$, expressing the result as an Infinite or Circulating Decimal.

10. A certain sum of money was put at simple interest at 9 per cent, December 21, 1790. At what date did it become tripled ?

XV.

1. What is the Greatest Common Divisor of two numbers? of 4760 and 3432 ?

2. Subtract $\frac{3}{4}$ of $\frac{8}{9}$ from $\frac{2}{3}$ of $\frac{3\frac{1}{2}}{4\frac{1}{5}}$, add to the remainder $\frac{5}{16}$, divide the result by $6\frac{7}{9}$, and change the quotient to a decimal.

3. Divide 0.000647808 by 6.72. Write the quotient in words.

4. I owe three notes bearing interest from date: the first, dated June 1, 1866, is for $450.00; the second, dated Dec. 17, 1866, is for $750.00; the third, dated March 15, 1867,

is for $600.00. I wish to substitute for these a single note for $1800.00: what should be the date of it?

5. Find the square root of 0.9.

6. Find the cube root of 751089.429.

7. Find the cube of 4; of 0.4; of 0.0004.

8. A sum of money was put at interest, at $7\frac{3}{10}$ per cent, October 30, 1866: at what date will it be tripled? (A year = 365 days.)

9. If 4 men dig a trench 84 feet long and 5 feet wide in 3 days of 8 hours each, how many men can dig a trench 420 feet long and 3 feet wide in 4 days of 9 hours each?

10. How many feet, board measure, in a plank 12 ft. 4 in. long, 2 ft. 5 in. wide at one end, 2 ft. 1 in. wide at the other, and 4 in. thick?

11. In what proportion shall sugars worth 7 and 12 cents a pound be taken to form a mixture worth $9\frac{1}{8}$ cents a pound?

XVI.

1. What is the Least Common Multiple of two or more numbers? of 48, 98, 21, and 27?

2. Add $\frac{4\frac{1}{5}}{6\frac{3}{10}}$ and $\frac{2\frac{1}{4}}{7\frac{7}{8}}$; divide the result by $7\frac{13}{21}$, and change the quotient to a decimal.

3. A certain bank declares a semiannual dividend of 4 per cent: what can I afford to pay for its shares if I wish to get 6 per cent a year for my money?

4. Reduce .445 of an acre to rods, feet, and inches.

5. Divide 0.0018891 by 3.75. Write the quotient in words.

6. Find the cube root of 748613.312; of 0.27.

7. Find the square of 0.9; of three millionths. Write the results in words.

8. How many feet, board measure, in a plank 16 ft. 4 in. long, 1 ft. 7 in. wide, and $4\frac{1}{2}$ in. thick?

9. A, B, and C hire a pasture for $92. A pastures 6 horses for 8 weeks, B 12 oxen for 10 weeks, and C 50 cows for 12 weeks. Now, if 5 cows are reckoned as 3 oxen, and 3 oxen as 2 horses, how much shall each man pay?

10. If 496 men, in 5 days of 12 h. 6 m. each, dig a trench of 9 degrees of hardness, 465 feet long, $3\frac{2}{3}$ feet wide, and $4\frac{2}{3}$ feet deep, how many men will be required to dig a trench 2 degrees of hardness, $168\frac{3}{4}$ feet long, $7\frac{1}{2}$ feet wide, and $2\frac{4}{5}$ feet deep, in 22 days of 9 hours each?

XVII.

1. What is the Least Common Multiple of two or more numbers? What is the Least Common Multiple of 3150 and 2310?

2. From $\frac{1}{4}$ of $1\frac{2}{7}$ take $\frac{\frac{3}{8}}{2\frac{1}{2}}$, add to the remainder $\frac{2}{3}$, and divide the result by $6\frac{2}{7}$.

3. Divide 0.00091471 by 9.43. Write the quotient in words.

4. How many yards of carpet which is $\frac{3}{4}$ of a yard wide does it require to cover a floor 17 feet long and 16 feet 6 inches wide?

5. Reduce 0.758762 acres to square rods, square feet, etc.

6. Find the square root of 0.002539 to five places of decimals.

7. Find the cube root of 0.15 to three places of decimals.

8. What is the interest of $875.26 from October 10, 1866, to July 10, 1868, at $7\frac{3}{10}$ per cent?

9. One metre (in Long Measure) = 39.37 inches. Express one foot in the metric system, both in Long Measure and in Square Measure.

XVIII.

1. Find the Greatest Common Divisor and the Least Common Multiple of 340200, 583200, and 2268000.

2. From $\frac{4}{21}$ of $2\frac{2}{5}$ subtract the product of 0.075 and $1\frac{1}{9}$, and divide the remainder by 12. Reduce the result to its lowest terms as a Vulgar Fraction, and also to a decimal form.

3. Divide 10 times $\left(\frac{7}{9} \text{ of } \frac{1\frac{2}{3}}{12\frac{1}{4}} \text{ of } 9\frac{9}{10} \right)$ by $\frac{2\frac{2}{4}}{7\frac{1}{2}}$

4. Divide 189695.4 by 2.708. What is the quotient of 0.01896954 divided by 2.708? Write the latter quotient in words.

5. Reduce to their lowest terms as Vulgar Fractions the Infinite or Circulating Decimals $0.003\dot{6}$ and $0.01\dot{3}\dot{6}$. Add $0.0\dot{7}$ to $0.3\dot{8}\dot{2}$, expressing the result as an Infinite or Circulating Decimal.

6. A certain square field contains 38.75 acres. Compute the length of one side of the field in metres. Given one square metre = 1550 square inches.

7. The sum £46 6s. 8d. was put at interest at 4 per cent on the 20th June, 1868. Required the amount on the 5th May, 1875.

8. Find the cubic root of 77869 to three places of decimals.

9. At what rate of compound interest will $2500 amount in 3 years to $4320? At what rate of simple interest?

XIX.

1. Reduce $\frac{184800}{1180410}$ to its lowest terms. What is a

Prime Number? When are two numbers said to be prime to each other? Reduce the numerator and denominator of the above fraction to their Prime Factors.

2. From $5\frac{1}{3}$ subtract $\frac{3\frac{7}{16}}{3\frac{1}{3}} \div \left(\frac{3}{10} \text{ of } \frac{4\frac{5}{7}}{2\frac{2}{3}} \text{ of } 4\frac{1}{6}\right)$.

3. Divide 33368949.63 by 0.007253. What is the quotient of 3336.894963 by 72530? What is the third power of 0.1? of 100? Write these answers in words.

4. Find the cube root of 0.0093 to five places of decimals. Find the square root of 531.5 to three places of decimals.

5. Reduce to their lowest terms as vulgar fractions the Infinite or Circulating Decimals 0.2̇2̇5̇, 0.002̇2̇5̇, and 0.25̇22̇5̇. Reduce $\frac{2}{7}$ to a Circulating Decimal.

6. From 1 sq. rod 5 sq. ft. subtract 7 sq. yd. 139 sq. in.

7. Find the amount of £50 12s. 5ds. at simple interest at 8 per cent, at the end of 5 yrs. 2 mos. and 3 ds.

8. One metre = 39.37 inches. Compute from this datum the value of 4 miles in kilometres.

XX.

1. Divide two thousand five hundred one and four tenths by four thousand one hundred twenty-five ten millionths. Divide 1.29136109 by 184.3, and write the quotient in words.

2. How do you divide one Vulgar Fraction by another? Give the rule and the reason of the rule. Illustrate by an example.

3. From the sum of $\frac{7\frac{1}{2}}{13\frac{1}{3}}$ and $\frac{3\frac{7\frac{1}{2}}{8}}{6\frac{2}{6\frac{2}{3}}}$ subtract $\frac{1\frac{9}{2}}{2\frac{1}{4}}$, and divide the result by the product of $3\frac{1}{6}$ and $2\frac{1}{3}$.

4. Find the cube root of 10 to four places of decimals.

5. Find the square root of 0.0000001.

6. A merchant sold a quantity of goods for $29900. He deducts five per cent from the amount of the bill for cash, and finds that he has made fifteen per cent. on the investment. What did he pay for the goods?

7. What is the compound interest on £47 13 s. 6 d. for 3 yrs. 4 mos. 15 ds., at $3\frac{1}{2}$ per cent?

8. How many feet of board in a plank 17 ft. long, 22 inches wide at one end, 13 inches wide at the other, and 4 inches thick?

9. Write the tables for Long Measure and Square Measure.

XXI.

1. Reduce 179487 to the product of its Prime Factors.

2. Find the Greatest Common Divisor of 13212 and 1851.

3. To divide by a Vulgar Fraction: give the rule and the reason of the rule.

4. Find the sum of the following numbers: fifty-seven and three thousandths; three hundred and sixty-four hundred thousandths; forty-seven thousand and eight thousand and seven hundred thousandths; eighty-seven hundred millionths; four hundred and twenty-seven ten thousandths.

5. Divide $(2\frac{1}{7} \times \frac{3}{16})$ by $(2\frac{1}{4} - 1\frac{5}{7})$, and reduce the result to a decimal.

6. What is the difference between Bank Discount and True Discount? Give an example.

7. Bought $1500 worth of goods, half on 6 months' and half on 9 months' credit. What sum at 7 per cent interest, paid down, would discharge the whole bill?

8. Find the cube root of 0.29 to three places of decimals.

9. The interest on £ 50 12 s. 6 d. for a year is £ 1 15 s. 5¼d. What is the rate per cent?

10. A cubical vat measures 9 feet in each direction: what is its capacity in *Litres?* (Given 1 metre = 39.37 inches.)

11. In the Metric System of Weights and Measures what is the unit of length? of surface? of volume? of weight? How are they related to each other?

XXII.

1. Divide four millionths by four millions, and write the quotient in words.

2. The metre = 39.371 inches: compare the kilometre with the mile.

3. Change $\frac{5}{7}$ to a decimal, and extract the cube root to four places.

4. Express 38 sq. rods, 21 sq. yards, 5 sq. feet, 108 sq. inches, in decimals of an acre.

5. The capital stock of a certain bank is $500,000, and directors have declared a dividend of 4 per cent. The sum set aside from the profits to meet this dividend is subject to a revenue tax of 5 per cent. What sum must be set aside in order that the stockholder may receive a dividend of 4 per cent on his stock?

6. From $\dfrac{4\frac{4}{15} \times 2\frac{5}{8}}{5\frac{1}{3} \quad 4\frac{1}{2}}$ subtract $\dfrac{7\frac{1}{3}}{2\frac{1}{4}}$.

7. A man has a bin 7 ft. long, 2½ ft. wide and 2 ft. deep, which contains 28 bushels of corn; how deep must he build another, which is to be 18 ft. long, 1 ft. 10½ in. wide, in order to contain 120 bushels? (Solve this question by analysis, and give your reasoning in full.)

ARITHMETIC.   129

8. What is the present worth of $10,000, due three years hence, at 7 per cent compound interest?

9. Find the Greatest Common Divisor of 1274, 2002, 2366, 7007, and 13013.

10. How do you verify your work when you have multiplied together two large numbers? Give an example to illustrate your method.

XXIII.

1. Find the Greatest Common Divisor and the Least Common Multiple of 13860 and 38500. What is the Least Common Multiple of 15, 18, and 35? When are two numbers said to be prime to each other?

2. Divide $\frac{4\frac{5}{7}}{2\frac{1}{12}}$ by $\frac{2}{5}$ of $\left(\frac{2}{1\frac{3}{4}} - \frac{1}{5}\right)$. Simplify by cancelling.

3. Reduce to its lowest terms as a Vulgar Fraction $0.05\dot{4}0\dot{5}$. Reduce $\frac{9}{28}$ to a Circulating Decimal.

4. Find the number of cubic inches (to the nearest tenth) in the British imperial gallon, which contains 10 pounds of water. Given 1 gramme = weight of 1 cubic centimetre of water, 1 cubic metre = 35.3 cubic feet, 1 kilogramme = 2.2 pounds.

5. Find the square root of 0.076 to six significant figures.

6. A rectangular field measures 30 rods and 6 feet by 21 rods and 11 feet. Find its area in acres, roods, rods, and feet.

7. Find the sum on which the interest at 9 per cent for 5 years 1 month and 18 days is $947.10.

8. Find the interest on one pound sterling at 5 per cent for one year; for one month.

XXIV.

1. What is the Greatest Common Divisor of 1872 and 432? Obtain the answer, if possible, by factoring.

2. What is the smallest sum of money that can be made up either of 2-cent, of 3-cent, of 5-cent, of 10-cent, or of 25-cent pieces?

3. Add $\frac{2}{5}$ to $\left(7\frac{1}{5} \div \frac{\frac{2}{3} \times 7\frac{1}{2}}{\frac{1}{3}}\right)$.

4. By a pipe of a certain capacity a cistern can be emptied in $3\frac{7}{15}$ hours; in what time can it be emptied by a pipe the capacity of which is $\frac{2}{5}$ greater?

5. Find the value of 7 acres 35 rods 127 feet of land, at \$108.15 per acre.

6. How many litres are there in a rectangular vat 2.8 m. long, 2 m. wide, 5 dcm. deep?

7. Find the square root of 0.9 to four places of decimals.

8. My agent sells for me 2000 yards of cloth at 24 cents a yard. He allows the purchaser 5 per cent discount for cash, and charges me $2\frac{1}{2}$ per cent on the cash receipts. How much money does he pay over to me?

XXV.

1. Find the Greatest Common Divisor of 187 and 153. Also the Least Common Multiple of the same two numbers.

2. Multiply 108 billionths by two thousand, and extract the cube root of the product.

3. Add $\frac{4}{1\frac{2}{7}}$ to $\frac{8\frac{1}{2} - 2\frac{2}{3}}{9}$.

4. A cellar is to be dug 30 feet long and 20 feet wide: at what depth will 50 cubic yards of earth have been removed?

5. What is the amount of \$340 at 8 per cent for 1 year 3 months, the interest being compounded semiannually?

6. A man receives \$18 for six days' work of 8 hours

each; what should he receive for 5 days' work of 9 hours each?

7. A cistern is 4 metres long, 24 decimetres wide, and 80 centimetres deep. How much water will it hold in cubic metres? In litres? In cubic centimetres? In grammes? In kilogrammes?

8. I have a rectangular lot of land, 64 rods long and 36 rods wide, and a square lot of the same area; how many more feet of fencing will be needed for the former lot than for the latter?

XXVI.

1. Add $\frac{17}{5}$ of $\frac{4}{1\frac{6}{11}}$ to $\frac{2\frac{2}{3}}{1\frac{2}{5}}$.

2. Multiply 0.14̇5̇ by 0.2̇97̇, and give the answer as a Circulating Decimal.

3. Find the Greatest Common Divisor of 43700 and 9430. Also obtain their Least Common Multiple.

4. I buy one fifth of an acre of land for $2178. For how much a square foot must I sell it, in order to gain twenty per cent of the cost?

5. The kilogramme equals 2 lb. 8 oz. 3 dwt. 2 gr. How many centigrammes equal one grain?

6. What is the present worth of $678.75, due 3 years 8 months hence, at 7 per cent compound interest?

7. Multiply the square root of 0.173056 by the cube root of $\frac{15625}{32768}$.

8. A can do a certain piece of work in 10 days, working 8 hours a day. B can do the same work in 9 days, working 12 hours a day. They decide to work together, and to finish the work in 6 days. How many hours a day must they work?

XXVII.

1. Divide 0.75 by $\frac{2\frac{7}{9}}{15} \times 0.081$.

2. Find the least common multiple of $\frac{1}{15}$, $\frac{1}{11}$, $2\frac{1}{2}$, 5, and $6\frac{1}{3}$.

3. A and B, 44 miles apart, travel towards each other. A travels $\frac{3}{11}$ of the whole distance, while B travels $\frac{4}{7}$ of the remainder. How far are they then apart?

4. In what time will $680, at 4 per cent simple interest, amount to $727.60?

5. How many cubic yards are there in a cistern the dimensions of which are 64 dcm., 225 cm., and 3.75 m.?

6. If 9 men build $247\frac{2}{13}$ rods of wall in 28 days, in how many days will 8 men build 51 rods?

7. What is the difference between the square root and the cube root of 1771561?

8. A can do a piece of work in 10 days, A and C can do it in 7 days, A and B can do it in 6 days; in how many days can B and C together do it?

XXVIII.

1. The sum of $\frac{\frac{3}{8} \times 0.83\frac{1}{3}}{0.5}$ and $\frac{2\frac{3}{4} \times \frac{3}{8}}{3\frac{1}{8}}$ is how many times the difference?

2. How many kilometres are there in 2 m. 6 fur. 39 rd. 5 yd.?

3. What common fraction equals the sum of $0.\dot{1}\dot{8}$ and $0.\dot{3}0769\dot{2}$?

4. A cube contains 79507 cubic inches. How many square inches does its surface contain?

5. Having purchased an acre of land, I sell from it a rectangular lot, 121 yds. long, and 25 yds. wide, for what the whole acre cost me. What per cent do I gain on the land thus sold?

6. A collector who charges 8 per cent commission on what he collects pays me $534.75 for a bill of $775. What amount of the bill does he collect?

7. A can travel around a certain island in $2\frac{2}{15}$ days, B in $3\frac{1}{5}$ days, C in $3\frac{1}{3}$ days. If they set out at the same time from the same point, and travel in the same direction, in how many days will they all come together at the starting-point, and how many times will each man have gone around the island?

ALGEBRA.

I.

1. MULTIPLY $a^3 + 2a^2x + 2ax^2 + x^3$ by $a^3 - 2a^2x + 2ax^2 - x^3$.
2. Divide 1 by $1 - m^2$, finding five terms of the series.
3. Divide $-6x^4 + 96$ by $-3x + 6$.
4. Divide $\dfrac{4(x^2 - y^2)(a + b)}{3m^2}$ by $2a(x + y)$.
5. Find the greatest common divisor of numerator and denominator, and reduce the fraction $\dfrac{x^4 - 1}{x^5 + x^3}$ to its lowest terms.
6. Divide $\dfrac{ab - bx}{a + p}$ by $\dfrac{ac - cx}{a + p}$.
7. Reduce $1 - \dfrac{a^2 - x^2}{a^2 + x^2}$ to the form of a fraction.
8. A farmer sells to one man 5 cows and 7 oxen for $370, and to another, at the same rate, 10 cows and 3 oxen for $355. Required, the price of a cow and that of an ox.
9. What is the fourth power of $-3p^2q^2$?
10. What is the third root of $-729\,a^6b^3c^{12}$?
11. Find $(2a - b)^4$ by the Binomial Theorem.

II.

1. Multiply $a^6 + 3a^4b^2 - 5a^2b^4$ by $7a^4 - 4a^2b^2 + b^4$.
2. What is the value of $-(10 - 3a^3)(10 + 3a^3)$?
3. Divide $12a^4b^9 - 14a^5b^6 + 6a^6b^3 - a^7$ by $2a^2b^3 - a^3$.

4. Reduce $\dfrac{x^4 - y^4}{(x^2 + y^2)(x^2 - 2xy + y^2)}$ to its lowest terms.

5. Reduce $\dfrac{a}{b} + \dfrac{a - 3b}{cd} - \dfrac{b^2 + ab - a^2}{bcd}$ to its simplest form as a single fraction.

6. Divide $\dfrac{10\,a^8 x^2}{9\,m^3 y^2}$ by $\dfrac{5\,a^7 x}{27\,m\,y^2}$, and reduce the answer to its lowest terms.

7. Find the value of x in terms of a, b, c, from the equation $\dfrac{2x - a}{b} = \dfrac{bc - cx}{a}$. What does this value become when $a = 2$, $b = -1$, and $c = 3$?

8. The sum of the distance passed over by two locomotives, the first running 6 hours and the second 4 hours, is 228 miles; but the second goes 24 miles more in 8 hours than the first goes in 12 hours. Find the distance each goes in an hour?

9. $(-7\,x^2 y^3)^2 =$ what? $\sqrt[3]{\left(-\dfrac{64\,a^3 b^9}{c^6}\right)} =$ what?

10. Find by the Binomial Theorem $(a - b)^6$ and $(1 - 3\,x^3)^6$.

III.

1. From $5\,a^2 b + 3\,b^2 c - 7\,c^3 de$ take $-6\,a^2 b - (4\,c^3 de - 4\,b^2 c)$.

2. Multiply $x^2 + xy + y^2$ by $x^2 - xy + y^2$.

3. Divide $x^4 - y^4$ by $x - y$.

4. Reduce $\dfrac{(x^2 - y^2)(x - y)}{(x + y)(x^2 - 2xy + y^2)}$ to its lowest terms.

5. Add together $3x + \dfrac{2xy}{5}$ and $x - \dfrac{8x}{9}$.

6. Multiply $x - \dfrac{x + 2}{a}$ by $x^2 - \dfrac{x^2 + 3}{b}$.

7. Divide $\dfrac{x(a+b)}{x-1}$ by $\dfrac{a^2-b^2}{x^2-2x+1}$.

8. Divide the number 75 into two such parts that three times the greater may exceed seven times the less by 15.

9. What is the fourth power of $\dfrac{-ax^2}{by}$?

10. What is the third root of $\dfrac{-27\,b^9}{x^8}$?

11. Find $(b - 2c^3)^4$ by the Binomial Theorem.

IV.

1. Reduce $a + b - (2a - 3b) - (5a + 7b) - (-13a + 2b)$ to its simplest form.

2. Multiply $a^3 + b^2 - c$ by $a^2 - b^3$.

3. Divide $-1 + a^3 n^3$ by $-1 + an$.

4. Reduce to one fraction $\dfrac{a}{a+2} + \dfrac{2}{a-2}$.

5. Multiply $a + \dfrac{ax}{a-x}$ by $x - \dfrac{ax}{a+x}$.

6. Divide $\dfrac{a^2 + 2ab + b^2}{x^4 - y^4}$ by $\dfrac{a^2 - b^2}{x^2 - y^2}$.

7. How much money have I when the fourth and fifth part amount together to $2.25.

8. Find the fifth power of $-2a^2$.

9. Find the fourth root of $\dfrac{a^8 b^{20} c^4}{16\, d^{12}\, z^{16}}$.

10. Find $(5a - 4x)^4$ by the Binomial Theorem.

V.

1. Reduce the following expression to its simplest form:
$ax + b(x+c) + c^2 - [(a-b)x - (b-c)(b+c)]$.

2. Multiply $x + 2y - 3z$ by $x - 2y + 3z$.

3. Divide $8a^2 - 12a^5 + 8a^4 + 18a^3 - 30$ by $6 - 4a^2$.

4. Combine $\dfrac{3}{1-2x} - \dfrac{8}{1+2x} - \dfrac{20x-4}{1-4x^2}$ in a single fraction, and reduce it to its lowest terms.

5. Divide $x^2 + \dfrac{1}{x^2} - 2$ by $x - \dfrac{1}{x}$.

6. Find $(a-b)^4$ and $\left(\dfrac{x}{y} - 2y^2\right)^4$ by the Binomial Theorem.

7. Solve the equation $\dfrac{x}{a} - \dfrac{a}{a+b} = \dfrac{x}{a-b}$, in which a and b denote known quantities. Find also the value of x when $a = -1$, $b = 3$.

8. Find a certain fraction which is such that if 3 be subtracted from both numerator and denominator, the value of the fraction becomes $\frac{1}{4}$, and that if 11 be added to both numerator and denominator, the value of the fraction becomes $\frac{3}{5}$.

9. Solve the equations $2x - y = 5$, $3y - 2z = -13$, $2z - 4x = 2$.

10. Verify the answers of Nos. 7, 8, and 9, by showing that they satisfy the original conditions.

VI.

1. From $6ac - 5ab + c^2$ subtract $3ac - [3ab - (c-c^2) + 7c]$.

2. Divide $28a^2 - 6a^3 - 6a^5 - 4a^4 - 96a + 264$ by $3a^2 - 4a + 11$.

3. Reduce $\dfrac{(a^2-b^2)(a+b)}{(a-b)(a^2+2ab+b^2)}$ to its lowest terms.

4. From $3x + \dfrac{x}{b}$ take $x - \dfrac{x-a}{c}$.

5. Divide $\dfrac{x^4 - b^4}{x^2 - 2bx + b^2}$ by $\dfrac{x^2 + bx}{x - b}$, and reduce the answer to its lowest terms.

6. Multiply $\left(-\dfrac{2a}{b^4 c^3}\right)^4$ by $\sqrt[3]{\left(-\dfrac{b^{15}}{8 a^{18} c^3}\right)}$.

7. Find $(x - y)^5$ and $(a^2 - 3b)^5$ by the Binomial Theorem.

8. Find a number from which if 5 be subtracted $\tfrac{2}{3}$ of the remainder will be 40.

9. Solve the equations $x - 6z = 6 - 2y$, $3x - 5y = 20$, $4z = 5x - 27$.

10. Verify the answers to Nos. 8, 9, by showing that they satisfy the original conditions of those problems.

VII.

1. From $4a^2 x - (2abc - 4bc + 8d)$ subtract $8abc - (4a^2 x - 2d) + abc$.

2. Multiply $x^2 + xy + y^2$ by $x^2 - xy + y^2$.

3. Divide $3a^4 - 8a^2 b^2 + 3a^2 c^2 + 5b^4 - 3b^2 c^2$ by $a^2 - b^2$.

4. Reduce $\dfrac{(a^4 - b^4)(a^2 + 2ab + b^2)}{(a^2 - 2ab + b^2)(a^2 - b^2)(a + b)}$ to its simplest form by inspection.

5. From $x - \dfrac{a+b}{2}$ take $2x + \dfrac{a-b}{c}$.

6. Divide $\dfrac{a^2 - b^2}{a^2 + 2ab + b^2}$ by $\dfrac{x(a-b)}{(a+b)^2}$.

7. Divide $\dfrac{\sqrt[3]{-64 a^8 b^2}}{c^6}$ by $(-2a^2 b c^3)^5$.

8. Subtract $(a - 2b)^5$ from $(a + 2b)^5$. Use the Binomial Theorem.

9. In a mixture of wine and cider one half the whole

plus twenty-five gallons was wine, and one third part minus five gallons was cider; how many gallons were there of each?

10. Solve the equations $\frac{x}{7} + 7y = 99$, $\frac{y}{7} + 7x = 51$.

VIII.

1. Reduce the following expression to its simplest form: $(a+b)x - (b-c)c - [(b-x)b - (b-c)(b+c)] - ax$.

2. Multiply $2x^3 - 3xy + 6y^2$ by $3x^2 + 3xy + 5y^2$.

3. Divide $40a + 8a^4 - 50a^2 - 8$ by $5a - 2a^2 - 2$.

4. Give the rule for multiplying different powers of the same quantity, and explain its reason. Example: $x^m \times x^n =$ what?

5. Reduce the following expression to a single fraction, having the least possible denominator: $\dfrac{1+x}{(1-x)^2} - \dfrac{4x}{1-x^2} - \dfrac{1-x}{(1+x)^2}$.

6. Divide $\dfrac{25 a^2 b^3 x^2}{18 c^3 y^2}$ by $-\dfrac{10 a b^3}{27 c x y^3}$.

7. Find by the Binomial Theorem the first four terms of $(a-b)^{20}$ and of $\left(1 - \dfrac{2x}{5y^2}\right)^{20}$.

8. Find the value of x in the equation $x - a = \dfrac{bc}{d} + \dfrac{c^2 x}{de}$, in which $a, b, c, d,$ and e denote known quantities. Find, also, what the value of x becomes when $a = -3, b = 0, c = -2, d = -2, e = 4$; and verify it for this case by showing that it satisfies the equation.

9. A and B have together $\frac{2}{3}$ as much money as C; B and

C have together 6 times as much as A; and B has $680 less than A and C together have: how much has each? Eliminate by comparison; and verify the answers by showing that they satisfy the given conditions.

IX.

1. Reduce the following expression to its simplest form: $ab - c(x - b) - [(x + c)(x - c) - c(b - \{c - x\}) - x^2]$.

2. Into what two factors can the following expressions be severally resolved: $(4x^6y^2 - 25x^{16})$; $(m^3 - n^3)$.

3. Multiply $6a^3 - 2a^2b + 4ab^2$ by $2a^2b - 5ab^2 - 3b^3$.

4. Divide $9x^2 - 6x^4 - 45x + 3x^3 + 54$ by $3x + 3x^2 - 9$.

5. State the rule for multiplying different powers of the same quantity, and give its reason. Examples: $x^m \times x^n =$ what? $(x^m \times x^n)^p =$ what?

6. Reduce to one fraction (with least possible denominator) $\dfrac{3}{(1-x)^2} + \dfrac{2}{1-x} + \dfrac{2}{1+x} - \dfrac{1}{1-x^2}$.

7. Divide $\dfrac{4a^2b^5}{21c^9d^8}$ by $-\dfrac{2cb^5}{3a^2d^{10}}$; and raise the quotient to the second power.

8. A and B are building a wall. A alone can build it in a days, and B alone in b days. In what time can both together build it?

9. Solve the equations $\frac{1}{2}x + \frac{1}{4}y = \frac{1}{5}z - 1$, $2z - \frac{1}{2}y = 28 - \frac{2}{3}x$, $\dfrac{4x - 3z}{2} = y + 5$.

10. Solve the equation $x^2 - 5x - 6 = 0$; and verify the answers by showing that they satisfy the equation.

11. Show that no binomial can be an exact second power.

X.

1. Reduce the following expression to its simplest form

$$x^3y^2 - \left(-xy^2 + x^3 - \frac{x^4}{y}\right)xy - x^2[-\{y^3 - y(xy - x^2)\}].$$

2. Reduce the following expression to its simplest form:
$(a + b)b + c - [(c + d)(a + d) - c(a + b - 1) - (a + c)(d - b)]$.

3. Multiply $15 a^2 + 18 ab - 14 b^2$ by $4 a^2 - 2 ab - b^2$.

4. Divide $43 x^2 y^2 - 22 x^3 y + 24 y^4 + 8 x^4 - 38 x y^3$ by $3xy - 2x^2 - 4y^2$.

5. From $\dfrac{1 + m^2}{1 - m^4}$ take $\dfrac{1 - m^2}{1 + m^2}$.

6. Divide $\dfrac{a + x}{a - x} + \dfrac{a - x}{a + x}$ by $\dfrac{a + x}{a - x} - \dfrac{a - x}{a + x}$; and reduce the quotient to its lowest terms.

7. Divide $\dfrac{6 a^3 c^7}{-45 b^8 c^6}$ by $\dfrac{9 a^9 b}{20 c^2}$; and find the second power and the third root of the quotient.

8. Reduce to one fraction $\dfrac{a + b}{(b - c)(c - a)} - \dfrac{b + c}{(a - c)(a - b)} + \dfrac{a + c}{(b - a)(c - b)}$. What is the Least Common Denominator in this example?

9. State the rule for multiplying different powers of the same quantity, and give its reason. $x^m \times x^n =$ what? $(x^m \times x^n)^3 =$ what? $(a^2 b)^m = ?$ $\sqrt{\left(\dfrac{a^{m+n}}{a^{m-n}}\right)} = ?$ What is denoted by $a^{\frac{4}{3}}$?

10. What is the reason that any term may be transposed from one member of an equation to the other, provided its sign is changed?

11. Solve the equation $\dfrac{x}{a} - \dfrac{a}{a+b} = \dfrac{x}{a-b}$. What is the value of x if $a = -2, b = 3$?

12. Out of a cask of wine from which a third part had leaked away, 21 gallons were afterwards drawn, and the cask was then half full. How much did it hold?

13. Solve the equations $3x - 5y = 63$, $\tfrac{1}{2}x + \tfrac{2}{3}y = -3$.

14. Solve the equations $x + y - z = 29$, $x - 2y + 3z = -46$, $\tfrac{1}{2}x - \tfrac{1}{3}y - \tfrac{1}{4}z = 4$.

15. Solve the equation $x^2 - 3x - 10 = 0$, and verify the answers by showing that they satisfy the equation.

16. Find four terms of $(a-b)^{25}$ and of $\left(\dfrac{x^2}{y} - xy^2\right)^{25}$ by the Binomial Theorem.

XI.

1. Reduce the following expression to its simplest form: $(a+b)x - (b-c)c - [(b-x)b - (b-c)(b+c)] - ax$.

2. Multiply $x^6 + 3x^4 y^2 - 5x^2 y^4$ by $7x^4 - 4x^2 y^2 + y^4$.

3. Divide $23a - 30 - 7a^3 + 6a^4$ by $3a - 2a^2 - 5$.

4. What is the reason that when different powers of the same quantity are multiplied, their exponents are added?

5. Reduce to one fraction, with least possible denominator, $\dfrac{1+x}{(1-x)^2} - \dfrac{4x}{1-x^2} - \dfrac{1-x}{(1+x)^2}$.

6. Divide $\dfrac{10 a^2 x^3}{9 b^2 y^3}$ by $\dfrac{5 a x^7}{27 b^2 y}$, and reduce the answer to its lowest terms.

7. A had twice as much money as B, A gained $30 and B lost $40, whereupon A gave B $\tfrac{3}{10}$ as much as B had left.

A then had what he had in the beginning and 20 per cent more. How much had each in the beginning?

8. Solve the equations $5y - 8x = -280 - 30z$, $x - 20 = z - y$, $20z - 4x = 5y$.

9. Solve the equation $2x^2 - 7x + 3 = 0$; and verify the answers by showing that they satisfy the equation.

XII.

1. Reduce the following expression to its simplest form: $(a^2 - b^2)c - (a - b)(a[b + c] - b[a - c])$.

2. Multiply $3x^5y^2 - 6x^2y^2z + y^3$ by $3x^5y^2 + 6x^2y^2z - y^3$.

3. Divide $9a^2 + 1 - 4a^4 - 6a$ by $1 + 2a^2 - 3a$.

4. What is the reason that when different powers of the same quantity are multiplied together, their exponents are added?

5. Reduce $\dfrac{x^4 - y^4}{(x^2 + y^2)(x^2 - 2xy + y^2)}$ to its lowest terms.

6. Reduce to one fraction with the least possible denominator $\dfrac{a}{b} \cdot \dfrac{b^2 - a^2 + ab}{bcd} - \dfrac{3b - a}{cd} + \dfrac{c}{bd}$.

7. Divide $\dfrac{6x^3y}{35m^5z^2}$ by $\dfrac{14y^3z^2}{15m^2x^8}$; and reduce the answer to its lowest terms.

8. Find the value of x, in terms of a, b, and c, in the equation $\dfrac{a - 2x}{b} = \dfrac{cx - bc}{a}$. What does this value become when $a = 2$, $b = -1$, $c = 3$?

9. Solve the equations $\frac{3}{4}x + 2y + 3\frac{1}{2}z = 80$, $4\frac{2}{3}y - z - \frac{2}{3}x = 66$, $5z + 18x - 7y = 140$.

10. Solve the equation $x^2 = 4x + 60$; and verify the answers by showing that they satisfy the original equation.

XIII.

1. Free the following expression from parentheses and reduce it to its simplest form: $(x + a) a + y - [(y + b) (x + b) - y (x + a - 1) - (x + y) (b - a)]$.

2. Divide $24 a^8 y^2 + 21 x^2 y^8 - 9 x^{10} + 4 x^4 y^6$ by $2 x^4 y^2 - 3 x^2 y^4 - 3 x^6$.

3. What is the reason that when different powers of the same letter are multiplied the exponents are added?

4. Resolve the following expression into a single fraction (finding the least common denominator, and reducing the answer to its lowest terms): $\dfrac{4a^2 + 3ab}{4a^2 - 3ab} - 1 - \dfrac{48 a^3 b}{16 a^4 - 9 a^2 b^2}$.

5. Divide $\dfrac{15 m^3 x^5}{14 y^4 z^7}$ by $-\dfrac{3 m^6 z^2}{8 x^8 y}$.

6. Having a certain sum of money in my pocket, I lost c dollars, afterwards spent one ath part of what remained, and then found that what I had left was one bth part of what I had had at the beginning. Find the original sum. What does the answer become if $a = 3, b = 9, c = 5$?

7. Solve the equations $8x + \frac{3}{4} y - 5z = 0, 12 z - 19 = 7 x, y - 7 = 12 - 8 z$. (If any answers are fractional, reduce them to their lowest terms.)

8. Solve the equation $2 x^2 - x - 21 = 0$. Verify each answer by substituting it in the original equation.

9. Find, by the Binomial Theorem, $(a - b)^5$, $(2 x - y^2)^5$.

XIV.

1. Free the following expression from parentheses, and reduce it to its simplest form: $(a - b + c)^2 - (a [c - a - b] - [b \{a + b + c\} - c \{a - b - c\}])$.

2. Divide $5 x y^3 - 7 x^3 y + 10 x^4 - 24 y^4$ by $xy - 3 y^2 - 2 x^2$.

3. What is the reason that when different powers of the same letter are multiplied the exponents are added?

4. Resolve the following expression into a single fraction (finding the least common denominator, and reducing the answer to its lowest terms): $\dfrac{3+2x}{2-x} - \left(\dfrac{2-3x}{2+x} - \dfrac{(16-x)x}{x^2-4}\right).$

5. Divide $\dfrac{4(a^2-ab)}{b(a+b)^2}$ by $\dfrac{6ab}{a^2-b^2}.$

6. Solve the equation $a - \dfrac{1+x}{1-x} = 0.$

7. A gentleman has two horses and one chaise. The first horse is worth a dollars less than the chaise, and the second horse b dollars less than the chaise. If $\tfrac{3}{5}$ of the value of the first horse be subtracted from that of the chaise, the remainder will be the same as if $\tfrac{7}{3}$ of the value of the second horse is subtracted from twice that of the chaise. Find the value of each horse and that of the chaise. What are the answers, if $a = -50$, $b = 50$?

8. Solve the equations $5y - 2x = 4z + 13\tfrac{1}{8}$, $\tfrac{1}{3}x = \dfrac{z-40}{4}$, $2x - y + 6z = 0.$ (If any answers are fractional, reduce them to their lowest terms.)

9. Solve the equation $18x^2 - 33x - 40 = 0.$ Verify each answer by substituting it in the original equation.

10. Find $(a-b)^7$ by the Binomial Theorem.

XV.

1. Reduce to its simplest form the expression $a - c - b - \dfrac{(c-d)e}{e}.$

2. Solve the equation — $3x^2 + 5x = 2$.

3. Find the values of the unknown quantities in the equations $x + 2y = 11$, $2x + 3z = 13$, $3y - 2z = 7$.

4. What are similar terms? What is the rule for multiplying together different powers of the same letter? For dividing? By the rule, what do you get for the exponent of a in the quotient of $a^5 \div a^3$, $a^4 \div a$, $a^2 \div a^2$, $a^3 \div a^5$? When is the square of a number larger than the number itself? How do you raise fractions to powers, $\left(\frac{a^2}{b}\right)^2$, $\left(\frac{a}{b^2}\right)^2$, for example?

5. Separate $x^8 - y^8$ into prime factors.

6. A can do a piece of work in a days, B in b days, C in c days. In how many days can A and B together do it? B and C together? A and C together? All three together?

7. Find the value of x in the equation

$$x = \frac{\dfrac{ab}{a^2 - b^2}}{\dfrac{a+b}{a-b} - \dfrac{a-b}{a+b}},$$ in its simplest form.

8. If I buy a certain number of pounds of beef at 25 cents a pound, I shall have 25 cents left; if I buy the same number of pounds of lard at 15 cents a pound, I shall have $1.25 left. How much money have I?

XVI.

1. Reduce to its simplest form the expression $\dfrac{1 + \dfrac{n-1}{n+1}}{1 - \dfrac{n-1}{n+1}}$.

2. Solve the equation — $2x^2 + 7x - 3 = 0$.

3. Find the values of the unknown quantities in the

equations $y - \dfrac{z}{3} = \dfrac{x}{5} + 5$, $\dfrac{x-1}{4} - \dfrac{y-2}{5} = \dfrac{3-z}{10}$, $x - \dfrac{2y-5}{3} = 1\tfrac{3}{4} + \dfrac{z}{12}$.

4. Separate $a^8 - b^8$ into prime factors.

5. A and B can do a piece of work in a days, A and C in b days, B and C in c days. In how many days could each person do it?

6. What is the rule for multiplying together different powers of the same letter? For dividing? Explain the reason. Multiply a^5 by a^2; a^m by a^n. Divide a^5 by a^2; a^2 by a^5; a^3 by a^3; a^m by a^n; $6\,a$ by $2\,a$.

7. Divide $x^5 - y^5$ by $x - y$.

8. Find the seventh power of $3\,a - 2\,b$ by the Binomial Theorem.

XVII.

1. Reduce the following expression to its simplest form: $(x^2 + y^2)\,z - (x + y)\,(x\,[z - y] - y\,[z - x])$.

2. What is the reason that, when different powers of the same quantity are multiplied together, the exponents are added? $x^{m+n} \times x^{m-n} =$ what? $x^{m+n} \div x^{m-n} =$ what? Give the square root of each of these results.

3. Resolve the following expression into a single fraction (finding the least common denominator, and reducing the answer to its lowest terms): $\dfrac{4\,x^2}{x^2 - y^2} - \dfrac{x-y}{x+y} - 1$. What is the most reduced value of $\dfrac{100\,a^4\,b^2}{25\,a^4\,b^2 - 9\,b^6} - \dfrac{5\,a^2\,b - 3\,b^3}{5\,a^2\,b + 3\,b^3} - 1$?

4. Divide $\dfrac{24\,a^3\,x^2\,y^3}{35\,b^4}$ by $\dfrac{32\,x^2\,y^2}{25\,a^2\,b}$.

5. The owners of a certain mill make a dollars a day each, sharing equally. If the number of owners were b less, they would make c dollars each. Required the number of owners and the total daily profit of the mill. What are the answers if $a = 80$, $b = -3$, $c = 50$?

6. Solve the equations $37 + \tfrac{1}{8}x - 12y = 8z + 55$, $\tfrac{3}{2}y = \dfrac{9-z}{9}$, $x = 4z$.

7. Solve the equation $\dfrac{2}{1-x} - 1 = \dfrac{2}{3x}$.

8. Find $(a-b)^5$ and $\left(\dfrac{2x^2}{y} - y\right)^5$ by the Binomial Theorem.

XVIII.

1. Reduce the following expression to its simplest form: $(a + b)a - ((a - b)^2 - (b - a)b)$.

2. Separate $n^5 - n$ into its prime factors.

3. From $\dfrac{1+x^2}{1-x^2}$ subtract $\dfrac{1-x^2}{1+x^2}$ and divide the result by $\dfrac{4x}{1+x^2}$.

4. "In multiplication and division, like signs give *plus* and unlike signs give *minus*." Explain fully why this is so.

5. A can perform a piece of work in a days, B can perform the same in b days, and C in c days. In how many days will the work be performed if they all labor together?

6. Solve the equations $y + \dfrac{z}{3} = \dfrac{x}{5} + 5$, $\dfrac{x-1}{4} - \dfrac{y-2}{5} = \dfrac{z+3}{10}$, $x - \dfrac{2y-5}{3} = 2$.

7. Solve the equation $\dfrac{90}{x} - \dfrac{90}{x+1} - \dfrac{27}{x+2} = 0$.

8. Find $(a+b)^5$ and $\left(1 - \dfrac{2}{3}x^3\right)^5$ by the Binomial Theorem.

XIX.

1. Reduce the following expression to its simplest form: $(9a^2b^2 - 4b^4)(a^2 - b^2) - (3ab - 2b^2)(3a[a^2+b^2] - 2b[b^2 + 3ab - a^2])b$.

2. Divide $36x^2 + 1 - 64x^4 - 12x$ by $6x - 1 - 8x^2$.

3. What is the reason that when different powers of the same quantity are multiplied together their exponents are added?

4. Reduce to one fraction with the lowest possible denominator $\dfrac{3a+2b}{a+b} - \dfrac{25a^2 - b^2}{a^2 - b^2} - \dfrac{a}{2b}$.

5. Divide $\dfrac{x+y}{x^2 - 2xy + y^2}$ by $\dfrac{x^2 + xy}{x-y}$, and reduce the answer to its lowest terms.

6. Find x in terms of a, b, and c, from the equation $\dfrac{a - 2x}{b} = \dfrac{cx - bc}{a}$. What is the value of x when $a = 2$, $b = -1$, $c = 3$?

7. A man bought a watch, a chain, and a locket for \$216. The watch and locket together cost three times as much as the chain, and the chain and locket together cost half as much as the watch. What was the price of each?

8. Solve the equation $\dfrac{5x}{x+12} - \dfrac{8-3x}{3x-1} = 1$.

9. Find $(a-b)^6$ and $\left(xy - \dfrac{x^3}{2y}\right)^6$ by the Binomial Theorem.

XX.

1. Separate into prime factors $x^5 - x$.

2. Reduce to its simplest form $3\,a^5 - 4\,a^3 + 2\,b - ca^3$
$(a^2 - 1) + \{2\,b - [7\,a^5 - a^3\,(4 - c) - a^5\,(4 + c)]\}$.

3. Divide $x^2 + \dfrac{x^4}{a^2 - x^2}$ by $\dfrac{a\,x}{a - x} - x$, and subtract the quotient from $\dfrac{a^2}{a - x}$.

4. It is said that when a term is transposed from one member of an equation to the other, its sign should be changed. Why is this so?

5. A reservoir is supplied by two pumps. Both pumps were worked three hours and the reservoir was found to be half full. On another occasion the larger pump was worked two hours and the smaller seven hours, when the reservoir was found to be two thirds full. How many hours required by either pump alone to fill the reservoir?

6. A laborer, having built 105 rods of stone fence, found that if he had built two rods less a day he would have been six days longer in completing the job. How many rods a day did he build?

7. What is Elimination? Describe fully the several processes by which it can be effected, and illustrate by examples of your own selection.

8. What is the Binomial Theorem? Find the seventh power of $\tfrac{1}{2}a - 4\,b\,c$ by aid of it.

XXI.

1. Reduce to its simplest form the following expression:
$(a + b)\,x - (b - c)\,c - [(b - x)\,b - (b - c)\,(b + c)] - a\,x$.

ALGEBRA. 151

2. Divide $\dfrac{25 a^2 b^3 x^2}{18 c^3 y^2}$ by $-\dfrac{10 a b^8}{27 c x y^8}$.

3. Divide $8 a^4 - 22 a^3 b + 43 a^2 b^2 - 38 a b^3 + 24 b^4$ by $2 a^2 - 3 a b + 4 b^2$.

4. Separate $a^8 - x^8$ into its prime factors.

5. Reduce to its simplest form the following expression:
$\left(\dfrac{1}{m} + \dfrac{1}{n}\right)(a + b) - \left(\dfrac{a+b}{m} - \dfrac{a-b}{n}\right)$.

6. Find, by the Binomial Theorem, the sixth term in the development of $(a - b)^{18}$; and the fourth term in the development of $\left(2 x - \dfrac{3 x^2}{4 y}\right)^7$.

7. Find the values of x, y, and z, from the equations
$\dfrac{3y-1}{4} = \dfrac{6z}{5} - \dfrac{x}{2} + 1\tfrac{4}{5},\ \dfrac{5x}{4} + \dfrac{4z}{3} = y + \dfrac{5}{6},\ \dfrac{3x+1}{7} - \dfrac{z}{14} + \dfrac{1}{6} = \dfrac{2z}{21} + \dfrac{y}{3}$.

8. A person performs a journey of 192 miles in a certain number of days; had he travelled 8 miles more a day he would have performed the journey in two days less time. Find how many days it took him to perform the journey.

9. Solve the equation $(x - 1)(x - 2) = 6$, and verify the results.

XXII.

1. Reduce to its simplest form the expression $a - (2 b + [3 c - 3 a - (a + b)] + 2 a - (b + 3 c))$.

2. Separate into its prime factors the expression $x^6 - y^6$.

3. Divide $(a^2 - b c)^3 + 8 b^3 c^3$ by $a^2 + b c$.

4. Solve the equation $(a + x)(b + x) = (c + x)(d + x)$.

5. A can build a wall in one half the time that B can; B can build it in two thirds of the time that C can; all to-

gether they can build it in 6 days: find the time it would take each alone.

6. Solve the equations $\dfrac{2}{x}+\dfrac{1}{y}=\dfrac{3}{z},\ \dfrac{3}{z}-\dfrac{2}{y}=2,\ \dfrac{1}{x}+\dfrac{1}{z}=\dfrac{4}{3}.$

7. Solve the equation $\dfrac{x+2}{x-1}-\dfrac{4-x}{2x}=\dfrac{7}{3}.$

8. The length of a rectangular field exceeds the breadth by one yard, and the area is three acres; find the dimensions.

9. Expand the expression $\left(2a+\dfrac{1}{b^2}\right)^7.$

10. What is Elimination? How many methods are you familiar with? Explain them in full.

XXIII.

1. Simplify $(a+b)(b+c)-(c+d)(d+a)-(a+c)(b-d).$

2. Reduce to its simplest form $\dfrac{a^4-x^4}{a^2-b^2}\times\dfrac{a+b}{a^2+x^2}\times\dfrac{a-b}{a-x}.$

3. Find the first four terms of $\left(\dfrac{a^2}{2x}-\dfrac{\sqrt{x}}{3}\right)^{10}.$

4. Find a number such that three times its square diminished by five times the number itself shall amount to 50. Solve completely.

5. What fraction is that which becomes equal to $\tfrac{3}{4}$ when 6 is added to its numerator, and equal to $\tfrac{1}{2}$ when 2 is subtracted from its denominator?

6. Solve the equation $\dfrac{2x-3}{3x-5}=\dfrac{5}{2}-\dfrac{3x-5}{2x-3}.$

7. A and B find a purse of dollars. A takes out 2 dol-

lars and $\frac{1}{6}$ of what remains; B takes out 3 dollars and $\frac{1}{6}$ of what then remains. They find that each has taken out the same amount. How many dollars were there in the purse?

8. Solve the equations $7x - 3y = a$, $5x - 11y = a$, $9y - 5z = a$.

XXIV.

1. Find the value of $a + 2x - \{b + y - [a - x - (b - 2y)]\}$ when $a = 2$, $b = 3$, $x = 6$, and $y = 5$.

2. Divide $\frac{1}{3} - 6a^2 + 27a^4$ by $\frac{1}{3} + 2a + 3a^2$.

3. Reduce to its lowest terms $\dfrac{-x^4 - a^4}{x^5 - a^2 x^3}$.

4. Find both roots of the equation $\dfrac{90}{x} - \dfrac{90}{x+1} - \dfrac{27}{x+2} = 0$.

5. Expand, by the Binomial Theorem, $(m - n)^5$ and $\left(\sqrt{a} - \dfrac{2b}{3}\right)^5$.

6. Solve the equations $y + \dfrac{z}{3} = \dfrac{x}{5} + 5$, $\dfrac{x-1}{4} - \dfrac{y-2}{5} = \dfrac{z+3}{10}$, $x - \dfrac{2y-5}{3} = \dfrac{7}{4} - \dfrac{z}{12}$.

7. A man hires a certain number of acres of land for $336. He cultivates 7 acres for himself, and lets the rest for $4 an acre more than he pays for it. He receives for the portion that he lets what he pays for the whole, or $336. Find the number of acres.

8. The value of a fraction, if its numerator is doubled and its denominator increased by 7, is $\frac{2}{3}$; while, if its denominator is doubled and its numerator increased by 2, its value is $\frac{3}{5}$. What is the fraction?

XXV.

1. A certain piece of work can be done by A and B working together in $3\frac{3}{4}$ days, by B and C in $4\frac{2}{7}$ days, and by C and A in 6 days. Required the time in which either can do it alone, and the time in which all can do it together.

2. Solve the equation $\dfrac{2+x}{2-x} - \dfrac{1-x}{1+x} = \dfrac{9}{5}$.

3. Solve the equation $x^2 - (a - b + c)x = (b - a)c$.

4. Divide $1 - \dfrac{ax + b^2}{a^2 + ax}$ by $\dfrac{a^2 - (b - 2x)b + 2b^2}{a+x} - 2b$, and reduce the result to its lowest terms.

5. Divide $9a^{2n} - a^{3n} - 27a^n + 27$ by $a^n - 3$.

6. Divide $\sqrt[4]{\dfrac{a}{b}}$ by $\sqrt{\dfrac{b}{a}}$.

7. What is the *reason* that $a^m a^n = a^{m+n}$?

XXVI.

1. Solve the equation $x - 3 = 4x - \dfrac{15 - x}{x}$.

2. What are the three methods of Elimination? Solve the following equations by any two of the three methods: $6x + \frac{1}{2}y = 0$, $2(4x - 1) = 3(y - 8)$.

3. M's age is to N's as a is to b; but c years ago M's age was to N's as a' to b'. Required the present ages of both.

4. Divide $1 - \dfrac{2}{x} \times \dfrac{1+x}{x-3}$ by $\dfrac{x^3 - 5x}{(x-3)(x+2)} - x$; and reduce the answer to its lowest terms. Simplify the division by cancelling.

5. Find the fourth term of $\left(a^2 b - \dfrac{\sqrt{b}}{2a}\right)^7$.

XXVII.

1. Solve the equation $\dfrac{ax}{c} - \dfrac{c}{bx - a} = \dfrac{b}{c}$.

2. What are eggs a dozen when two more in a shilling's worth lowers the price one penny per dozen?

3. A merchant adds yearly to his capital one third of it, but takes from it at the end of each year $5,000 for his expenses. At the end of the third year, after deducting the last $5,000, he finds himself in possession of twice the sum he had at first. How much was his original capital?

4. Divide $\sqrt[3]{a^3}\cdot\sqrt{b}$ by $\dfrac{\sqrt[3]{a^5\cdot c}}{\sqrt[4]{b^3}}$.

5. Find x from the proportion $6a^{m-2}b : x = 15a^3b^5 : 40a^{-(m-1)}$.

6. Divide $a^2 - \dfrac{a^4}{a^2 - b^2}$ by $\dfrac{ab}{b - a} - a$.

7. What is the rule for transposing a term from one side of an equation to the other; and what is the reason of the rule?

8. Solve the equations $4x + 3y + 2z = 40$, $5x - 9y - 7z = 47$, $9x - 8y - 3z = 97$.

9. Find $(a - b)^7$ by the Binomial Theorem.

XXVIII.

1. A certain sum of money at simple interest will amount to a dollars in m months, and to b dollars in n months. Find the principal and the rate of interest. Find the answers when $a = 1837.50$, $b = 1890.00$, $m = 10$, $n = 16$.

2. Solve the equation $\dfrac{27}{x - 2} - \dfrac{90}{x} = \dfrac{90}{1 - x}$.

3. Simplify $\dfrac{a + \dfrac{b}{1 + \dfrac{a}{b}}}{a - \dfrac{b}{1 - \dfrac{a}{b}}}$ $(a^6 - b^6)$.

4. Find $(x - y)^5$ and $\left(\dfrac{3a^2}{b} - \dfrac{\sqrt{b^3}}{2a}\right)^5$ by the Binomial Theorem.

5. Divide $13a^2x^2 - 5a^4 - 13ax^3 + 6x^4 - 13a^3x$ by $3ax + a^2 - 2x^2$.

6. Find two numbers of which the sum is a and difference b. State a *rule* for finding two numbers when their sum and difference are given.

7. Find the greatest common divisor and the least common multiple of $12a^3bc^4$ and $27abc^7d$.

ADVANCED ALGEBRA.

COURSE II.

I.

1. FIND the greatest common divisor of $a^6 - a^2 x^4$ and $a^6 + a^5 x - a^4 x^2 - a^3 x^3$.

2. Divide $a^{\frac{3n}{2}} - a^{\frac{-3n}{2}}$ by $a^{\frac{n}{2}} - a^{\frac{-n}{2}}$

3. Multiply $\frac{a}{b}\sqrt{\frac{c}{d}}$ by $\frac{x}{y}\sqrt[3]{\frac{d^2}{c^2}}$.

4. Divide 14 into two parts such that the quotient of the greater divided by the less shall be to the quotient of the less divided by the greater as 16 to 9.

5. Solve the equation $\sqrt{x+8} + \sqrt{x+3} = \sqrt{x}$.

6. The sum of two numbers is 17; and twice the square of the first, increased by 30, is equal to 3 times the square of the second. Find the numbers.

7. Explain the method of inserting a given number of arithmetical means between two given terms.

8. Find the sum of an infinite number of terms of the series 4, $\frac{12}{5}$, $\frac{36}{25}$, &c.

9. What is the seventh term in the expansion of $(a-x)^{10}$?

10. A and B have the same number of horses. A can make up twice as many teams, taking 3 horses at a time, as B can make up, taking 2 at a time. Find the number of horses.

II.

1. Find the least common multiple of $x^3 - x$, $x^3 - 1$, and $x^3 + 1$. Obtain the result, if possible, by factoring.

2. Simplify $(a^{\frac{1}{3}} \times a^{\frac{5}{7}})^{\frac{7}{11}}$.

3. Add together $\sqrt[3]{40}$, $\sqrt[3]{135}$, $\sqrt[3]{625}$.

4. Find both roots of the equation $2x + \sqrt{5x + 10} = 11$.

5. What two numbers are those whose difference is to the less as 4 to 3, and whose product multiplied by the less is 504?

6. What is the 4th term in the expansion of $\left(c - \dfrac{d}{4}\right)^9$?

7. The difference of two numbers is 3, and the difference of their cubes is 63. What are the numbers?

8. Obtain the formula for the sum of the terms of an Arithmetical Progression.

9. Find the sum of the series $2, \dfrac{2}{3}, \dfrac{2}{9}$, to infinity.

10. How many arrangements can be made of the letters in the word *Richmond*, taking four letters in a set?

III.

1. Reduce the following expression to its simplest form:
$a^2 - [2ab - \{bc - (a + b - c)(a - (b - c))\} + 3ab] - (b + c)^2$.

2. State and prove the rule for the *sign* of a power and of a root. How do *imaginary* quantities arise?

3. What is denoted by a^0? by a^{-3}? by $a^{\frac{2}{5}}$?

4. Reduce $1 - \dfrac{\dfrac{x}{x-1} - 1}{1 - \dfrac{x}{x+1}}$ to its simplest form.

5. Solve the equation $ax^2 + 2hx + b = 0$; and prove that the product of the roots $= \dfrac{b}{a}$.

6. There are seven numbers in Arithmetical Progression such that the sum of the 1st and 5th is 16, and the product of the 4th and 7th is 160. Find the numbers. (This question admits two solutions. Both are required.)

7. Multiply $1 - 5\sqrt{7}$ by $-2 - 3\sqrt{7}$. Divide $\dfrac{\sqrt{a}}{d^3\sqrt{c}} \cdot \sqrt[5]{\dfrac{b}{d^2}}$ by $\dfrac{a\sqrt{d^5}}{cd\sqrt[4]{c}}$.

8. Find the sixth term of $\left(\dfrac{2a}{b^2} - \dfrac{1}{3}b\sqrt{a}\right)$.

9. Find the greatest common divisor and the least common multiple of $6x^3 - 6x^2 - 72x$ and $4x^4 - 16x^3 - 84x^2$.

IV.

1. Extract the cube root of $64 - 96x - x^6 + 40x^3 - 6x^5$.

2. Solve the equation $\dfrac{x+2}{x-1} - \dfrac{4-x}{2x} - 3\tfrac{1}{2} = 0$.

3. Multiply together $2 + 3\sqrt{-1}$, $3 - 2\sqrt{-1}$, and $12 - 5\sqrt{-1}$.

4. Three times the product of two numbers, diminished by the square of the first, equals the square of the second plus one. Also the first number is greater by one than twice the second. Find the numbers. (Give both solutions.)

5. Solve the equation $ax^2 + bx + c = 0$, and state

what *relative* values of a, b, and c will make the roots equal, and what values will make them imaginary.

6. In an Arithmetical Progression, given the number of terms, the common difference, and the sum of the terms; —obtain formulas for the first term and the last.

7. In a Geometrical Progression the first term is $2\frac{1}{4}$, and the fifth term is $\frac{4}{9}$. Find the sum of the series to infinity.

8. Find the sixth term of $\left(\sqrt{\dfrac{a}{bc}} - \dfrac{\sqrt{c}}{3ab}\right)^7$.

9. How many whole numbers of four figures each can you form, each number either beginning or ending with 5, and no number containing the same figure twice?

V.

1. What are eggs a dozen when two more in a shilling's worth lowers the price one penny per dozen?

2. Solve the equations $x^3 - y^3 = 63$, $x^2y - xy^2 = 12$.

3. Multiply $\frac{3}{2} + \frac{3}{2}\sqrt{\frac{1}{2}}$ by $\frac{1}{2} - 7\sqrt{\frac{1}{2}}$.

Divide $\dfrac{\sqrt[3]{b^2}}{a^4\sqrt{a}.\sqrt[6]{c}.d^3}$ by $\dfrac{\sqrt{a^3}.d^2\sqrt[3]{d^5}}{a^8\sqrt[9]{b^8}.c}$.

4. Solve the equation $\sqrt{(21 + 4x)} + \sqrt{(x+3)} - \sqrt{(x+8)} = 0$.

5. From the letters $a\ b\ c\ d\ e$, how many combinations of 2 letters can be taken? how many of 3? how many of 4? Give the reasons.

6. Prove that the sum of any number of antecedents of a continued proportion is to the sum of the corresponding consequents as any one antecedent is to its consequent.

7. Find the greatest common divisor of $27x^5 + 3x^3 - 10x^2$ and $162x^5 - 32x$.

8. For what values of a, b, and c is $\dfrac{a-b}{b-c}$ positive, and for what values negative? For what values is it 0? ∞? indeterminate?

9. Find r and n in an arithmetical progression when a, l, and S are known.

VI.

1. A certain sum of money at simple interest will amount to a dollars in m months, and to b dollars in n months. Find the principal and the rate of interest. Find the answers when $a = 1837.50$, $b = 1890.00$, $m = 10$, $n = 16$.

2. There are three numbers in geometric progression of which the continued product is 64 and the sum of their cubes 584. Find the numbers.

3. Simplify $\dfrac{a + \dfrac{b}{1 + \dfrac{a}{b}}}{a - \dfrac{b}{1 - \dfrac{a}{b}}} (a^6 - b^6)$.

4. Find the greatest common divisor of $24x^7 + 6x^3 - 30x$ and $4x^{10} - 4x^2$.

5. Find the square root of $25x^6 - 20x^5y - 6x^4y^2 + 34x^3y^3 - 11x^2y^4 - 6xy^5 + 9y^6$.

6. Solve the equation $2\sqrt{x} - \sqrt{4x + \sqrt{7x + 2}} = 1$.

7. To find two numbers when their sum and product are given. In what case are the answers imaginary? How must a given number be divided in order that the product of its parts shall be as great as possible?

8. State and prove the Rule of Three.

PLANE GEOMETRY.

I.

1. DEFINE a Surface, a Plane, a Plane Figure, a Polygon. Mention all the different kinds of quadrilaterals.

2. Prove that if two angles of a triangle are equal, the sides opposite these angles are also equal.

3. How many degrees in each interior angle of a regular decagon? State and prove the proposition which enables you to answer this question.

4. What is the measure of an angle made by two tangents? by two chords which intersect? by two chords which do not intersect? by a tangent and a chord drawn through the point of contact? Draw a figure for each case.

5. What is the length of the longest line that can be drawn through a rectangular block of marble 12 feet long, 4 feet wide, and 3 feet thick?

6. On a given line as chord, to construct an arc of a given number of degrees.

7. Two tangents drawn to a circle make with each other an angle of 60 degrees; how many degrees of arc between the two points of contact?

8. What is meant by the equation $\pi = 3.1416$? Calculate the difference in area between a circle whose diameter is 20, and the square inscribed in it.

9. Construct a triangle, having given the base, an adjacent angle, and the altitude.

II.

1. Define a Point; a Surface; a Plane; an Angle. What is assumed as the measure of angles?

2. Prove that when two oblique lines are drawn at unequal distances from the perpendicular, the more remote is the greater.

3. Prove that when the opposite sides of a quadrilateral are equal, the figure is a parallelogram.

4. Two angles of a triangle being given, to find the third by geometric construction.

5. What is the measure of an inscribed angle? State and prove.

6. Two tangents drawn to a circle make with each other an angle of 20°; how many degrees of arc between the two points of contact?

7. The side of an equilateral triangle is 12; what is its altitude?

8. Construct a triangle, having given the base and adjacent angle, and the altitude.

III.

1. Define a Right Angle, a Perpendicular, Parallel Lines. On what does the magnitude of an angle depend? What arc is assumed as the usual measure of an angle? Why?

2. To inscribe a circle in a given triangle.

3. Prove that two triangles are equal if the three sides of one are equal respectively to the three sides of the other.

4. Define Similar Polygons.

5. To find a mean proportional between two given lines. Prove the theorem on which your solution depends.

6. Prove that every equilateral polygon inscribed in a circle is regular.

7. The ratio of the squares described on the two legs of a right triangle is equal to the ratio of what two lines?

8. To construct a square which shall be to a given square in a given ratio. Take for the given ratio 2 : 3.

9. What are the expressions for the circumference and area of a circle in terms of π and the radius?

IV.

1. Define a Plane, a Plane Figure, a Parallelogram.

2. Prove that, if in a triangle two angles are equal, the opposite sides are also equal and the triangle is isosceles.

3. What is the measure of an inscribed angle? State and prove.

4. Upon a given straight line to construct a segment such that any angle inscribed in it shall have a given magnitude.

5. To find a fourth proportional to three given lines.

6. Define Similar Polygons. Draw two polygons mutually equiangular, but not similar; also two polygons having proportional sides, but not similar. In what cases are triangles similar.

7. Prove that any two parallelograms of the same base and altitude are equivalent.

8. Prove: (*a.*) That similar triangles are to each other as the squares of their homologous sides. (*b.*) Prove that of similar polygons.

V.

1. Prove that the perpendicular from the centre of a circle upon a chord bisects the chord and the arc subtended by the chord.

2. To circumscribe a circle about a given triangle.

3. Prove that two angles are to each other in the ratio of two arcs described from their vertices as centres with equal radii.

4. Prove that a line drawn through two sides of a triangle parallel to the third side divides those two sides into proportional parts.

5. State and prove the proportion which exists between the parts of two chords which cut each other in a circle. State what proportion exists when two secants are drawn from a point without the circle.

6. Prove that two regular polygons of the same number of sides are similar.

7. Prove that similar triangles are to each other as the squares of their homologous sides.

8. Show how the area of a polygon circumscribed about a circle may be found; then how the area of a circle may be found; then prove that circles are to each other as the squares of their radii.

VI.

1. Prove that if two opposite sides of a quadrilateral are equal and parallel, the other two sides are also equal and parallel.

2. To describe a circle of which the circumference shall pass through three given points not in a straight line.

3. To find a fourth proportional to three given lines by a geometrical construction.

4. Prove that a perpendicular dropped in a right triangle from the vertex of the right angle to the hypotheuuse divides the triangle into two triangles which are similar to each other and to the whole triangle.

5. To find a mean proportional between two given lines.

6. To circumscribe about a circle a regular polygon similar to a given inscribed regular polygon.

7. Similar polygons are to each other as the squares of their homologous sides. What is the ratio between the areas of two circles?

8. Prove that the area of a circle of which r is the radius is equal to πr^2.

VII.

1. Prove that if two triangles have two sides of the one respectively equal to two sides of the other, while the included angles are unequal, the third sides will be unequal, and the greater third side will belong to that triangle which has the greater included angle.

2. Prove that the greater of two chords in a circle is subtended by the greater arc; and the converse.

3. Find the common measure of these two lines, and express their ratio in numbers:

―――――――――――――――――――――
―――――――――――――――――――――――――――――

4. To divide one side of a triangle into two parts proportional to the other two sides. (Solve and prove.)

5. The perimeters of similar polygons are to each other in what ratio? (State and prove.)

6. To circumscribe a circle about a given regular polygon. (Solve and prove.)

7. Prove that the line which joins the middle points of the two sides of a trapezoid which are not parallel is parallel to the two parallel sides and equal to half their sum. What is the area of a trapezoid?

8. To construct a parallelogram equivalent to a given square and having the sum of its base and altitude equal to a given line. (Solve and prove.)

VIII.

1. Prove that only one perpendicular can be drawn from a point to a straight line.

2. Prove that of two sides of a triangle that is the greater which is opposite the greater angle. State and prove the converse.

3. Through a given point to draw a tangent to a given circle.

4. Prove that if a line be drawn so as to divide two sides of a triangle into proportional parts, this line is parallel to the third side.

5. To inscribe in a circle a regular decagon.

6. Prove that a triangle is equivalent to half of any parallelogram of the same base and altitude.

7. To find a triangle equivalent to a given polygon.

8. To construct a parallelogram equivalent to a given square, and having the difference of its base and altitude equal to a given line.

IX.

1. Prove that when oblique lines are drawn from a point in a perpendicular to points unequally distant from the foot of the perpendicular, the more remote line is the longer.

2. To bisect a given angle.

3. Draw a number of lines radiating from a point, and

then draw two parallel lines intersecting them: prove that the parts of these parallels are proportional.

4. A tangent and a secant being drawn from a point outside a circle, prove that the tangent is a mean proportional between the entire secant and its exterior part.

5. What is the centre of a regular polygon? Prove that the sides of a regular polygon are equally distant from the centre.

6. The circumference of a circle is 341.8 feet; what is the circumference of another circle having twice the area of the former? (If you have not time to perform the computation, you can explain how to do it.)

X.

1. In what three cases is it proved that two triangles are equal? In what three cases, that they are similar? Define similar polygons.

2. Prove that if two opposite sides of a quadrilateral are equal and parallel, the other two sides are also equal and parallel. Define a Trapezoid.

3. Prove that if two polygons are composed of the same number of triangles which are respectively similar and similarly disposed, the polygons are similar.

4. State and prove the theorem concerning the ratio between the areas of two similar triangles.

5. Prove that two regular polygons of the same number of sides are similar.

6. Find the formula for the area of a circle in terms of the radius and the ratio of the circumference to the diameter.

XI.

1. To how many right angles is the sum of all the interior angles of any polygon equal? State and prove; and then state and draw the figure for the theorem on which this one immediately depends.

2. What is the measure of the angle formed by two chords which cut each other between the centre and the circumference? by two chords which meet at the circumference? by two secants which meet without the circumference? Draw the figure for each case, and prove the last one.

3. To describe a circle through three given points.

4. Prove that two regular polygons of the same number of sides are similar.

5. The area of a trapezoid is half the product of its altitude by the sum of its parallel sides.

6. The perimeter of a regular hexagon is 18. Find
   (*a.*) The area of the circumscribed circle;
   (*b.*) The area of the square inscribed in this circle.

7. Prove the proportion that exists between the parts of two intersecting chords.

XII.

1. Two parallel lines are cut by a third line. Prove what angles formed by these lines are equal, and also what angles are supplements of each other.

2. Obtain the value of any interior angle of a regular octagon.

3. An angle inscribed in a circle is measured by half the arc intercepted by its sides. Prove this proposition for each of the three cases which may arise.

4. State and prove the method of finding the centre of a given circle or arc.

5. State and prove the method of finding a mean proportional between two given straight lines.

6. From a point without a circle secants are drawn to the circle. Prove the proportion existing between the entire secants and the parts lying outside the circle.

What corollary results when one of these secant lines becomes a tangent.

7. Show how a square may be constructed equal in area to any given polygon.

XIII.

1. The perimeters of similar polygons are to each other in what ratio? The areas of similar polygons are to each other in what ratio? Proof in both cases.

2. To make a square which is to a given square in a given ratio.

3. Prove that two rectangles are to each other as the products of their bases by their altitudes. What follows if we suppose one of the rectangles to be the unit of surface?

4. Prove that two similar polygons may be divided into the same number of triangles, that are similar each to each and similarly placed.

5. To divide this line into three parts proportional to the numbers 2, 4, and 3, and prove the principle involved.

6. Prove that a line which divides two sides of a triangle proportionally is parallel to the third side.

7. Prove that a tangent to a circle is perpendicular to the radius drawn to the point of contact.

8. Prove that parallel chords intercept upon the circumference equal arcs.

XIV.

1. Prove that two triangles are equal when a side and the two adjacent angles of the one are respectively equal to a side and the two adjacent angles of the other. Under what other conditions are two triangles equal to each other?

2. Prove that the diagonals of a parallelogram mutually bisect each other. Prove at what angle the diagonals of a rhombus bisect each other.

3. Given the circumference of a circle, show how to find the centre. Show also how to draw a tangent to the circumference, either from a point on the circumference or from one without it. Give the proof in the last case.

4. Prove that the area of any circumscribed polygon is half the product of its perimeter by the radius of the inscribed circle.

5. Show how a regular hexagon may be inscribed in a circle; also an equilateral triangle. Find the ratio of the side of the inscribed equilateral triangle to the radius of the circle.

6. Prove that similar triangles are to each other as the squares of their homologous sides.

7. Show how to find a triangle equivalent to a given polygon.

SOLID GEOMETRY.
COURSE II.
I.

1. PROVE that two parallel lines are always in the same plane.

2. Prove that the sum of the plane angles, which form a solid angle, is always less than four right angles. (This theorem is sometimes stated thus: The sum of the face angles of a polyhedral angle is less than four right angles.)

3. Prove that parallel sections of a pyramid are similar polygons. What proposition relating to the volumes of pyramids is proved by aid of this proposition? (State, but do not prove.)

4. Prove that the sum of the angles of a spherical triangle is greater than two right angles.

5. A spherical triangle has angles of 75°, 94°, and 91°; what is its area in degrees? How large a portion of the surface of the sphere does it cover?

6. The surface of a sphere is 31.17 square feet; what is the surface of another sphere having three times the volume of the former?

II.

1. Define a Plane, a Prism, a Great Circle. How many faces has a parallelopiped? How many edges? How is the angle between two planes measured?

2. Prove that if two planes are perpendicular to a third plane, their line of intersection is also perpendicular to the third plane.

SOLID GEOMETRY.

3. Prove that the section of a pyramid made by a plane parallel to the base is a polygon similar to the base.

4. Prove that a triangular pyramid is a third part of a triangular prism of the same base and altitude.

5. Prove that the sum of the angles of a spherical triangle is greater than two right angles.

6. Given the radius of a sphere = 2 inches. Compute the volume and convex surface.

III.

1. If two planes are perpendicular to each other, the line drawn in one plane perpendicular to the common intersection is also perpendicular to the second plane.

2. The sum of all the plane angles which form a solid angle is always less than four right angles.

3. The solidity of a triangular prism is the product of its base by its altitude. Prove; and then show briefly how this theorem is made use of in finding the volume of a cylinder. Give the formula to express that volume.

4. Define similar polyhedrons. Prove that similar prisms, or pyramids, are to each other as the cubes of their altitudes.

5. Prove that if two spherical triangles on the same sphere, or on equal spheres, are equilateral with respect to each other, they are also equiangular with respect to each other.

6. The length of a perfectly round log of wood is 20 feet, and the diameter of each end is 12 feet. Find: (*a.*) Its convex surface. (*b.*) The surface of the greatest sphere which can be cut out of it. (*c.*) The volume of this sphere.

IV.

1. Prove that the intersections of two parallel planes with any third plane are parallel lines. Define parallel planes.

2. Planes are passed through a pyramid parallel to its base; prove that the sections formed are similar polygons, and that these polygons are to each other as the squares of their distances from the vertex.

3. What are the regular polyhedrons? How many faces has each? how many vertices? how many edges? What are the faces in each case?

4. A spherical triangle being given, to construct its polar. Prove the relations that exist between the sides and angles of a spherical triangle and those of the polar triangle.

5. The surface of a sphere is given, to find the surface of a sphere whose volume is five times as great.

6. A right cylinder and a right cone have the same circular base and the same altitude; compare their volumes. Compare with these the volume of a sphere having the same radius as the base of the cone.

V.

1. Prove that oblique lines drawn from a point to a plane, at equal distances from the perpendicular, are equal; and that of two oblique lines unequally distant from the perpendicular the more remote is the greater. As a corollary to this theorem, show how a perpendicular may be drawn to a plane from a given point without the plane.

2. Prove that two straight lines, comprehended between

three parallel planes, are divided into parts which are proportional to each other.

3. Prove that the sum of any two of the face angles of a triedral angle is greater than the third.

4. By what may a right cone be considered to be generated? To what is the area of its convex surface equal? To what is its solidity equal? Compare the solidity of a right cone with that of a right cylinder, when both solids have the same altitude, and the radius of the base of the cylinder is double that of the base of the cone.

5. Prove that the sum of the sides of a spherical triangle is less than four right angles, and that the sum of the angles is greater than two right angles.

6. Prove that every triangular pyramid is one third of a triangular prism having the same base and altitude.

ANALYTIC GEOMETRY.

I.

1. How do you find the co-ordinates of the point where two given lines intersect?

2. Find the vertices of a triangle of which the sides are $2x + 4y + 7 = 0$, $2x + y - 2 = 0$, $2x - 2y + 1 = 0$.

3. Draw the lines just given and find the angles of the triangle they form.

4. What curve is represented by each of the following equations? (i.) $x^2 + y^2 + 4y = 0$. (ii.) $9x^2 + 25y^2 = 400$. (iii.) $y^2 - 7x$. (iv.) $16y^2 - 9x^2 + 36 = 0$. Find the points at which each of these curves cuts the axes of co-ordinates.

5. Explain briefly how to construct a conic section when you have given the eccentricity (Boscovich's ratio), and the distance from the directrix to the focus. Take, for example, the eccentricity $= \frac{4}{5}$, and the distance from the directrix to the focus $= 2\frac{1}{4}$.

6. Find the equation of a conic section when the directrix is the axis of ordinates, and a perpendicular from the focus on the directrix is the axis of abscissas. Take, for example, the same data as are given in the preceding question.

Find what this equation becomes if transformed to a new set of axes parallel to the former and passing through the centre of the curve.

7. What is the locus of a point whose distance from a

fixed line is equal to its distance from a fixed point? Find the equation.

8. Construct a hyperbola whose transverse axis is 6 and less focal distance 2. Find also the conjugate axis, and the foci and directices of the conjugate hyperbola.

LOGARITHMS AND TRIGONOMETRY.

I.

1. Find the value of the following fraction by logarithms:
$$\left(\frac{0.010006}{1.4 \times \sqrt[4]{0.325062}}\right)^3.$$

2. Find the value of the following fraction by logarithms:
$$\left(\frac{(0.050395)^2}{3.2 \times \sqrt[3]{0.546781}}\right).$$

3. Find the value of the following fraction by logarithms, using arithmetical complements: $\left(\dfrac{0.00101904 \times 0.99992}{760 \times \sqrt[3]{(0.0275142)}}\right)^4.$

4. Define a logarithm.

5. Find, by logarithms, the value of the following quantities to six significant figures: $\sqrt[3]{0.0117283}$; $(0.50396)^2$; $\left(\dfrac{1}{0.50396}\right)^2$; $\dfrac{\sqrt[3]{0.0117283}}{2.4 \times (0.50396)^2}$: use arithmetical complements in dividing.

6. Solve the equation $32^x = 8$ by logarithms.

7. Prove that the sum of the logarithms of several numbers is equal to the logarithm of their product.

8. Find, by logarithms, the values of the following quantities to six significant figures: $\sqrt[5]{(0.62394)}$; $(0.00102173)^2$; $\sqrt[5]{\dfrac{1}{0.62394}}$; $\dfrac{(0.0012173)^2}{3.1 \times \sqrt[5]{(0.62394)}}.$

II.

1. In a system of which the base is 9, what is the logarithm of 81? of 3? of 27? of 9? of 1? of $\frac{1}{9}$? of $\frac{1}{81}$? of 0?

2. Find, by common logarithms, the values of the following quantities (to five significant figures): $\sqrt[3]{0.492162}$; $(0.011009)^5$; $\dfrac{1}{(0.011009)^5}$; $\dfrac{\sqrt[3]{0.492162}}{9.8 \times (0.011009)^5}$. Use arithmetical complements in dividing.

3. Solve the equation $2048^x = 16$, by logarithms.

4. Express in a decimal form the numbers which have the following logarithms in a system of which the base is $16:2$; -2; -0.25; 2.75; 0.

5. Find, by common logarithms, the values of the following quantities (to five significant figures): $\sqrt[3]{0.485463}$; $(0.00130106)^2$; $\dfrac{1}{(0.00130106)^2}$; $\dfrac{\sqrt[3]{0.485463}}{2.7 \times (0.00130106)^2}$. Use arithmetical complements in dividing.

6. Prove that the logarithm of the product of two numbers is equal to the sum of the logarithms of the numbers.

7. Find, by logarithms, the values of the following quantities (to six significant figures): $\sqrt[3]{0.0126534}$; $\left(\dfrac{1}{0.56036}\right)^{\frac{1}{2}}$; $\dfrac{\sqrt[3]{0.0126534}}{.204 \times (0.56036)^2}$. Use arithmetical complements in dividing.

8. Solve the equation $243^x = 81$ by logarithms.

9. What is the characteristic of a logarithm?

10. What is the logarithm of 1.? of .1? of 1000.? of .00001? of one hundred billionth?

11. Find, by logarithms, the value of the following quantities (to six significant figures): $\left(\dfrac{1}{.0126534}\right)^{\frac{1}{2}}$; $\dfrac{\sqrt{0.0357635}}{(\sqrt{2.04}+\sqrt{1.2036})^2}$.

12. Solve the equation $1024^x = 64$.

III.

1. Prove that the logarithm of a quotient is equal to the logarithm of the dividend diminished by the logarithm of the divisor.

2. Find, by logarithms, the values of the following quantities: $\sqrt[3]{0.03478}$, $\dfrac{(0.278)^2 \times (0.008)^{\frac{5}{2}}}{0.03478 \times (0.7)^3}$.

3. Prove the formula $(\sin A)^2 + (\cos A)^2 = 1$. What is the formula for the cosine of the sum of two angles?

4. Solve the oblique triangle in which $a = 50$, $A = 45°$, $B = 60°$. N. B.—a, b, c denote the sides; A, B, C the angles respectively opposite to a, b, c.

5. In a system of logarithms, of which 4 is the base, determine the logarithms of the following numbers: 4; 16; 2; 8; 32; 1; $\frac{1}{2} = 0.5$; $\frac{1}{4} = 0.25$; $\frac{1}{8} = 0.125$; $\frac{1}{16} = 0.0625$; 0. What is the base of the common system of logarithms?

6. Find, by logarithms, using arithmetical complements, the values of the expressions: $(0.001109)^2$; $\dfrac{1}{(0.001109)^2}$; $\sqrt[5]{\dfrac{\sqrt[3]{(0.492)} \times 560}{9 \times (0.001109)^2}}$.

7. What single function of any angle $A = \dfrac{\sec A}{\tan A}$? What function is the reciprocal of the secant.

8. Give the formulas for the sine and cosine of the sum and of the difference of two angles; and deduce from these the formulas for the sine and cosine of the double of an angle and of the half of an angle.

9. What is the sine and cosine of $0°$, $90°$, $180°$, $270°$, $360°$. Work out the formulas for the trigonometric functions of $(270° - N)$.

10. Solve the triangle in which $b = 0.007625$, $c = 0.015$, $B = 29°$. Find both solutions. N.B.—A, B, C denote the angles respectively opposite the sides a, b, c.

IV.

1. What is the logarithm of 1 in any system? of any number in a system of which that number is the base? In a system of which the base is 4, what is the logarithm of 64? of 2? of 8? of $\tfrac{1}{2}$?

2. Find by logarithms, using arithmetical complements, the value of the fraction $\dfrac{(0.02183)^2 \times (7)^{\frac{1}{3}}}{\sqrt{(0.0046)} \times 23.309}$.

3. Prove the formula for the cosine of the sum of two angles; and deduce the formulas for the cosine of the double of an angle and the cosine of the half of an angle.

4. In what quadrants is the cosine positive, and in what quadrant is it negative? Prove the values of the cosine of $0°, 90°, 180°, 270°$.

5. Given in an oblique triangle $b = 0.254$, $c = 0.317$, $B = 46°$. Solve completely.

V.

1. Prove that the logarithm of the product of several factors is equal to the sum of the logarithms of the factors.

2. Prove that the logarithm of the nth root of a number is $\tfrac{1}{n}$th of the logarithm of the number.

3. Work the following examples: $0.01706 \times 8.7634 \times 0.001 = ?$; $\dfrac{1}{0.01706} = ?$; $\sqrt{4.9} = ?$; $\sqrt[3]{0.29} = ?$; $\dfrac{\sqrt[5]{(8.7634)^3} \times 100}{9 \times \sqrt{0.1109} \times (4.9)}?$ Use arithmetical complements in working the last.

4. Which of the trigonometric functions are always less than unity? which always greater? which sometimes greater and sometimes less?

5. Write down the formulas for the sine and cosine of the sum, and the sine and cosine of the difference of two angles.

6. Prove the formula $\sin^2 a + \cos^2 a = 1$.

7. From the formulas of the two preceding questions deduce formulas for the sine, cosine, and tangent of twice an angle, and of half an angle.

8. To solve a triangle in which two sides and an angle opposite one of them are given. Example: one side = 47.6, another side = 32.9, and the angle opposite the latter side = 53° 24′.

VI.

1. Prove that the logarithm of the product of several factors is equal to the sum of the logarithms of the factors.

2. Work the following examples: (*a.*) $0.01706 \times 8.7634 \times 0.001 = ?$ (*b.*) $\frac{1}{0.01706} = ?$ (*c.*) $\sqrt{4.9} = ?$ $\sqrt[3]{0.29} = ?$ (*d.*) $\sqrt[5]{\frac{(8.7634)^3 \times 100}{9 \times \sqrt{0.1109} \times (4.9)^{\frac{2}{3}}}} = ?$ Use arithmetical complements in working the last.

3. Find the sines, cosines, and tangents, both natural and logarithmic, of the following angles: (*A.*) 24° 47′ 22″. (*B.*) 56° 23′ 14″. (*C.*) 134° 28′. Find the angles which correspond to the following functions: log sin A = 9.94325. nat cos B = − 0.57832. nat tan C = 1.473.

4. Prove the formula $a^2 = b^2 + c^2 - 2bc \cos A$.

5. Prove the formulas $1 + \cos A = 2 (\cos \frac{1}{2} A)^2$. $1 - \cos A = 2 (\sin \frac{1}{2} A)^2$.

LOGARITHMS AND TRIGONOMETRY. 183

6. From the formulas of the last two questions deduce the formula $\sin \frac{1}{2} A = \sqrt{\frac{(s-b)(s-c)}{bc}}$.

7. The sides of a triangle are 37, 41, and 48; what are the angles?

8. To solve a triangle when two sides and the included angle are given. *Example:* Given the sides 47.6 and 58.4, the included angle 52° 24'.

VII.

1. In a system of logarithms of which the base is 16, what is the number of which the logarithm is — 1.25? In the system of which 10 is the base, why do the logarithms of two numbers composed of the same series of significant figures differ only in their characteristics?

2. Prove that the logarithm of the continued product of several numbers is equal to the sum of their logarithms.

3. Write (without proving) the formulas for the sine and cosine of the sum and of the difference of two angles; and prove the formula $\cos A + \cos B = 2 \cos \frac{1}{2}(A+B) \cos \frac{1}{2}(A-B)$.

4. Give the values of the sine, cosine, and tangent of 0°, 90°, 180°, 270°, 360°. Find the formula for $\cos(270° - \varphi)$.

5. Given in a triangle $b = 0.1072$, $c = 0.0625$, $C = 20° 17'$. Solve completely.

6. Find by logarithms the value of $\dfrac{\sqrt[3]{(0.07323)^2}}{0.35308 \times 3700}$.

7. Given the cotangent of an angle equal to $2\sqrt{2}$; find the other trigonometric functions, by computation.

VIII.

1. What is the reason that, in the common system, the logarithms of two numbers consisting of the same series of significant figures differ only in their characteristics?

2. Write (without proving) the formulas for the sine and cosine of the sum and of the difference of two angles; and deduce those for the sine and cosine of the double of an angle and of the half of an angle.

3. Find, by means of formulas, the trigonometric functions of 30° and 60°.

4. Prove that, in any triangle, $\dfrac{a+b}{a-b} = \dfrac{\tan \frac{1}{2}(A+B)}{\tan \frac{1}{2}(A-B)}$.

5. Solve the triangle in which $a = 110.6$, $b = 56.7$, $C = 108° \, 24'$.

6. Find, by logarithms, the value of the fraction $\dfrac{\sqrt[5]{(0.027919)^3}}{(0.0010708)^2 \times 7.9}$.

IX.

1. Obtain a formula by which, when the sine of an angle is known, its cosine may be found. Also formulas for finding the tangent and cotangent of an angle, when the sine and cosine are given.

2. Obtain, by the formulas of the previous question, the trigonometric functions of 45°.

3. Prove that, in any triangle, the sines of any two angles are proportional to the opposite sides.

4. Solve the triangle in which two sides are 32.64 and 25.14, and the angle opposite the second side is $32° \, 48'$. Are there two solutions to this problem? Why?

5. Find, by logarithms, the value of $\sqrt[4]{\dfrac{32.85 \times (.0146)^2}{(23.9)^3}}$.

6. State the process and give the formulas by which, when two sides and the included angle of a triangle are known, the remaining parts can be obtained.

X.

1. In the system of logarithms with six for its base, of what numbers will 3 and -3 be the logarithms? What will be the index of the logarithm of 2000?

2. Find, by logarithms, the value of $\frac{4}{7}\sqrt[3]{\frac{84.9 \times .001}{(.4286)^2}}$.

3. Show, by means of a diagram, what lines may be taken to represent the sine and the cosine of angles in each of the four quadrants of a circle, the radius of the circle being unity. Show also what are the algebraic signs of these same functions in the different quadrants.

4. Obtain formulas for the trigonometric functions of a negative angle.

5. In a right plane triangle, one side is 0.1426 and the opposite angle is 47° 29'. Solve the triangle.

6. Write the formulas for the sine and the cosine of the sum of any two angles; and obtain from them formulas for the sine and the cosine of the double angle. The sine of a certain angle is $\frac{7}{25}$. Find the trigonometric functions of double that angle.

7. Two sides of a plane oblique triangle are 16.49 and 21.37, and the included angle is 129° 37'. Find the other two angles. State the method of finding the remaining side.

8. One angle of a plane triangle is 30°, and an adjacent side is 12. What values of the side opposite the given angle will give two solutions to the triangle? What values will give only one? What values will give no solution?

XI.

1. Between what two integers does the common logarithm of 327.8 lie? Give the reason for your answer.

2. Find, by logarithms, the value of $\frac{1}{.04682} \times (\frac{4}{9})^3 \times \sqrt[3]{824.7}$.

3. In what quadrants may an angle be taken whose secant is 1.25? Obtain the corresponding values of the sine.

4. Find all the functions of $(180° + y)$.

5. The hypothenuse of a right triangle is 0.3287, and one side is 0.1938. Solve the triangle.

6. By means of the formulas for the sine and the cosine of the sum of two angles, obtain the formula, $\tan(x+y) = \frac{\tan x + \tan y}{1 - \tan x \tan y}$.

7. The three sides of a triangle are 1.328, 1.416, and 0.9388. Find the angles.

XII.

1. In a certain system of logarithms the logarithm of 0.125 is -1.5. What is the base?

2. Find, by logarithms, the value of $\sqrt{\frac{2}{3}} \div \frac{3}{7} (.0048659)^{\frac{2}{3}}$.

3. Of the following angles, which have a cosine equal to -0.5? a tangent equal to 1? a cosecant equal to $-\sqrt{2}$? 45°; 120°; 225°; 240°; 315°; $-240°$; $-315°$; 600°.

4. If $\sin \phi = m$, obtain the values of $\sin 2\phi$ and $\cos 2\phi$.

5. In any triangle ABC, prove that $a^2 = b^2 + c^2 - 2bc \cos A$.

6. Solve the right triangle, given an angle 47° 48′ 13″, and the opposite side 0.043629.

7. Find all the trigonometric functions of (180° + y).

8. Give the formulas and state the process by which an oblique triangle is solved when two sides and the included angle are given.

PHYSICS.
COURSE II.
I.

1. DEFINE the terms Force, Weight, Mass.

2. If two forces acting perpendicularly on a straight lever in opposite directions and on the same side of the fulcrum balance each other, they are inversely as their distances from the fulcrum; and the pressure on the fulcrum is equal to the difference of the forces.

3. The pressure upon any particle of a fluid of uniform density is proportional to its depth below the surface of the fluid.

4. In 50 cubic yards of rock, whose average specific gravity is 142, there enter 32 cubic yards of a substance whose specific gravity is 124. Find the specific gravity of the remainder of the rock.

5. How would you graduate a hydrometer for ascertaining the strength of alcohol.

6. How do you change from Fahrenheit to Centigrade?

II.

1. How does the weight of a body differ from the mass? How are forces represented? If it be stated that two forces of 5 lbs. and 10 lbs. act upon a body, what more is wanting to enable us to determine the result?

2. Prove the proposition, "If two forces, acting at any angles on the arm of any lever, balance each other, they are

inversely as the perpendiculars drawn from the fulcrum to the directions in which the forces act."

3. The direction of two forces, P and Q, which act on a bent lever and keep it at rest, make equal angles with the arms of the lever, which are at 6 and 8 inches respectively. Find the ratio of Q to P.

4. Find the centre of gravity of a triangle. One half of a given triangle is cut off by a straight line parallel to the base: find the centre of gravity of the remaining trapezium.

5. Prove the proposition, "When a body of uniform density floats on a fluid, the part immersed: the whole body :: the specific gravity of the body: the specific gravity of the fluid."

6. If the difference of readings of a thermometer, which is graduated both according to Fahrenheit's and the Centigrade scale, be 40, find the temperature in each scale.

MECHANICS.

COURSE II.

I.

1. Prove the law of the parallelogram of forces.

2. Find the centre of gravity of any number of heavy points.

3. Deduce the law of the equilibrium of movable pulleys, taking the weight of the pulleys into account.

4. Find the relation of P's velocity to that of W on an inclined plane.

5. Prove the equality of fluid pressures. Explain the Hydrostatic Paradox.

6. If the volume of the receiver of a condensing pump is five times that of the barrel, find the pressure on the valve after ten strokes.

II.

1. Define Force; Weight; Mass; and Density. How are forces represented?

2. Give the axioms of the lever. Assuming the properties of the straight lever, prove the laws of the bent lever.

3. Can the resultant of two forces, in any case, be equal to one of the components? If so, what are the conditions?

4. A string passing round a smooth peg is pulled at each end by a force equal to the strain upon the peg. Find the angle between the two parts of the string.

5. Deduce the laws of the inclined plane, both when the body on the plane is at rest, and when it is in motion.

6. Prove that when a body of uniform density floats on a fluid, the part immersed : the whole body : : the specific gravity of the body : the specific gravity of the fluid.

7. Prove that the elastic force of air at a given temperature varies as the density. A barometer is sunk to the depth of twenty feet in a lake, find the consequent rise in the mercurial column. (Specific gravity of mercury $= 13.57$.)

III.

1. Describe the different kinds of levers, giving examples of each kind.

2. Enunciate the *Parallelogram of Forces*. Assuming it to be true for the *direction* of the resultant, prove it for the *magnitude* of the resultant.

3. A string passing around a smooth peg is pulled at each end by a force equal to the strain on the peg. Find the angle between the two parts of the string.

4. On the *inclined plane* when the power acts parallel to the plane, prove that the power : the weight : : height of the plane : length of the plane.

5. In the leaning tower of Pisa the top overhangs the base by 12 feet; why does it not fall?

6. Prove that when a body of uniform density floats on a fluid, the part immersed : the whole body : : the specific gravity of the body : the specific gravity of the fluid.

7. A piece of iron weighs 12 pounds in water; and when a piece of wood which weighs 5 pounds is attached to it, the two together weigh 9 pounds in water. Find the specific gravity of the wood.

8. Explain why a balloon rises, and why the higher it gets the slower it rises. Why does it ever cease to rise?

9. Describe the construction of the common suction pump and its operation. (Draw a diagram of the pump.)

10. A piece of wood floats in a cup of water under the receiver of an air-pump. Will it sink deeper or rise higher when the air is exhausted? Why?

IV.

1. If two weights, P and Q, acting perpendicularly on a straight *Lever* on opposite sides of the fulcrum balance each other, determine the position of the fulcrum and the pressure on it.

The scale-pans of a *Balance* are of unequal weight, and its arms consequently also of unequal length; find the true weight of any substance from its apparent weights, when placed in the two scale-pans respectively.

2. If two forces, acting at any angles on the arms of any *Lever*, balance each other, they are inversely as the perpendiculars drawn from the fulcrum to the directions in which the forces act.

3. If three forces, represented in direction and magnitude by the sides of a triangle taken in order, act on a point, they will produce equilibrium.

Two forces whose magnitudes are $\sqrt{3} \times P$ and P, respectively, act at a point in directions at right angles to each other; find the magnitude and direction of the force which will balance them.

4. In that system of *Pulleys*, in which the same string passes round any number of pulleys, and the parts of it

MECHANICS. 193

between the pulleys are parallel, there is equilibrium (neglecting the weights of the pulleys) when $P : W :: 1 :$ the number of strings (n) at the lower block.

5. Prove that when a body is suspended from a point, it will rest with its *Centre of Gravity* in the vertical line passing through the point of suspension. Hence show how the *Centre of Gravity* of any plane figure of irregular outline may practically be determined.

6. Describe an experimental proof, that, if the pressure at any point of a fluid be increased, the pressure at all other points will be equally increased. By what short form of words is this property of fluid pressure sometimes described?

In the common *Hydraulic Press*, are the fluid pressures and tendency to break uniform throughout the cylinders?

7. Prove that if a body floats in a fluid, it displaces as much of the fluid as is equal in weight to the weight of the body; and it presses downwards, and is pressed upwards, with a force equal to the weight of the fluid displaced.

A uniform cylinder, when floating vertically in water, sinks to a depth of 4 inches; to what depth will it sink in alcohol of specific gravity 0.79?

8. Describe the construction of the *Condenser*, and the mode of its operation.

A cylinder, filled with atmospheric air, and closed by an air-tight piston, is sunk to the depth of 500 fathoms in the sea; required the compression of the air (assume specific gravity of sea water to be 1.027, specific gravity of mercury 13.57, and height of *Barometer* 30 inches).

EXAMINATION PAPERS
OF JUNE, 1874.

ANCIENT HISTORY AND GEOGRAPHY.

[Take the *first three*, and *one* other; *four* in all.]

1. Name in the order of time the successive conquests made by the Romans, and note distinctly the position of each conquered state or district.

2. By a map or by words represent or describe Sicily. Point out its place in Grecian and in Roman history.

3. Name *eight* places that were noted in ancient times: four Greek, and four Roman. Give their situation, and show their importance in history.

4. What objects would a Roman be sure to point out to a stranger visiting Rome in the time of Augustus? Describe some of them. Show, by a rough plan, their position relatively to each other, and connect them with events in Roman history.

5. The legislation of Solon.

6. The Gracchi and the Agrarian Laws. State precisely the character of these laws.

MODERN AND PHYSICAL GEOGRAPHY.

1. Upon what principle is Mercator's map constructed? How do the parallels and meridians appear upon it? What distortion is produced in the forms of the countries?

2. Draw an outline map of Africa and put upon it, in their proper positions, the equator, and the meridian of Greenwich. Give also the names of the bodies of water surrounding the continent, and the positions of important islands near the coast.

3. What is shown by a profile of a country? Draw a profile of South America, from the mouth of the Amazon to the Pacific Ocean.

4. Describe the southern coast of Europe, giving the names of countries, bodies of water, important islands, principal seaports, and largest rivers.

5. What time is it at Madras when it is eight o'clock in the morning at Boston? Longitude of Madras, 80° E.; of Boston, 71° W.

6. Where is the Great Bear Lake? Why was it so named? What other large lakes are near it? Which continent has the smallest number of lakes? Where are the principal salt lakes, and why are they salt?

7. Where does the Colorado River rise and empty? What are the most striking physical features of the country through which it flows?

Answer the same questions for each of the following rivers: — Columbia; Niagara; Hudson; Seine; Ganges.

8. What cities of Europe are in nearly the same latitude as New York?

9. Describe two water routes between Marseilles and Hong Kong.

GREEK COMPOSITION.

TRANSLATE INTO GREEK:—

When these ten thousand Greeks had come in their march to the great river Euphrates, they found a barbarian soldier who told them that the great king with all his army was only two stages (day's march) distant, and that if they should go forward during all that night and the following (ἐπιέναι) day, they would see the king's forces before the time for supper came. When the generals heard this, they determined (it seemed good to them) not to remain where they were, but to cross (διαβαίνειν) the river and send Xenophon with a hundred hoplites so that they might know whether the man had spoken the truth.

GREEK PROSE.

☞ Read the following notice before doing any of the paper:—
[Those offering Greek Reader, take 2, 4, 5. Those offering Anabasis, four books, and 7th book of Herodotus, take 1, 2, 5. Those offering the whole of Anabasis, take 1, 2, 3.]

1. (Anab. II. v. 10, and part of 11.) εἰ δὲ δὴ καὶ μανέντες σε κατακτείναιμεν, ἄλλο τι ἂν ἢ τὸν εὐεργέτην κατακτείναντες πρὸς βασιλέα τὸν μέγιστον ἔφεδρον ἀγωνιζοίμεθα; ὅσων δὲ δὴ καὶ οἵων ἂν ἐλπίδων ἐμαυτὸν στερήσαιμι, εἰ σέ τι κακὸν ἐπιχειρήσαιμι ποιεῖν, ταῦτα λέξω. ἐγὼ γὰρ Κῦρον ἐπεθύμησά μοι φίλον γενέσθαι, νομίζων τῶν τότε ἱκανώτατον εἶναι εὖ ποιεῖν ὃν βούλοιτο. From what and where is μανέντες?

2. (Anab. IV. I. 23, 24.) Καὶ εὐθὺς ἀγαγόντες τοὺς ἀνθρώπους ἤλεγχον διαλαβόντες εἴ τινα εἰδεῖεν ἄλλην ὁδὸν ἢ τὴν φανεράν. ὁ μὲν οὖν ἕτερος οὐκ ἔφη, μάλα πολλῶν φόβων προσαγομένων· ἐπεὶ δὲ οὐδὲν ὠφέλιμον ἔλεγεν, ὁρῶντος τοῦ ἑτέρου κατεσφάγη. ὁ δὲ λοιπὸς ἔλεξεν ὅτι οὗτος μὲν οὐ φαίη διὰ ταῦτα εἰδέναι, ὅτι αὐτῷ ἐτύγχανε θυγάτηρ ἐκεῖ

παρ' ἀνδρὶ ἐκδεδομένη· αὐτὸς δ' ἔφη ἡγήσεσθαι δυνατὴν καὶ ὑποζυγίοις πορεύεσθαι ὁδόν. Explain mood of εἰδεῖεν.

3. (Anab. VI. IV. 20, 21.) Καὶ πάλιν τῇ ὑστεραίᾳ ἐθύετο, καὶ σχεδόν τι πᾶσα ἡ στρατιὰ διὰ τὸ μέλειν ἅπασιν ἐκυκλοῦντο περὶ τὰ ἱερά· τὰ δὲ θύματα ἐπελελοίπει. οἱ δὲ στρατηγοὶ ἐξῆγον μὲν οὔ, συνεκάλεσαν δέ. εἶπεν οὖν Ξενοφῶν, Ἴσως οἱ πολέμιοι συνειλεγμένοι εἰσὶ καὶ ἀνάγκη μάχεσθαι· εἰ οὖν καταλιπόντες τὰ σκεύη ἐν τῷ ἐρυμνῷ χωρίῳ ὡς εἰς μάχην παρεσκευασμένοι ἴοιμεν, ἴσως ἂν τὰ ἱερὰ προχωροίη ἡμῖν.

4. (Phaedo, p. 109 of Reader, s. 24.) Καὶ ὁ Κρίτων ἀκούσας ἔνευσε τῷ παιδὶ πλησίον ἑστῶτι, καὶ ὁ παῖς ἐξελθών, καὶ συχνὸν χρόνον διατρίψας, ἧκεν ἄγων τὸν μέλλοντα διδόναι τὸ φάρμακον, ἐν κύλικι φέροντα τετριμμένον· ἰδὼν δὲ ὁ Σωκράτης τὸν ἄνθρωπον, Εἶεν, ἔφη, ὦ βέλτιστε, σὺ γὰρ τούτων ἐπιστήμων τί χρὴ ποιεῖν; Οὐδὲν ἄλλο, ἔφη, ἢ πιόντα περιιέναι, ἕως ἄν σου βάρος ἐν τοῖς σκέλεσι γένηται, ἔπειτα κατακεῖσθαι· καὶ οὕτως αὐτὸ ποιήσει. καὶ ἅμα ὤρεξε τὴν κύλικα τῷ Σωκράτει. From what and where is πιόντα?

5. (Herod. VII. 234; Reader, p. 155, § 57.) Οἱ μὲν δὴ περὶ Θερμοπύλας Ἕλληνες οὕτω ἠγωνίσαντο· Ξέρξης δὲ καλέσας Δημάρητον εἰρώτα ἀρξάμενος ἐνθένδε. Δημάρητε, ἀνὴρ εἶς ἀγαθός. τεκμαίρομαι δὲ τῇ ἀληθείῃ· ὅσα γὰρ εἶπας, ἅπαντα ἀπέβη οὕτω. νῦν δέ μοι εἰπέ, κόσοι τινές εἰσι οἱ λοιποὶ Λακεδαιμόνιοι, καὶ τούτων ὁκόσοι τοιοῦτοι τὰ πολέμια, εἴτε καὶ ἅπαντες. ὁ δ' εἶπε· Ὦ βασιλεῦ, πλῆθος μὲν Λακεδαιμονίων πολλόν, καὶ πόλιες πολλαί· τὸ δὲ ἐθέλεις ἐκμαθεῖν, εἰδήσεις.

GREEK POETRY.

1. Τώ γ' ὡς βουλεύσαντε διέτμαγεν· ἡ μὲν ἔπειτα
Εἰς ἅλα ἆλτο βαθεῖαν ἀπ' αἰγλήεντος Ὀλύμπου,
Ζεὺς δὲ ἑὸν πρὸς δῶμα. θεοὶ δ' ἅμα πάντες ἀνέσταν
Ἐξ ἑδέων, σφοῦ πατρὸς ἐναντίον· οὐδέ τις ἔτλη

Μεῖναι ἐπερχόμενον, ἀλλ' ἀντίοι ἔσταν ἅπαντες.
Ὣς ὁ μὲν ἔνθα καθέζετ' ἐπὶ θρόνου· οὐδέ μιν Ἥρη
Ἠγνοίησεν ἰδοῦσ' ὅτι οἱ συμφράσσατο βουλάς
Ἀργυρόπεζα Θέτις, θυγάτηρ ἁλίοιο γέροντος.
Iliad, I. 531–539.

Where is διέτμαγεν found? Attic for σφοῦ, ἔσταν.

2. Δεύτερον αὖτ' Ὀδυσῆα ἰδὼν ἐρέειν' ὁ γεραιός·
"Εἴπ' ἄγε μοι καὶ τόνδε, φίλον τέκος, ὅς τις ὅδ' ἐστὶν
Μείων μὲν κεφαλῇ Ἀγαμέμνονος Ἀτρείδαο,
Εὐρύτερος δ' ὤμοισιν ἰδὲ στέρνοισιν ἰδέσθαι.
Τεύχεα μέν οἱ κεῖται ἐπὶ χθονὶ πουλυβοτείρῃ,
Αὐτὸς δὲ κτίλος ὣς ἐπιπωλεῖται στίχας ἀνδρῶν.
Ἀρνειῷ μιν ἔγωγε ἐΐσκω πηγεσιμάλλῳ,
Ὅς τ' ὀΐων μέγα πῶϋ διέρχεται ἀργεννάων."
Iliad, III. 191–198.

Divide two first verses into feet.

GREEK GRAMMAR.

[All Greek words must be written with their accents.]

1.* Decline πολίτης, ἄνθρωπος, and ἀληθής in the *singular*; θής, ἀνήρ, and λύων in the *plural*.

2. Decline ναῦς, μείων, σύ, ὅς, and τίς (interrogative) throughout.

3.* Compare σοφός, φίλος, μέγας, and ῥᾴδιος.

4. Inflect the present optative and imperfect indicative of ὁράω; the imperfect of δείκνυμι; and the present indicative of εἰμί and εἶμι, with the meaning of each.

5.* Give the principal parts of τυγχάνω, θνῄσκω, δράω, λείπω, and ἵημι.

6. Translate οἶδα τοῦτον γράφοντα and οἶδα τοῦτο γράφειν. Translate ὁ αὐτὸς ἀποτέμνεται τὴν κεφαλήν, and explain the accusative.

7. Translate οἶκος δ' αὐτός, εἰ φθογγὴν λάβοι, σαφέστατ' ἂν λέξειεν, and explain the optatives.

8. Explain the subjunctive in ἐφοβούμην μὴ τοῦτο γένηται. Could it be changed to the optative?

9. Explain the optative in εἶπεν ὅτι γράφοι. Could you have any other mood than the optative in this case?

10. What is an iambus? a spondee? an anapæst? What is a dactylic hexameter, and what substitutions are allowed in it?

* Candidates for ADVANCED STANDING will omit 1, 3, and 5, and answer the following questions.

11. Translate τί μ' οὐκ ἔκτεινας εὐθὺς ἵνα μήποτε εἶδον τὸ φῶς, and explain ἵνα εἶδον.

12. Translate εἰ αὐτοὺς ἴδοιεν ἂν ἔφυγον. What is the construction of ἴδοιεν? of ἄν?

13. How would you express in Greek: *Would that Cyrus were alive! He said that he would do it, He said that he would have done it, He said that he did it?*

14. What is an iambic trimeter of tragedy, and what substitutions are allowed in it?

LATIN COMPOSITION.

☞ Candidates for the Freshman Class are required to translate the whole of I. and in II. only to 2, "He said." Candidates for Advanced Standing will translate the whole of I. and II.

I.

TRANSLATE INTO ENGLISH: —

Restat ut doceam omnia, quæ sint in hoc mundo, quibus utantur homines, hominum causa facta esse et parata. Principio ipse mundus deorum hominumque causa factus est, quæque in eo sunt, ea parata ad fructum hominum et inventa sunt. Est enim mundus quasi communis deorum atque hominum domus aut urbs utrorumque. Ut igitur Athenas et Lacedæmonem Atheniensium Lacedæmoniorumque causa putandum est conditas esse, omniaque, quæ sint in his urbibus, eorum populorum recte esse dicuntur, sic quæcumque sunt in omni mundo deorum atque hominum putanda sunt.

II.

TRANSLATE INTO LATIN: —

1. It was the custom[1] in old times for senators at Rome to enter[2] the senate-house[3] attended-by[4] their young[5] sons. The mother of Papirius asked[6] her son what-in-the-world[7] the fathers had been doing[8] in the senate. The boy answered that it must be-kept-secret.[9] The woman gets[10] more eager[11] to hear. Then the boy resorts-to[12] an ingenious[13] lie.[14]

[1] mos. [2] introire. [3] curia. [4] cum. [5] prætextatus. [6] percontari. [7] quisnam. [8] agere. [9] tacere. [10] fieri. [11] cupidus. [12] consilium capere (with the genitive). [13] festivus. [14] mendacium.

2. He said that the-discussion-had-been[1] whether it was more expedient[2] for one man to have two wives or for one woman to have two husbands.[3] The-next-day[4] the matrons beg[5] the senate that one woman might rather[6] be-married-to[7] two men than that two women might have one husband.

[1] agere (passive). [2] utilis. [3] maritus. [4] postridie. [5] obsecrare. [6] potius. [7] nubere.

LATIN GRAMMAR.

Mark the quantity of the penults and last syllables of the following words: *custodis, arbores, frigora, gladiolus, infamis* (nom.), *victricis* (acc. plur.), *inopis, petitur, perivit, periit, peritus, ambitus, apices.*

Decline *decus, locus, specus, celeber, quivis;* compare *inferus, humilis.* Form and compare *loquax, sanctus.*

Form derivatives with the terminations *-tas, -tor, -ensis, -olus, -sco,* and give their meaning.

Give the principal parts of *sumo, sentio, libet, pateo, patior, spondeo, adjuvo, tollo, disco, vereor, facio* with *con, eo* and *do* with *re.*

Give a synopsis of the Subjunctive Active and Passive (first Person) of two of these verbs not of the same conjugation. Give a complete synopsis of one other. Inflect the Imperative of *patior.* Give all the Participles and Infinitives of *sentio.*

Explain the formation of the presents *gigno* and *frango,* of the perfects *didici* and *dixi,* and of the participle *natus.*

What case or cases (separately or together) follow *persuadeo, moneo, obliviscor, solvo, vereor, præ, sub*?

TRANSLATE into Latin, with gerundive (participle in *-dus*), The city must be spared, I must go.

What construction is used in clauses (or verbs) after *timeo, gaudeo, dico, audeo*?

How are future conditions expressed in Latin? Express in Latin, in as many ways as you can, "Antony came to bury (*sepelio*) Cæsar."

What difference in meaning between *utinam sim* and *utinam essem*?

LATIN. — *Course I.*

CÆSAR AND SALLUST.

TRANSLATE *two* passages, — the *first* and one other.

I. Quo prœlio bellum Venetorum totiusque oræ maritimæ confectum est. Nam quum omnis juventus, omnes etiam gravioris ætatis, in quibus aliquid consilii aut dignitatis fuit, eo convenerant, tum navium quod ubique fuerat in unum locum coegerant; quibus amissis reliqui neque quo se reciperent neque quemadmodum oppida defenderent habebant. Itaque se suaque omnia Cæsari dediderunt. In quos eo gravius Cæsar vindicandum statuit, quo diligentius in reliquum tempus a barbaris jus legatorum conservaretur. Itaque omni senatu necato reliquos sub corona vendidit. — CÆSAR, B. G. III.

II. His rebus permotus Q. Titurius, quum procul Ambiorigem suos cohortantem conspexisset, interpretem suum Cn. Pompeium ad eum mittit rogatum ut sibi militibusque parcat. Ille appellatus respondit: Si velit secum colloqui, licere; sperare a multitudine impetrari posse quod ad militum salutem pertineat; ipsi vero nihil nocitum iri, inque eam rem se suam fidem interponere. Ille cum Cotta saucio communicat, si videatur, pugna ut excedant et cum Ambiorige una colloquantur; sperare ab eo de sua ac militum salute impetrare posse. Cotta se ad armatum hostem iturum negat atque in eo constitit. — CÆSAR, B. G. V.

III. Atheniensium res gestæ, sicut ego æstumo, satis amplæ magnificæque fuere, verum aliquanto minores tamen quam fama feruntur. Sed quia provenere ibi scriptorum magna ingenia, per terrarum orbem Atheniensium facta pro maxumis celebrantur. Ita eorum qui ea fecere virtus tanta habetur, quantum ea verbis potuere extollere præclara ingenia. At populo Romano numquam ea copia fuit, quia prudentissumus quisque maxume negotiosus erat; ingenium nemo sine corpore exercebat: optumus quisque facere quam dicere, sua ab aliis bene facta laudari quam ipse aliorum narrare malebat. — SALLUST, CAT. viii.

IV. Patres conscripti. Micipsa pater meus moriens mihi

præcepit, uti regni Numidiæ tantummodo procurationem existumarem meam, ceterum jus et imperium ejus penes vos esse; simul eniterer domi militiæque quam maxumo usui esse populo Romano, vos mihi cognatorum, vos affinium loco ducerem; si ea fecissem, in vostra amicitia exercitum, divitias, munimenta regni me habiturum. Quae quum præcepta parentis mei agitarem, Jugurtha, homo omnium quos terra sustinet sceleratissumus, contempto imperio vostro, Masinissæ me nepotem et jam ab stirpe socium atque amicum populi Romani regno fortunisque omnibus expulit. — SALLUST, JUG. xiv.

OVID.

TRANSLATE *any* ONE *of the following passages:* —

V. Inde loco medius rerum novitate paventem
    Sol oculis juvenem, quibus adspicit omnia, vidit,
    'Quæque viæ tibi causa? Quid hac,' ait, 'arce petîsti,
    Progenies, Phaëthon, haud infitianda parenti?'
    Ille refert: 'O lux immensi publica mundi,
    Phœbe pater, si das hujus mihi nominis usum,
    Pignora da, genitor, per quæ tua vera propago
    Credar, et hunc animis errorem detrahe nostris.'
    Dixerat. At genitor circum caput omne micantes
    Deposuit radios, propiùsque accedere jussit,
    Amplexuque dato, 'Nec tu meus esse negari
    Dignus es, et Clymene veros,' ait, 'edidit ortus.'—METT. II.

VI. Psittacus, Eois imitatrix ales ab Indis,
    Occidit: exsequias ite frequenter aves.
    Ite, piæ volucres; et plangite pectora pennis;
    Et rigido teneras ungue notate genas.
    Horrida pro mæstis lanietur pluma capillis:
    Pro longa resonent carmina vestra tuba.
    Quid scelus Ismarii quereris, Philomela, tyranni?
    Expleta est annis ista querela suis.
    Alitis in raræ miserum devertite funus.
    Magna, sed antiqui causi doloris Itys.
    Omnes quae liquido libratis in aëre cursus;
    Tu tamen ante alias, turtur amice, dole. — AM. II.

VII. Sin autem ad pugnam exierint — nam sæpe duobus
Regibus incessit magno discordia motu,
Continuoque animos volgi et trepidantia bello
Corda licet longe præsciscere; namque morantis
Martius ille æris rauci canor increpat, et vox
Auditur fractos sonitus imitata tubarum;
Tum trepidæ inter se cœunt, pennisque coruscant,
Spiculaque exacuunt rostris, aptantque lacertos,
Et circa regem atque ipsa ad prætoria densæ
Miscentur, magnisque vocant clamoribus hostem.
Virg. Georg. IV.

VIII. Ipse, caput nivei fultum Pallantis et ora
Ut vidit levique patens in pectore volnus
Cuspidis Ausoniæ, lacrimis ita fatur obortis:
Tene, inquit, miserande puer, cum læta veniret,
Invidit Fortuna mihi, ne regna videres
Nostra, neque ad sedes victor veherere paternas?
Non hæc Euandro de te promissa parenti
Discedens dederam, cum me complexus euntem
Mitteret in magnum imperium, metuensque moneret
Acris esse viros, cum dura prœlia gente.—Virg. Æn. XI.

LATIN.—*Course I.*

CICERO.

Translate *two* passages. [*If you have read the Cato Major, translate I. and either III. or IV.; if not, translate II. and either III. or IV. Answer all the questions.*]

I. An ne eas quidem vires senectuti relinquemus ut adolescentulos doceat, instituat, ad omne officii munus instruat? Quo quidem opere quid potest esse præclarius? Mihi vero Cn. et P. Scipiones et avi tui duo, L. Æmilius et P. Africanus, comitatu nobilium juvenum fortunati videbantur; nec ulli bonarum artium magistri non beati putandi, quamvis consenuerint vires atque defecerint. — De Senectute, ix. 29.

II. Quid autem aliud egimus, Tubero, nisi ut quod hic potest nos possemus? Quorum igitur impunitas, Caesar, tuae clementiae laus est, eorum ipsorum ad crudelitatem te acuit oratio. Atque in hac causa non nihil equidem, Tubero, etiam tuam, sed multo magis patris tui prudentiam desidero, quod homo cum ingenio tum etiam doctrina excellens genus hoc causae quod esset non viderit; nam si vidisset, quovis profecto quam isto modo a te agi maluisset. — Pro Ligario, iv.

III. Tertium genus est aetate jam affectum, sed tamen exercitatione robustum, quo ex genere iste est Manlius, cui nunc Catilina succedit : sunt homines ex eis coloniis, quas Sulla constituit; quas ego universas civium esse optimorum et fortissimorum virorum sentio, sed tamen ii sunt coloni, qui se in insperatis ac repentinis pecuniis sumptuosius insolentiusque jactarunt. Hi dum aedificant tamquam beati, dum praediis lectis, familiis magnis, conviviis apparatis delectantur, in tantum aes alienum inciderunt, ut, si salvi esse velint, Sulla sit eis ab inferis excitandus. — In Catilinam, II. ix.

IV. Quare quis tandem me reprehendat aut quis mihi jure succenseat, si, quantum ceteris ad suas res obeundas, quantum ad festos dies ludorum celebrandos, quantum ad alias voluptates et ad ipsam requiem animi et corporis conceditur temporum, quantum alii tribuunt tempestivis conviviis, quantum denique alveolo, quantum pilae, tantum mihi egomet ad haec studia recolenda sumpsero? Atque hoc ideo mihi concedendum est magis, quod ex his studiis haec quoque crescit oratio et facultas, quae quantacumque in me est, numquam amicorum periculis defuit. — Pro Archia, vi.

1. What offices did the Romans generally go through before their consulship?

2. What is the difference between *ne* and *ut non* followed by the Subjunctive?

3. What was the fate of Catiline's fellow-conspirators, and what complaint was made of it?

VIRGIL.

TRANSLATE *two* passages, — *I. and either II. or III.* Answer all the questions.

I. Pauca tamen suberunt priscae vestigia fraudis,
    Quae temptare Thetim ratibus, quae cingere muris
    Oppida, quae iubeant telluri infindere sulcos.
    Alter erit tum Tiphys, et altera quae vehat Argo
    Delectos heroas; erunt etiam altera bella,
    Atque iterum ad Troiam magnus mittetur Achilles.
    Hinc, ubi iam firmata virum te fecerit aetas,
    Cedet et ipse mari vector, nec nautica pinus
    Mutabit merces: omnis feret omnia tellus. — ECL. IV.

II. Postera iamque dies primo surgebat Eoo,
    Humentemque Aurora polo dimoverat umbram:
    Cum subito e silvis, macie confecta suprema,
    Ignoti nova forma viri miserandaque cultu
    Procedit supplexque manus ad litora tendit.
    Respicimus. Dira inluvies inmissaque barba,
    Consertum tegumen spinis; at cetera Graius,
    Et quondam patriis ad Troiam missus in armis. — ÆN. III.

III. Primus equum phaleris insignem victor habeto,
    Alter Amazoniam pharetram plenamque sagittis
    Threiciis, lato quam circumplectitur auro
    Balteus, et tereti subnectit fibula gemma;
    Tertius Argolica hac galea contentus abito.
    Haec ubi dicta, locum capiunt, signoque repente
    Corripiunt spatia audito, limenque relinquunt,
    Effusi nimbo similes, simul ultima signant. — ÆN. V.

1. Give a brief summary of the events in Æneid IV.

2. Divide into feet, marking quantities and ictus (or verse accent), the fifth line in I.

3. How does the metre help to determine the meaning of the fifth line in II.?

ARITHMETIC AND LOGARITHMS.

[Give the work in full, and arrange it in an orderly manner. Reduce each answer to its simplest form.]

LOGARITHMS.

1. Find, by logarithms, the value of $\dfrac{0.9 \times 147.2}{5.047}$.

2. Find, by logarithms, the value of $\left(\dfrac{(134.9)^2 \times \sqrt[5]{.16}}{10000 \times 46.49}\right)^3$.

3. Give a proof of the process of finding any root of a quantity by logarithms. If the characteristic of the logarithm of the given quantity is negative, how is the characteristic of the logarithm of the root obtained?

ARITHMETIC.

4. What part of $2\frac{1}{3}$ is $\dfrac{7\frac{2}{3}}{31\frac{8}{9}} \times \dfrac{7\frac{2}{3} - 4\frac{1}{4}}{\frac{1}{2} \times 3\frac{5}{6}}$?

5. A carriage, at the rate of $8\frac{1}{2}$ miles an hour, completes $\frac{2}{3}$ of a certain distance in $3\frac{1}{3}$ days; in how many days will it complete $\frac{4}{7}$ of the same distance, going at the rate of 10 miles an hour?

6. A merchant buys $2\frac{2}{3}$ hectometres of silk for $480, and sells the silk at $1.95 a yard. Does he gain or lose, and how much?

7. Find the cube root of 0.083453453.

8. Thirty-six persons buy 2766 *A.* 3 *R.* 12 *P.* of land on equal shares. What does one man receive, who sells $\frac{2}{3}$ of his share at 1 *s.* 9 *d.* 2 *f.* per square rod? [Give the answer in pounds and decimals of a pound.]

9. What is gold quoted at, when one dollar in currency is worth only seventy-five cents?

ALGEBRA. — *Course I.*

[Give the whole work clearly, and reduce each answer to its simplest form.]

1. Divide
$$\frac{a-1}{a} + \frac{b-1}{b} + \frac{c-1}{c} - 1 \text{ by } 2 - \left(\frac{1}{a} + \frac{1}{b} + \frac{1}{c}\right).$$

2. A can do a piece of work in half the time in which B can do it, B can do it in two thirds the time in which C can do it, and all three, working together, can do it in 6 days. Find the time in which each can do it alone.

3. Find the two middle terms in the expansion of $(a-x)^9$. What is the *reason* that one of these terms is negative, and the other is positive?

4. Find the fourth root of $\sqrt[3]{a^2c^2}$. [Fractional exponents may be used if desired.]

5. One number is $\frac{10}{3}$ of another, and the product of these two numbers is 750. What are the numbers?

6. Solve the equations $ax + by = c,$
$mx - ny = d.$

7. I bought a certain number of oxen for £80. Had I bought four more with the same money, each ox would have cost £1 less. How many did I buy, and what did I pay for each?

8. Find the square root of
$$a^{4m} + 6a^{3m}c^n + 11a^{2m}c^{2n} + 6a^m c^{3n} + c^{4n}.$$

ALGEBRA. — *Course II. and Advanced Standing.*

[Give the whole work clearly, and reduce each answer to its simplest form.]

1. Simplify $\dfrac{\dfrac{a+1}{b} - 2 + \dfrac{b-1}{a}}{\dfrac{a-1}{b} - 2 + \dfrac{b+1}{a}}$.

2. A man rides a certain distance at the rate of 8 miles an hour, and walks back to his starting-point at the rate of 4 miles an hour. The time employed in going and returning is 6 hours. How far does he walk?

3. Divide $\dfrac{\sqrt{c}}{d^{\frac{1}{3}}}$ by $c^{\frac{1}{3}} d^{-\frac{2}{3}}$.

4. Solve the equation $x^2 + 2ax = b$. What will the roots be if $a = 2$, $b = -4$? If $a = 4$, $b = -20$?

5. What is the 4th term of $(a - x)^{n+1}$?

6. The greater of two numbers is a^2 times the less; the product of these two numbers is b^2. Find the numbers.

7. There are 3 numbers in arithmetical progression: the sum of these numbers is 18, and the sum of their squares is 158. Find the numbers.

8. I have 4 single books and a set of 3 books. In how many ways can I arrange these 7 books on a shelf, provided the books which make the set cannot be separated?

PLANE GEOMETRY.—*Courses I. and II.*

1. In a triangle ABC the angle A is greater than the angle B, and B is greater than C; what is true of the sides? State and prove.

State and prove the converse.

2. Prove that two triangles are equal if the sides of one are respectively equal to the sides of the other.

3. Prove that when two circumferences touch each other the point of contact and the centres lie in one straight line.

4. Draw two circles touching each other, and through the point of contact draw a straight line forming a chord in each circle: prove that these chords are proportional to the diameters of the circles.

5. To draw the circumference of a circle through three given points. Solve and prove. When would the problem be impossible? Why?

Given any curve, to ascertain whether it is the arc of a circle or not.

6. Prove that the perimeters of regular polygons of the same number of sides are proportional to the diameters of their inscribed or circumscribed circles. Go on to prove that the ratio of the circumference to the diameter is the same in all circles.

7. Draw, in your book, a regular hexagon of which each side shall be of this length ——————————
Explain how you do it. Now draw another having *half the area* of the first. Solve and prove.

SOLID GEOMETRY. — *Course II.*

1. Prove that the intersections of two parallel planes with a third plane are parallel planes.

2. Prove that the sum of the line angles that compose a solid angle is less than four right angles.

3. What is the frustum of a pyramid? Show how to find the convex surface of a regular pyramid. Prove that the surface of a right circular cone is equal to the product of the slant height multiplied by the circumference of a section drawn midway between the bases.

4. Given the radius of a sphere: write a formula for its surface and one for its volume.

5. What is the segment of a sphere? Explain how to find the volume of a segment of a sphere having two bases, one each side of the centre.

6. Given a spherical triangle, to draw its polar triangle. What relations exist between the sides and angles of a spherical triangle and those of its polar triangle? State and prove.

7. Given a spherical triangle, to draw another symmetrical with it on the same sphere. Prove that two symmetrical triangles on the same sphere have the same area.

8. What is a regular polyhedron? How many are there? Give their names, and a brief description of each.

ANALYTIC GEOMETRY.

Course II. and Advanced Standing.

[Ask for a Table of Natural Cosines.]

1. To find the equation of a straight line that passes through two given points.

2. Find the equation of a line that passes through the origin and the point $(-3, 2)$.

3. Find the equation of a line which passes through the point $(2, -1)$ and makes an angle of $45°$ with the line $x - 2y + 3 = 0$.

4. Establish formulas for changing rectangular into polar coordinates.

5. Write down the equation of a circle having a radius $= 7$ and its centre at $(3, -4)$.

6. What curves do these equations represent?

$$9x^2 + 16y^2 = 144, \qquad 9x^2 - 16y^2 = 144.$$

What are the polar equations of these curves? Sketch one of these curves from its rectangular equation, and the other from its polar equation. Find the *foci*. Find the *parameter* of each curve, and draw it.

7. Which of the points $(4, 2\frac{1}{2})$, $(3, -3\frac{1}{5})$, $(3, 3\frac{2}{5})$, is on the curve $\dfrac{x^2}{25} + \dfrac{y^2}{16} = 1$. Find the equation of the *tangent* and that of the *normal* at this point. Find also the lengths of the subtangent and subnormal.

8. How do you find the points where two curves intersect? As an example take these two curves: $y^2 = 4x$ and $x^2 + 6x + y^2 = 24$. What are these curves? Draw them.

PLANE TRIGONOMETRY.
Course II. and Advanced Standing.

1. The sine of an angle x is greater than the sine of another angle y, both angles being in the second quadrant. Compa͏ the other trigonometric functions of these angles (cosine with cosine, etc.), stating which in each set is numerically the larger. Prove your results, either by formulæ or by a diagram.

2. Obtain, from fundamental formulæ, the trigonometric functions of $(360° - y)$. *Given* the functions of $(180° - y)$, how can those of $(180° + y)$ be obtained?

3. Solve the right triangle in which one angle is 74° 18′, and the hypothenuse is $\sqrt[3]{.01}$.

4. What angle in the third quadrant has a cosine equal to the sine of 330°?

5. Obtain, from fundamental formulæ,
$$\frac{\cos(x+y)}{\cos(x-y)} = \frac{1 - \tan x \tan y}{1 + \tan x \tan y}.$$

6. Obtain, from the second member of the equation in the previous question, an equally simple expression in terms of the cotangents of x and y.

7. Find the smallest angle in the triangle whose sides are 1236, 1342, 1729.

8. Obtain the formulæ necessary for the complete solution of an oblique triangle, in which are given two sides and the included angle.

ENGLISH COMPOSITION.

A short English composition is required, correct in spelling, punctuation, grammar, and expression. Thirty lines will be sufficient. Make at least two paragraphs.

SUBJECT : —

    The story of the Caskets, in the Merchant of Venice;

    Or, The story of Shakespeare's Tempest;

    Or, The story of Rebecca, in Scott's Ivanhoe.

EXAMINATION PAPERS
OF OCTOBER, 1874.

ANCIENT GEOGRAPHY AND HISTORY.

N. B. — When you name a place or country, state its position. You may omit one of the first three subjects given below, and one of the last three.

1. Point out some of the causes of the greatness of Sparta and of Athens.

2. Write in the order of time (with such dates as you remember) the principal events in the Peloponnesian War, and show the chief results of that war.

3. Amphictyonic Council, Ephors, Archons; Areopagus, Pnyx, Agora. Define or describe these.

4. The death of Demosthenes and the death of Cicero.

5. The life of C. Julius Cæsar.

6. Laws that are landmarks in Roman history.

MODERN AND PHYSICAL GEOGRAPHY.

1. State, in detail, what you know about the form and dimensions of the earth. Define the mathematical and geographical terms which occur in your statement.

2. What is the length in miles of a degree of latitude? Where are the degrees of latitude and of longtitude equal in length? How do the degrees of longitude differ in length among themselves?

3. State accurately the zone or zones in which each of the six continents lies.

4. Name eight of the most important of the West India Islands, and draw a map to show their relative position. To what country does each belong?

5. Describe as fully and precisely as possible the position of the following cities, stating in what part of the state or country, and near what river or other body of water, each one lies: Belgrade, Bogota, Bombay, Brest, Carlsruhe, Dantzic, Frankfort (in Europe), Montevideo, Montreal, Odessa, Singapore, Tripoli. Which of these names suggests some physical feature of the neighboring region, or some fact of historical interest connected with the settlement of the city?

6. What strait or channel lies between Wales and Ireland? Wales and the southern part of England? Ireland and Scotland? Borneo and Celebes? Patagonia and Terra del Fuego? Labrador and Greenland? Labrador and Newfoundland?

7. Why are there large cities at higher latitudes in Europe than in America?

8. In sailing from New York to Liverpool, at what season of the year would you expect to see icebergs? How far south are icebergs ever seen in the North Atlantic?

9. Describe the drainage systems of North America, and name the highlands which bound each of its important river basins. Are there any portions of this continent which have no outlet for their waters to the sea?

GREEK COMPOSITION.

After the death of Cyrus, the Greeks being despondent (ἀπορέω), Xenophon called together (συγκαλέω) the soldiers, and told them that he had seen a vision (ἐνύπνιον); in order that he might encourage (θαρρύνω) them and cause them to cease (παύω) thinking (ἐνθυμέομαι) what things they had already suffered (πάσχω) and were still to suffer, he told them that if they would obey (πείθομαι) him, he would bring them all through in safety (διασώζω) to their native land.

GREEK GRAMMAR.

[All Greek words must be written with the accents.]

1. Give the general rule for accenting nouns (the accent of the nominative singular being known). How is accent affected by the quantity of the penultimate and final syllables?

2. Decline the nouns μοῦσα, νῆσος, and ἐλπίς in the *singular;* and βασιλεύς, in the *plural.*

3. Compare the adjectives ἄξιος, ἀληθής, μικρός, ἀγαθός.

4. Decline the pronouns ἐγώ and ὅστις throughout.

5. Give the *principal parts* of γράφω, ἵστημι, λαμβάνω, ὁράω, τίθημι.

6. Inflect the imperfect active of τιμάω and the present optative passive of φιλέω (in the contract forms). Inflect the second aorist optative active of ἵστημι.

7.* What uses of the article ὁ are found in Homer which are not found in Attic Greek?

8.* Explain the *genitive absolute* and the *accusative absolute,* and give an example of the correct use of each.

9.* Explain the difference in the meanings of ποιῆσαι in βούλεται τοῦτο ποιῆσαι and φησὶ τοῦτο ποιῆσαι.

10.* Give the names of the most common metrical feet of two and of three syllables, and show the quantity of the syllables in each (by — and ᴗ). Explain the Elegiac distich.

* Candidates for ADVANCED STANDING will omit 7, 8, 9, and 10, and will answer the following: —

11. Explain the Attic use of the substantive pronoun of the third person (οὗ, οἷ, etc.), and give an example.

12. Explain the regular use of the future infinitive. What objection can you make to any of the following expressions: βούλεται τοῦτο ποιήσειν, — ἐλπίζει τοῦτο ποιήσειν, — ὑπέσχετο τοῦτο ποιήσειν, — δεῖ τοῦτο ποιήσειν?

13. Why is εἰ τοῦτο ποιήσοι, ἔλθοιμι ἄν incorrect? Write a sentence in which εἰ ποιήσαι shall be correctly used.

14. Explain the Iambic trimeter of Comedy, showing how it differs from that of Tragedy. Explain also the Trochaic tetrameter catalectic and the Anapæstic System.

GREEK PROSE.

[Those offering Greek Reader, take 2, 4, 5. Those offering Anabasis (four Books), and Herodotus (Book 7th), take 1, 2, 5. Those offering the whole Anabasis, take 1, 2, 3.]

TRANSLATE : —

1. (Anab. II. III. 11.) Καὶ ἐνταῦθα ἦν Κλέαρχον καταμαθεῖν ὡς ἐπεστάτει, ἐν μὲν τῇ ἀριστερᾷ χειρὶ τὸ δόρυ ἔχων, ἐν δὲ τῇ δεξιᾷ βακτηρίαν· καὶ εἴ τις αὐτῷ δοκοίη τῶν πρὸς τοῦτο τεταγμένων βλακεύειν, ἐκλεγόμενος τὸν ἐπιτήδειον ἔπαισεν ἄν, καὶ ἅμα αὐτὸς προσελάμβανεν εἰς τὸν πηλὸν ἐμβαίνων· ὥστε πᾶσιν αἰσχύνην εἶναι μὴ οὐ συσπουδάζειν.

Explain the mood of δοκοίη.

EXAMINATION PAPERS. 221

2. (Anab. III. v. 8, 9; Reader, p. 26.) Ἀπορουμένοις δ' αὐτοῖς προσελθών τις ἀνὴρ Ῥόδιος εἶπεν. Ἐγὼ θέλω, ὦ ἄνδρες, διαβιβάσαι ὑμᾶς κατὰ τετρακισχιλίους ὁπλίτας, ἂν ἐμοὶ ὧν δέομαι ὑπηρετήσητε καὶ τάλαντον μισθὸν πορίσητε. Ἐρωτώμενος δὲ ὅτου δέοιτο· Ἀσκῶν, ἔφη, δισχιλίων δεήσομαι· πολλὰ δὲ ὁρῶ ταῦτα πρόβατα καὶ αἶγας καὶ βοῦς καὶ ὄνους, ἃ ἀποδαρέντα καὶ φυσηθέντα ῥᾳδίως ἂν παρέχοι τὴν διάβασιν.

Explain the mood of δέοιτο. ἀποδαμέντα, in what voice, mood, tense, and from what verb?

3. (Anab. VII. II. 18.) Ἐπεὶ δ' ἐγγὺς ἦσαν αὐτοῦ, ἐπιτυγχάνει πυροῖς ἐρήμοις· καὶ τὸ μὲν πρῶτον ᾤετο μετακεχωρηκέναι ποι τὸν Σεύθην. Ἐπεὶ δὲ θορύβου τε ᾔσθετο καὶ σημαινόντων ἀλλήλοις τῶν περὶ Σεύθην, κατέμαθεν ὅτι τούτου ἕνεκα τὰ πυρὰ προκεκαυμένα εἴη τῷ Σεύθῃ πρὸ τῶν νυκτοφυλάκων, ὅπως οἱ μὲν φύλακες μὴ ὁρῷντο, ἐν τῷ σκότει ὄντες, μηδ' ὅπου εἶεν, οἱ δὲ προσιόντες μὴ λανθάνοιεν, ἀλλὰ διὰ τὸ φῶς καταφανεῖς εἶεν.

4. (Reader, p. 99, 15; Plato, Apol.) ἐγὼ μὲν γὰρ πολλάκις ἐθέλω τεθνάναι, εἰ ταῦτ' ἐστὶν ἀληθῆ· ἐπεὶ ἔμοιγε καὶ αὐτῷ θαυμαστὴ ἂν εἴη ἡ διατριβὴ αὐτόθι, ὁπότε ἐντύχοιμι Παλαμήδει καὶ Αἴαντι τῷ Τελαμῶνος καὶ εἴ τις ἄλλος τῶν παλαιῶν διὰ κρίσιν ἄδικον τέθνηκεν, ἀντιπαραβάλλοντι τὰ ἐμαυτοῦ πάθη πρὸς τὰ ἐκείνων, ὡς ἐγὼ οἶμαι, οὐκ ἂν ἀηδὲς εἴη. Καὶ δὴ τὸ μέγιστον, τοὺς ἐκεῖ ἐξετάζοντα καὶ ἐρευνῶντα ὥσπερ τοὺς ἐνταῦθα διάγειν, τίς αὐτῶν σοφός ἐστι, καὶ τίς οἴεται μὲν ἔστι δ' οὔ.

5. (Herod. VII. 37; Reader, p. 124, 12.) ὡρμημένῳ δέ οἱ ὁ ἥλιος ἐκλιπὼν τὴν ἐκ τοῦ οὐρανοῦ ἕδρην ἀφανὴς ἦν, οὔτ' ἐπινεφέλων ἐόντων, αἰθρίης τε τὰ μάλιστα, ἀντὶ ἡμέρης τε νὺξ ἐγένετο. ἰδόντι δὲ καὶ μαθόντι τοῦτο τῷ Ξέρξῃ ἐπιμελὲς ἐγένετο, καὶ εἴρετο τοὺς Μάγους, τὸ ἐθέλοι τροφαίνειν τὸ φάσμα. οἱ δὲ ἔφραζον, ὡς Ἕλλησι προδεικνύει ὁ θεὸς ἔκλειψιν τῶν πολίων, λέγοντες ἥλιον εἶναι Ἑλλήνων προδέκτορα, σελήνην δὲ σφέων. πυθόμενος δὲ ταῦτα ὁ Ξέρξης περιχαρὴς ἐὼν ἐποιέετο τὴν ἔλασιν.

GREEK POETRY.

Translate: —

1. Iliad I. 511–516.

Ὣς φάτο· τὴν δ' οὔτι προσέφη νεφεληγερέτα Ζεύς,
ἀλλ' ἀκέων δὴν ἧστο· Θέτις δ', ὡς ἥψατο γούνων,
513 ὣς ἔχετ' ἐμπεφυυῖα, καὶ εἴρετο δεύτερον αὖτις·
Νημερτὲς μὲν δή μοι ὑπόσχεο καὶ κατάνευσον,
ἢ ἀπόειπ'· ἐπεὶ οὔ τοι ἔπι δέος· ὄφρ' εὖ εἰδῶ,
ὅσσον ἐγὼ μετὰ πᾶσιν ἀτιμοτάτη θεός εἰμι.

Divide into feet vss. 513, 514. ὑπόσχεο, in what tense, mood, voice, and from what verb?

2. Iliad II. 308–316.

ἔνθ' ἐφάνη μέγα σῆμα· δράκων ἐπὶ νῶτα δαφοινός,
σμερδαλέος, τόν ῥ' αὐτὸς Ὀλύμπιος ἧκε φόωσδε,
βωμοῦ ὑπαΐξας, πρός ῥα πλατάνιστον ὄρουσεν.
ἔνθα δ' ἔσαν στρουθοῖο νεοσσοί, νήπια τέκνα,
ὄζῳ ἐπ' ἀκροτάτῳ, πετάλοις ὑποπεπτηῶτες.
ὀκτώ, ἀτὰρ μήτηρ ἐνάτη ἦν, ἣ τέκε τέκνα.
ἔνθ' ὅγε τοὺς ἐλεεινὰ κατήσθιε τετριγῶτας·
μήτηρ δ' ἀμφεποτᾶτο ὀδυρομένη φίλα τέκνα·
τὴν δ' ἐλελιξάμενος πτέρυγος λάβεν ἀμφιαχυῖαν.

3. Iliad III. 351–354.

Ζεῦ ἄνα, δὸς τίσασθαι, ὅ με πρότερος κάκ' ἔοργεν,
δῖον Ἀλέξανδρον, καὶ ἐμῇς ὑπὸ χερσὶ δάμασσον·
ὄφρα τις ἐρρίγῃσι καὶ ὀψιγόνων ἀνθρώπων,
ξεινοδόκον κακὰ ῥέξαι, ὅ κεν φιλότητα παράσχῃ.

LATIN COMPOSITION.

TRANSLATE INTO LATIN:—

While[1] this was done[2] where[3]-Cæsar-was, Labienus, leaving[4] the reinforcements[5] which had lately[6] come from Italy, at Agedicum, to serve[7] as a guard[8] for the baggage,[9] marches[10] to Lutetia with four legions. This is a town of the Parisii, which is situated[11] on an island[12] of the river Sequana. His arrival[13] being-known[14] by the enemy, large[15] forces[16] assembled[17] from the neighboring[18] states.[19] The chief-command[20] is given[21] to Camulogenus, who, almost[22] disabled[23] by years, nevertheless for[24] his unequalled[25] knowledge[26] of the art[27] military was detailed[28] for[29] this honorable-position.[30] Since[31] he had observed[32] that there was a marsh[33] which emptied[34] into the Sequana, he took-his-position[35] here, and began[36] to prevent[37] our men from crossing.[38]

[1] Dum. [2] gerere. [3] *simply with* apud. [4] relinquere. [5] supplementum. [6] nuper. [7] esse. [8] præsidium. [9] impedimenta. [10] proficisci. [11] ponere. [12] insula. [13] adventus. [14] cognoscere. [15] magnus. [16] copiæ. [17] convenire. [18] finitimus. [19] civitas. [20] summa imperii. [21] tradere. [22] prope. [23] confectus. [24] propter. [25] singularis. [26] scientia. [27] res. [28] evocare. [29] ad. [30] honos. [31] cum. [32] animadvertere. [33] palus. [34] influere. [35] considere. [36] instituere. [37] prohibere. [38] transitus (*substantive*).

TRANSLATE INTO ENGLISH:—

Parente P. Sestius natus est, judices, homine, ut plerique meministis, et sapiente et sancto et severo; qui cum tribunus plebis primus inter homines nobilissimos temporibus optimis factus esset, reliquis honoribus non tam uti voluit quam dignus videri. Eo auctore duxit honestissimi et spectatissimi viri, C. Albini filiam, ex qua hic est puer et nupta jam filia.

LATIN GRAMMAR.

Mark the quantity of the penults and last syllables of the following: *perbrevis* (nom. sing.), *fidei, arietis, cadaver, colloquor, molimen, peregre, cornicis, idus aprilis* (acc. plur.), *sentitis, ducitis.*

Give the vocative singular of *Marcus Tullius Cicero.*

Decline *manus* (*tener*) in the proper gender in the plural. Decline *collis* (*silvester*) in the proper gender. Give the principal parts of *pango, explico, vincio, vinco, tego, texo, sancio, tondeo, voveo, oportet.* Give three ways of forming the perfect stem in Latin. Give three ways of forming the present stem in the third conjugation.

Inflect the future indicative and the present subjunctive active of *volo, eo, domo, sumo.* Give the perfect subjunctive active of *surgo, censeo;* the imperative of *ordior;* the participles and infinitives of *veto, aperio, obliviscor.*

Compare *idoneus, tenax.* Compare *sæpe.* Form a word meaning "more watchfully" from *vigilo,* to watch. Form words meaning "belonging to Athens," "horned," "oaken," "an effort" (*conor*), "hardness," "seizure" (*rapio*). What two constructions follow the comparative degree? What is the rule for their use? How is the degree of difference expressed? What is the construction of *sententiam* in *Rogatus est sententiam?* What case or cases follow *proprius, adimo, obsto, ad, in, infra, sub, ante, pro?* Give, with examples, three uses of the *subjunctive* in independent clauses. State some cases in which there can be an apodosis without any accompanying conditional clause. Turn into direct discourse, *nisi jurasset, scelus se facturum arbitrabatur.* Explain the mood of *jurasset.* Why is it not either of the other tenses?

LATIN.

CÆSAR AND SALLUST.

Translate *two* passages, — the *first* and one other.

I. Milites non longiore oratione cohortatus quam uti suæ pristinæ virtutis memoriam retinerent neu perturbarentur animo hostiumque impetum fortiter sustinerent, quod non longius hostes aberant quam quo telum adjici posset, prœlii committendi signum dedit. Atque in alteram partem item cohortandi causa profectus pugnantibus occurrit. Temporis tanta fuit exiguitas hostiumque tam paratus ad dimicandum animus, ut non modo ad insignia accommodanda, sed etiam ad galeas induendas scutisque tegimenta detrudenda tempus defuerit. — Cæsar, B. G. II.

II. Primum omnium, qui ubique probro atque petulantia maxume præstabant, item alii per dedecora patrimoniis amissis, postremo omnes, quos flagitium aut facinus domo expulerat, ii Romam sicut in sentinam confluxerant. Deinde multi memores Sullanæ victoriæ, quod ex gregariis militibus alios senatores videbant, alios ita divites ut regio victu atque cultu ætatem agerent, sibi quisque si in armis foret ex victoria talia sperabat. Præterea juventus, quæ in agris manuum mercede inopiam toleraverat, privatis atque publicis largitionibus excita urbanum otium ingrato labori prætulerat; eos atque alios omnes malum publicum alebat. — Sallust, Cat. xxxvii.

III. Civitatibus maxima laus est quam latissime circum se vastatis finibus solitudines habere. Hoc proprium virtutis existimant, expulsos agris finitimos cedere neque quemquam prope audere consistere: simul hoc se fore tutiores arbitrantur repentinæ incursionis timore sublato. Quum bellum civitas aut illatum defendit aut infert, magistratus qui ei bello præsint ut vitæ necisque habeant potestatem deliguntur. In pace nullus est communis magistratus, sed principes regionum atque pagorum inter suos jus dicunt controversiasque minuunt. — Cæsar, B. G. VI.

OVID.

Translate *one* passage.

IV. Haud procul· Hennæis lacus est a mœnibus altæ,
Nomine Pergus, aquæ. Non illo plura Caystros
Carmina cygnorum labentibus audit in undis.
Silva coronat aquas, cingens latus omne, suisque
Frondibus, ut velo, Phœbeos submovet ignes.
Frigora dant rami, varios humus humida flores;
Perpetuum ver est. Quo dum Proserpina luco
Ludit, et aut violas aut candida lilia carpit;
Dumque puellari studio calathosque sinumque
Implet, et æquales certat superare legendo;
Pæne simul visa est, dilectaque, raptaque Diti:
Usque adeo est properatus amor. — METT. V.

V. Ille inter cædem Rutulorum elapsus in agros
Confugere, et Turni defendier hospitis armis.
Ergo omnis furiis surrexit Etruria iustis;
Regem ad supplicium præsenti Marte reposcunt.
His ego te, Ænea, ductorem milibus addam.
Toto namque fremunt condensæ litore puppes,
Signaque ferre iubent; retinet longævus haruspex
Fata canens: O Mæoniæ delecta iuventus,
Flos veterum virtusque virum, quos iustus in hostem
Fert dolor et merita accendit Mezentius ira,
Nulli fas Italo tantam subiungere gentem:
Externos optate duces. — VIRG. ÆN. VIII.

LATIN. — *Course I.*

CICERO.

Translate *two* passages. [*If you have read the Cato Major, translate I. and either III. or IV.; if not, translate II. and either III. or IV. Answer all the questions.*]

I. Vixerat M'. Curius cum P. Decio, qui quinquennio ante eum consulem se pro re publica quarto consulatu devoverat:

norat eumdem Fabricius, norat Coruncanius : qui quum ex sua vita tum ex eius, quem dico, Decii facto iudicabant esse profecto aliquid natura pulcrum atque præclarum quod sua sponte peteretur quodque spreta et contempta voluptate optimus quisque sequeretur. — CATO MAJOR XIII. 43.

II. Res erat minimè obscura : etenim palàm dictitabat, consulatum Miloni eripi non posse, vitam posse. Significavit hoc sæpe in senatu : dixit in contione. Quinetiam Favonio, fortissimo viro, quærenti ex eo, quâ spe fureret, Milone vivo, respondit, triduo illum, ad summum quatriduo, periturum : quam vocem ejus ad hunc M. Catonem statim Favonius detulit. — PRO MILONE IX. 26.

III. Quare, cùm et bellum ita necessarium sit, ut negligi non possit; ita magnum, ut accuratissimè sit administrandum; et cùm ei imperatorem præficere possitis, in quo sit eximia belli scientia, singularis virtus, clarissima auctoritas, egregia fortuna; dubitabitis, Quirites, quin hoc tantum boni, quod vobis a Diis mortalibus oblatum et datum est, in rempublicam conservandam atque amplificandam conferatis? — PRO LEGE MANILIA XVI. 49.

IV. Cùm facilè exorari, Cæsar, tum semel exorari, soles. Nemo unquam te placavit inimicus, qui ullas resedisse in te simultatis reliquias senserit. Quanquam cui sunt inauditæ cum Deiotaro querelæ tuæ? Nunquam tu illum accusavisti, ut hostem, sed ut amicum officio parum functum, quòd propensior in Cn. Pompeii amicitiam fuisset, quàm in tuam. Cui tamen ipsi rei veniam te daturum fuisse dicebas, si tantùm auxilia Pompeio, vel si etiam filium misisset, ipse excusatione ætatis usus esset. — PRO REGE DEIOTARO III. 9.

1. State concisely the circumstances and subject of any one of Cicero's Orations against Catiline.
2. Explain the use of the moods in indirect discourse.
3. What does Cicero usually mean by *Asia* and *Gallia*?

VIRGIL.

TRANSLATE *two* passages, — *II*. and either *I*. or *III*. Answer all the questions.

I.   *C.* Muscosi fontes et somno mollior herba,
Et quae vos rara viridis tegit arbutus umbra,
Solstitium pecori defendite; iam venit aestas
Torrida, iam laeto turgent in palmite gemmae.
    T. Hic focus et taedae pingues, hic plurimus ignis
Semper, et adsidua postes fuligine nigri;
Hic tantum Boreae curamus frigora, quantum
Aut numerum lupus, aut torrentia flumina ripas.

ECL. VII. 45–52.

II.   Anna, vides toto properari litore? Circum
Undique convenere; vocat iam carbasus auras,
Puppibus et laeti nautae inposuere coronas.
Hunc ego si potui tantum sperare dolorem,
Et perferre, soror, potero. Miserae hoc tamen unum
Exsequere, Anna, mihi; solam nam perfidus ille
Te colere, arcanos etiam tibi credere sensus;
Sola viri mollis aditus et tempora noras;
I, soror, atque hostem supplex adfare superbum:

ÆN. IV. 416–424.

III.   Tum contra Iuno; Terrorum et fraudis abunde est:
Stant belli caussae: pugnatur comminus armis;
Quae fors prima dedit, sanguis novus imbuit arma.
Talia coniugia et talis celebrent hymenaeos
Egregium Veneris genus et rex ipse Latinus.
Te super aetherias errare licentius auras
Haud Pater ille velit, summi regnator Olympi.
Cede locis. Ego, si qua super fortuna laborum est,
Ipsa regam. Talis dederat Saturnia voces.

ÆN. VII. 552–560.

1. Give a brief summary of the events in Æneid III.

2. Divide into feet, marking the quantities and ictus of every foot, the second and fifth lines in II.

3. How does the metre help to determine the meaning of the second line in I.?

ARITHMETIC AND LOGARITHMS.

LOGARITHMS.

1. Find, by logarithms, the value of $\sqrt{(1.06)^5}$.

2. Find, by logarithms, the value of $\dfrac{\frac{2}{3} \text{ of } 444.4}{0.864 \div 0.0001}$.

3. If the base of a system of logarithms is 8, between what integers does the logarithm of 9 lie? of 90? of 900?

ARITHMETIC.

4. Find the sum of $\dfrac{0.5 \times 0.006}{\frac{2}{3} \times \frac{4}{5} \times (\frac{1}{4})^2}$ and $\dfrac{\frac{1}{5} \text{ of } 1\frac{5}{6} \text{ of } (\frac{2}{3})^3}{1.6 + 0.625}$.

5. Obtain the answer to the first question on this paper *without using logarithms*.

6. Three men contract to do a piece of work for $8,775. The first man employs 20 men, 24 days, 10 hours a day; the second 25 men, 20 days, 12 hours a day; the third 30 men, 25 days, 9 hours a day. How much should each of the three contractors receive?

7. What circulating decimal is equivalent to the sum of $\frac{1}{3}$, $\frac{1}{7}$, and $\frac{1}{11}$?

8. A man buys 454 bushels of wheat for $3 a bushel, and sells the wheat at $8.75 a hectolitre. How much does he gain?

(Litre = 0.908 quart, dry measure.)

9. If 2 A. 3 R. 4 P. be multiplied by $2\frac{3}{4}$, what part is the product of 15 A. 1 R. 2 P.?

10. If a grocer's scales give only 15 oz. 4 dr. for a pound, out of how much money is a customer cheated who buys sugar to the amount of $55.04?

ALGEBRA.

1. Find the greatest common divisor of $2x^2 + x - 1$, $x^2 + 5x + 4$, and $x^3 + 1$. (Obtain the result, if possible, by separating each polynomial into its prime factors.)

2. Find the simplest expression for
$$\frac{1+x}{1+x+x^2} + \frac{1-x}{1-x+x^2} - \frac{2}{1+x^2+x^4}.$$

3. A number consists of two digits. If 9 be added to the number, the digits are inverted; and the sum of the number thus formed and of the original number is 33. Find the digits.

4. If n be divided into two parts, prove that the difference of the squares of the parts equals n times the difference of the parts.

5. Find the square root of $x^4 + 2x^3 - x + \frac{1}{4}$.

6. Given $\dfrac{2a+n}{3n+69a} = \frac{1}{33}$, a is $\frac{1}{3}$. Find the value of n.

7. Solve the equation $\dfrac{x+2}{x-1} = \dfrac{4-x}{2x} + \dfrac{7}{2}$.

8. A cistern is filled by two pipes in 2 h. 55 m. The larger pipe will fill the cistern, by itself, in two hours less time than the smaller pipe will fill it. In what time will each pipe fill the cistern?

9. The cube root of a number is twice the square root. Find the number.

ALGEBRA. — *Course II. and Advanced Standing.*

1. The sum of the two digits which form a number is 9, and if the number be divided by the sum of the digits the quotient is 5. Find the number.

2. Solve the equation $\dfrac{1}{x-2} - \dfrac{2}{x+2} = \dfrac{3}{5}$.

3. A merchant bought a certain number of pieces of silk for £180. Had he received three more pieces for the same money, each piece would have cost £3 less. How many pieces did he buy?

4. Obtain the equation whose roots are $m+n$ and $m-n$. What form will the equation take if $m=n$?

5. The first term of an Arithmetical Progression is 5, the last term is 302, the common difference is 3. Find the number of terms.

6. Solve the equation $\sqrt{x+4} - \sqrt{x} = \sqrt{x+\frac{3}{2}}$.

7. Find the first five terms of $\sqrt{1+x}$ by the Binomial Theorem.

8. In the Geometric Progression, $a, b, \ldots$ find the sum of an nfinite number of terms.

9. Out of 12 consonants and 5 vowels how many words can be formed, each containing 3 consonants and 2 vowels?

PLANE GEOMETRY.

1. Two sides of one triangle are respectively equal to two sides of another triangle, but the angles included by these sides are not equal. What is true of the third sides? State and prove. State the converse theorem. Is it true?

2. The area of a triangle. State and prove.

3. Prove that the areas of two rectangles are proportional to the products of their bases by their altitudes.

4. The radius of a given circle is ten inches; what is the radius of a circle having twice the area of the given circle? of a circle having one half the area of the given circle?

5. State and prove the Pythagorean theorem.

6. Given the base, the altitude, and one of the angles at the base of a triangle, to construct the triangle.

7. Prove that two triangles are similar, if an angle of one equals an angle of the other, and the sides which include these angles are proportional.

8. A perpendicular drawn from any point of a semi-circumference upon the diameter is a mean proportional between what? State and prove.

SOLID GEOMETRY.

1. Two planes are perpendicular to each other, and a straight line is drawn in one of them perpendicular to their intersection; prove that this straight line is perpendicular to the other plane.

2. Two planes are perpendicular to each other, and through any point of one is drawn a straight line perpendicular to the other: prove that this straight line lies wholly in the first plane.

3. Prove that if a solid angle is formed by three plane angles, the sum of either two of these angles is greater than the third.

4. Prove that sections of a pyramid made by parallel planes are similar polygons whose areas are proportional to the squares of their distances from the vertex.

5. Prove that two pyramids which have equal bases and altitudes are equivalent. Why not say *equal*?

6. Prove that a triangular pyramid is a third part of a triangular prism of the same base and altitude. Deduce from this a rule for finding the volume of any pyramid or cone.

7. How large a part of the surface of a sphere is covered by a spherical triangle whose angles are 90°, 150°, 132°?

8. What is a regular polyhedron? How many are there? Give their names and a brief description of each.

ANALYTIC GEOMETRY.

[Ask for Trigonometric Tables.]

1. What are *Rectangular Co-ordinates*? *Polar Co-ordinates*?

2. Lay down a few points of, and then draw the curves represented by, these equations: —

(i.) $7x^2 - 16y^2 = 112$,

(ii.) $\phi = \dfrac{1\frac{3}{4}}{1 - \frac{3}{4}\cos\theta}$.

What are these curves?

3. The centre of a circle is at the point $(-2, 0)$ and its radius $= 5$; what is its equation?

4. Define the *Ellipse, Parabola, Hyperbola*.

5. From its definition deduce the rectangular equation of the parabola.

6. Given the equation of a parabola $y^2 = 6x$; what is the distance from the origin to the focus? Transform this equation to a set of axes through the focus. What does the new equation represent? Transform it to polar co-ordinates. Illustrate by a diagram.

7. Is the point (2, 1) on the straight line $x - 3y + 1 = 0$? Why?

8. Find the equation of a straight line passing through (2, 1) and perpendicular to the line $x - 3y + 1 = 0$. Draw both lines.

9. In what point do the straight lines $x - 3y + 1 = 0$ and $x + 7y + 11 = 0$ intersect?

10. Find the angle between the two straight lines given in the last question.

PLANE TRIGONOMETRY.

Course II. and Advanced Standing.

1. The cosine of an angle in the first quadrant is 0.7. Find, either by formulæ or by tables, the sine of half that angle.

2. What is the sine of 240°? The cosine of 300°? The tangent of 225°? The secant of 150°?

3. One angle of a plane triangle is 64° 18′, and the other angles are equal. The greatest side is 10. Solve the triangle.

4. Find the trigonometric functions of $(270° - y)$.

5. Prove that the sides of a plane triangle are proportional to the sines of the opposite angles.

6. Obtain, from fundamental formulæ,

$$\cot(x - y) = \frac{\cot y \cot x + 1}{\cot y - \cot x}.$$

7. Two sides of a plane triangle are 4, 6, and the included angle is 38° 54′. Solve the triangle.

8. One side of a plane triangle is double another, and the third side equals one half the sum of the other two. Find the largest angle.

ENGLISH COMPOSITION.

A short English composition is required, correct in spelling, punctuation, grammar, and expression. Thirty lines will be sufficient. Make at least two paragraphs.

SUBJECT : —

    The Trial Scene, in the Merchant of Venice;
Or, The Story of Brutus, in Shakespeare's Julius Cæsar;
Or, The Passage of Arms at Ashby, in Ivanhoe.

HARVARD EXAMINATION PAPERS.

JULY, 1875.

LATIN GRAMMAR.

WHAT are the stem and root, respectively, of *donum?*

What is the gender of *nihil*, and why?

Decline *orbis* with the adjective *totus* in the proper gender, *formido* with *dirus.*

What is the vocative of *Gaius Julius Caesar?*

The genitive plural of *senex* and *dies*, the dative plural of *vir*, and *vis*, and the nominative plural of *calcar.*

Decline *quis* and *qualis.* Give the future second person singular active of *prosum, quaero, vinco, venio, creo.*

Give the perfect subjunctive active, first person, of *gaudeo, habeo, surgo, possum.*

Give the present subjunctive, first person, of *conor, gradior, adipiscor.*

Principal parts of *alo, pario, pareo, paro, venio, vincio, vinco, miseret*, and of *tango* compounded with *con.*

Explain the formation from the root of the present and perfect stem of *gigno, nosco, tango.*

Form and compare adverbs from *miser, bonus, dexter.*

Give the meaning of the following derivative terminations, with an example: *-bilis, -idus, -tura, -ades, -osus, -brum, -urio.*

What is the Latin for: *the rest of the army; a longing*

(desiderium) *for rest* (otium); *it is my interest; I envy you; I am persuaded; we must use diligence; freed from the laws; he was killed with a sword by Milo; he sold this for ten denarii.*

LATIN COMPOSITION.

TRANSLATE INTO LATIN:—

Sweet is the name of peace: but[1] the thing itself is not only pleasant[2] but salutary.[3] For that-man[4] seems not to hold[5] private hearths[6] nor public laws, nor the rights[7] of freedom[8] dear,[9] who loves[10] discords and murders[11] of his[12] countrymen[13] and civil war, and I think he should be cast[14] out[15] of the number of men, banished[16] from the bounds[17] of human nature. Nothing is more loathsome[18] than this citizen, than this man: if he is to be considered a citizen or a man who hankers-after[19] civil war.

[1] vero. [2] iucundus. [3] salutaris. [4] is. [5] habere. [6] focus. [7] ius. [8] libertas [9] carus. [10] *with* delectare. [11] caedis. [12] *omit*. [13] civis. [14] eicere. [15] ex. [16] ex terminare. [17] finis. [18] taeter. [19] concupiscere.

TRANSLATE INTO ENGLISH:—

Illud vereor, ne ignorans verum iter gloriae, gloriosum putes, plus te unum posse, quam omnes, et metui a civibus tuis, quam diligi malis. Quod si ita putas, totam ignoras viam gloriae. Carum esse civem, bene de re publica mereri, laudari, coli, diligi, gloriosum est: metui vero, et in odio esse, invidiosum, detestabile, imbecillum, caducum. Quod videmus etiam in fabulis, ipsi illi, qui "*Oderint, dum metuant,*" dixerit, perniciosum fuisse.

LATIN.

CAESAR, SALLUST, AND OVID.

[*N. B.—Translate one piece of Caesar, the piece of Sallust, and two pieces of Ovid. The order in which they are done is unimportant. The third piece of Cæsar and the piece of Virgil are only as substitutes for Sallust and Ovid, by those who have not read those authors. Answer all the questions.*]

I. Palus erat non magna inter nostrum atque hostium exercitum. Hanc si nostri transirent, hostes exspectabant; nostri autem, si ab illis initium transeundi fieret, ut impeditos aggrederentur, parati in armis erant. Interim proelio equestri inter duas acies contendebatur. Ubi neutri transeundi initium faciunt, secundiore equitum proelio nostris, Cæsar suos in castra reduxit. Hostes protinus ex eo loco ad flumen Axonam contenderunt, quod esse post nostra castra demonstratum est. Ibi vadis repertis partem suarum copiarum transducere conati sunt, eo consilio, ut, si possent, castellum, cui praeerat Quintus Titurius legatus, expugarent, pontemque interscinderent. — CAESAR, *Bell. Gall.*, II.

II. [*Only for those who do not offer Sallust.*]

Contra ea Titurius sero facturos clamitabat, cum majores hostium manus adjunctis Germanis convenissent; aut cum aliquid calamitatis in proximis hibernis esset acceptum; brevem consulendi esse occasionem: Caesarem arbitrari profectum in Italiam; neque aliter Carnutes interficiendi Tasgetii consilium fuisse capturos, neque Eburones, si ille adesset, tanta contemptione nostri ad castra venturos esse; non hostem auctorem, sed rem spectare; subesse Rhenum;

magno esse Germanis dolori Ariovisti mortem et superiores nostras victorias: ardere Galliam tot contumeliis acceptis sub populi Romani imperium redactam, superiore gloria rei militaris exstincta. — CAESAR, *Bell. Gall.*, V.

III. Sed postquam Cn. Pompeius ad bellum maritumum atque Mithridaticum missus est, plebis opes imminutae, paucorum potentia crevit. Hi magistratus, provincias, aliaque omnia tenere; ipsi innoxii, florentes, sine metu aetatem agere, ceteros judiciis terrere, quo plebem in magistratu placidius tractarent. Sed ubi primum dubiis rebus novandi spes oblata est, vetus certamen animos eorum arrexit, Quodsi primo proelio Catilina superior aut aequa manu discessisset, profecto magna clades atque calamitas rem publicam oppressisset; neque illis, qui victoriam adepti forent, diutius ea uti licuisset quin defessis et exsanguibus qui plus posset imperium atque libertatem extorqueret. — SALLUST *Bell. Cat.*, 39.

(*a*) Give the divisions of Gaul according to Caesar.

(*b*) Give the reason for any one of the subjunctives in the piece of Caesar translated.

(*c*) Give the date B. C. of Catiline's conspiracy, and the consuls of that year.

TRANSLATE: —

I. Serius egressus vestigia vidit in alto
   Pulvere certa ferae, totoque expalluit ore
Pyramus. Ut vero vestem quoque sanguine tinctam
Repperit, "Una duos" inquit "nox perdet amantes:
E quibus illa fuit longa dignissima vita,
   Nostra nocens anima est. Ego te, miseranda, peremi,
In loca plena metus qui iussi nocte venires,
Nec prior huc veni. Nostrum divellite corpus,
Et scelerata fero consumite viscera morsu,
   O quicumque sub hac habitatis rupe, leones.

OVID, *Metam.*, IV.

II. Inde per immensum croceo velatus amictu
 Aethera digreditur, Ciconumque Hymenaeus ad oras
 Tendit, et Orphea nequiquam voce vocatur.
 Affuit ille quidem. Sed nec sollemnia verba,
 Nec laetos vultus, nec felix attulit omen.
 Fax quoque, quam tenuit, lacrimoso stridula fumo
 Usque fuit, nullosque invenit motibus ignes.
 Exitus auspicio gravior, nam nupta, per herbas
 Dum nova naïadum turba comitata vagatur,
 Occidit, in talum serpentis dente recepto.
Ovid, *Metam.*, X.

III. Haec mea, si casu miraris, epistola quare
  Alterius digitis scripta sit, aeger eram.
 Aeger in extremis ignoti partibus orbis,
  Incertusque meae paene salutis eram.
 Quid mihi nunc animi dira regione iacenti
  Inter Sauromatas esse Getasque putes ?
 Nec caelum patior, nec aquis adsuevimus istis,
  Terraque nescio quo non placet ipsa modo.
 Non domus apta satis, non hic cibus utilis aegro,
  Nullus, Apollinea qui levet arte malum.
Ovid, *Trist.*, III.

IV. [*Only for such as do not offer Ovid.*]
 Interea pavidam volitans pennata per urbem
 Nuntia fama ruit, matrisque allabitur aures
 Euryali. At subitus miserae calor ossa reliquit;
 Excussi manibus radii, revolutaque pensa.
 Evolat infelix, et femineo ululatu,
 Scissa comam, muros amens atque agmina cursu
 Prima petit, non illa virûm, non illa pericli
 Telorumque memor ; caelum dehinc questibus implet :
 " Hunc ego te, Euryale, aspicio ? tune ille senectae
 Sera meae requies, potuisti linquere solam,

Crudelis ? nec te, sub tanta pericula missum,
Affari extremum miserae data copia matri ?
Heu, terra ignota canibus data praeda Latinis
Alitibusque jaces! nec te, tua funera, mater
Produxi, pressive oculos, aut vulnera lavi,
Veste tegens, tibi quam noctes festina diesque
Urgebam et tela curas solabar aniles.

VIRG., *Aen.*, IX.

(*a*) Write out, dividing into feet, and marking the quantity of every syllable, and the ictus or verse accent of every foot, the first two lines of each piece of Ovid which you translate, or, if you translate the Virgil, the first four lines.

(*b*) Point out any three words in the above pieces where the rules of metre will help you to distinguish their meanings in translation.

CICERO.

For all Candidates.

[*If you have read Cato Major, do I. and one other ; if not, do II. and one other. State clearly the principles of syntax that determine the forms in I. or II., printed at the end of each passage.*]

I. Audire te arbitror, Scipio, hospes tuus avitus Masinissa quae faciat hodie nonaginta natus annos; cum ingressus iter pedibus sit, in equum omnino non ascendere; cum autem equo, ex equo non descendere; nullo imbri, nullo frigore adduci ut capite operto sit; summam esse in eo corporis siccitatem ; itaque omnia exsequi regis officia et munera. Potest igitur exercitatio et temperantia etiam in senectute conservare aliquid pristini roboris. — CATO MAJOR, X.

Faciat, annos, ingressus sit, capite, sit, exsequi. Who was Masinissa? Why called *avitus hospes* of Scipio?

II. Quod si omnis impetus domesticorum hostium, depulsus a vobis, se in me unum convertit, vobis erit videndum, Quirites, qua condicione posthac eos esse velitis, qui se pro salute vestra obtulerint invidiae periculisque omnibus. mihi quidem ipsi quid est quod jam ad vitae fructum possit adquiri, cum praesertim neque in honore vestro neque in gloria virtutis quicquam videam altius, quo mihi libeat ascendere? — IN CATILINAM, III. xii.

Vobis, condicione, velitis, obtulerint, possit, videam. What does Cicero mean by *Neque in honore vestro quicquam videam altius?*

III. Ac primum quanta innocentia debent esse imperatores, quanta deinde in omnibus rebus temperantia, quanta fide, quanta facilitate, quanto ingenio, quanta humanitate? Quae breviter qualia sint in Cn. Pompeio consideremus; summa enim omnia sunt, Quirites, sed ea magis ex aliorum contentione quam ipsa per sese cognosci atque intelligi possunt. Quem enim imperatorem possumus ullo in numero putare, cujus in exercitu centuriatus veneant atque venierint? — DE IMP. CN. POMPEI, xiii.

State what you know of Cicero's relations with Pompeius.

IV. Quidam enim non modo armatis, sed interdum etiam otiosis minabantur, nec quid quisque sensisset, sed ubi fuisset cogitandum esse dicebant; ut mihi quidem videantur di immortales, etiam si poenas a populo Romano ob aliquod delictum expetiverunt, qui civile bellum tantum et tam luctuosum excitaverunt, vel placati jam vel satiati aliquando omnem spem salutis ad clementiam victoris et sapientiam contulisse. — PRO MARCELLO, vi.

Say what you know of Cicero's relations with Julius Caesar.

CAESAR.

For Course II.

Ea re constituta, secunda vigilia magno cum strepitu ac tumultu castris egressi, nullo certo ordine neque imperio, cum sibi quisque primum itineris locum peteret et domum pervenire properaret, fecerunt ut consimilis fugae profectio videretur. Hac re statim Caesar per speculatores cognita, insidias veritus, quod, qua de causa discederent, nondum perspexerat, exercitum equitatumque castris continuit.— B. G., II.

VIRGIL.

[*Course I. omit either II. or III. Course II. omit both.*]

I. Tum virgam capit; . . . . .
   illa fretus agit ventos, et turbida tranat
   nubila; jamque volans apicem et latera ardua cernit
   Atlantis duri, caelum qui vertice fulcit,
   Atlantis, cinctum adsidue cui nubibus atris
   piniferum caput et vento pulsatur et imbri;
   nix humeros infusa tegit; tum flumina mento
   praecipitant senis, et glacie riget horrida barba.
   hic primum paribus nitens Cyllenius alis
   constitit; hinc toto praeceps se corpore ad undas
   misit, avi similis, quae circum litora, circum
   piscosos scopulos humilis volat aequora juxta.
   haud aliter terras inter caelumque volabat,
   litus arenosum Libyae ventosque secabat
   materno veniens ab avo Cyllenia proles.— ÆN., IV.

Explain the mythological allusions in the last line. Write out the three lines beginning "piniferum," mark all the quantities, divide into feet, and mark the *ictus* of every foot.

II. Sic tua Cyrneas fugiant examina taxos;
sic cytiso pastae distendant ubera vaccae!
incipe, si quid habes: et me fecere poëtam
Pierides; sunt et mihi carmina; me quoque dicunt
vatem pastores; sed non ego credulus illis.
nam neque adhuc Vario videor, nec dicere Cinna
digna, sed argutos inter strepere anser olores.— ECL., IX.

III. Flectere si nequeo Superos, Acheronta movebo.
Non dabitur regnis, esto, prohibere Latinis,
Atque immota manet fatis Lavinia conjunx:
At trahere, atque moras tantis licet addere rebus,
At licet amborum populos exscindere regum.
Hac gener atque socer coëant mercede suorum.
Sanguine Trojano et Rutulo dotabere, virgo,
Et Bellona manet te pronuba. — ÆN., VII.

IV. Give a brief account of the life of Virgil.

GREEK GRAMMAR.

[Greek words must be written with accents.]

1. DECLINE Ἀτρείδης, λέων, and ἐλπίς in the singular; and τριήρης and γένος in the plural. Decline τιθείς throughout.

2. Decline ἐγώ, εἷς, ὅστις, and the Comparative of μέγας.

3. Form and compare Adverbs from σοφός, ἡδύς, πολύς.

4. Inflect the Present Optative Middle of νικάω. Give all the Active Infinitives of λείπω. Translate each of these

Infinitives with φησί. Which could regularly depend on βούλεται? Translate them.

5. Where are these words made, and from what Present Indicatives: ἔπαθεν, ἐπιέναι, ἐφιέναι, λέλυσαι, εἶπες, εἰδείην?

6. What is a Cognate Accusative? Give an example and translate it.

7. Ταῦτα βούλομαι. Show how these words would be quoted directly after ἔλεγεν, and also indirectly.

8. Explain the uses of the Article in Herodotus which differ from the Attic.

9. Explain the euphonic changes which occur in the following words: λέλειμμαι (λειπ-), σώμασι (σωματ-), πέπεισται (πειθ-), ἐτέθην (θε-).

10. Give the metrical feet of two syllables, showing the quantity of the syllables in each.

GREEK COMPOSITION.

[*Do either A or B, but not both. B consists of sentences from Jones'. Greek Composition.*]

A.

AND after the battle Clearchus called all the generals and captains together to inform (φράζειν) them that those messengers whom he had sent had come back, saying that the king had already marched off in the night and was now more than four stages (days'-march) distant (ἀπέχειν). If however, said he, I can trust the zeal (προθυμία) of your soldiers, I think that we shall cross (διαβαίνειν) the Euphrates River, before the army of the great king arrives (ἥκειν) there.

B.

1. Let us conquer those who have been drawn up before the king.

2. Cyrus feared that the king would come on the following day.

3. If you were willing to conquer some and save others, it would be well.

4. He will need not only soldiers, but also arms and chariots, if he attack this city.

5. May you be worthy of all the good things which you possess.

GREEK PROSE.

[*Those offering the Greek Reader will take* 2, 3, 4. *Those offering four books of the Anabasis and the seventh book of Herodotus will take* 1, 2, 5. *Those offering the whole Anabasis will take* 1, 2, 6.]

TRANSLATE : —

1. Οὐ μὲν δὴ οὐδὲ τοῦτ' ἄν τις εἴποι ὡς τοὺς κακούργους καὶ ἀδίκους εἴα καταγελᾶν, ἀλλὰ ἀφειδέστατα πάντων ἐτιμωρεῖτο. Πολλάκις δ' ἦν ἰδεῖν παρὰ τὰς στιβομένας ὁδοὺς καὶ ποδῶν καὶ χειρῶν καὶ ὀφθαλμῶν· στερουμένους ἀνθρώπους· ὥστε ἐν τῇ Κύρου ἀρχῇ ἐγένετο καὶ Ἕλληνι καὶ βαρβάρῳ μηδὲν ἀδικοῦντι ἀδεῶς πορεύεσθαι ὅπῃ τις ἤθελεν, ἔχοντι ὅ τι προχωροίη. — ANAB., I. IX. 13.

Explain the case of πάντων and the mood of προχωροίη. Where is εἴα made, and from what verb?

2. Ἐγὼ μὲν τοίνυν, ἔφη ὁ Ξενοφῶν, ἕτοιμός εἰμι τοὺς ὀπισθοφύλακας ἔχων, ἐπειδὰν δειπνήσωμεν, ἰέναι καταληψόμενος τὸ ὄρος. ἔχω δὲ καὶ ἡγεμόνας· οἱ γὰρ γυμνῆτες τῶν ἑπομένων ἡμῖν κλωπῶν ἔλαβόν

τινας ἐνεδρεύσαντες· τούτων καὶ πυνθάνομαι ὅτι οὐκ ἄβατόν ἐστι τὸ ὄρος, ἀλλὰ νέμεται αἰξὶ καὶ βουσίν· ὥστε ἐάνπερ ἅπαξ λάβωμέν τι τοῦ ὄρους, βατὰ καὶ τοῖς ὑποζυγίοις ἔσται. ἐλπίζω δὲ οὐδὲ τοὺς πολεμίους μενεῖν ἔτι, ἐπειδὰν ἴδωσιν ἡμᾶς ἐν τῷ ὁμοίῳ ἐπὶ τῶν ἄκρων· οὐδὲ γὰρ νῦν ἐθέλουσι καταβαίνειν ἡμῖν εἰς τὸ ἴσον. — ANAB., IV. VI. 17 and 18.

Where are ἰέναι, καταληψόμενος, and μενεῖν made, and from what verbs? Explain tense of μενεῖν.

3. Ἐκ δὲ τούτου οἱ τριάκοντα, οὐκέτι νομίζοντες ἀσφαλῆ σφίσι τὰ πράγματα, ἐβουλήθησαν Ἐλευσῖνα ἐξιδιώσασθαι, ὥστε εἶναι σφίσι καταφυγήν, εἰ δεήσειε. καὶ παραγγείλαντες τοῖς ἱππεῦσιν, ἦλθον εἰς Ἐλευσῖνα Κριτίας τε καὶ οἱ ἄλλοι τῶν τριάκοντα· ἐξέτασίν τε ποιήσαντες ἐν τοῖς ἱππεῦσι, φάσκοντες εἰδέναι βούλεσθαι πόσοι εἶεν καὶ πόσης φυλακῆς προσδεήσοιντο, ἐκέλευον ἀπογράφεσθαι πάντας· τὸν δὲ ἀπογραψάμενον ἀεὶ διὰ τῆς πυλίδος ἐπὶ τὴν θάλατταν ἐξιέναι. — HEL., II. IV. 8.

Explain the mood and tense of εἶεν and προσδεήσοιντο. What is the Attic use of σφίσι?

ARITHMETIC.

1. FIND, by logarithms, the cube root of $\dfrac{(1.469)^2 \times 0.001}{0.02584}$.

2. Reduce $\dfrac{12\frac{1}{4} \left(\frac{17}{18} + \frac{13}{10} \right)}{\frac{2}{3} \div 1\frac{3}{7}}$ to a repeating decimal.

3. Explain in full the method of finding the greatest common divisor of $3\frac{1}{3}$, $2\frac{1}{2}$, and $\frac{5}{6}$.

4. If $\frac{5}{8}$ of a bushel of corn be worth $\frac{3}{7}$ of a bushel of wheat, and wheat be worth $1.40 a bushel, how many bushels of corn can be bought for $27?

5. When $1 in gold is worth $1.595 in currency, how

many gold dollars and how much fractional currency ought 1 to receive for a ten-dollar U. S. note?

6. What part of 12 yds. 1 ft. 6 in. is $\frac{1}{2112}$ of a mile?

7. The gramme contains 15.4327 gr. Troy. How many pounds avoirdupois make a myriagramme?

8. A can do $\frac{1}{3}$ of a piece of work in 4 days; B $\frac{1}{4}$ in 5 days; C $\frac{1}{8}$ in 3 days; D $\frac{1}{8}$ in 1$\frac{1}{2}$ days. How long will it take them all to do it?

9. Extract the square root of 0.05331481. Verify the answer, as nearly as possible, by logarithms.

10. Six men, working 9 hours a day, can do a piece of work in 15 days. In how many days will a party of men, working 10 hours a day, do the work, the number of men being equal to the number of days?

ALGEBRA.

1. Find the simplest expression for
$$\frac{1}{a(a-b)(a-c)} + \frac{1}{b(b-c)(b-a)} + \frac{1}{c(c-a)(c-b)}.$$

2. Separate $4a^2b^2 - (a^2+b^2-c^2)^2$ into four trinomial factors.

3. Find two numbers such that the sum of $\frac{1}{2}$ of the first and $\frac{3}{4}$ of the second equals 11, and equals also three times the first diminished by the second.

4. Prove that $(a^m) = a^{mn}$. What is the relation between a, a^0, a^{-1}?

5. Solve the equation $\dfrac{2}{2+y} - \dfrac{1}{y-2} + \dfrac{3}{5} = 0.$

6. Give the first three and the last three terms of $(2a - \dfrac{b}{3})^6$.

7. A banker has two kinds of coin. It takes a pieces of the first, or b pieces of the second, to make a dollar. If a dollar is offered for c pieces, how many of each kind must be given?

8. Divide 16 into two parts such that their product added to the sum of their squares shall be 208.

9. Which is the larger: $\sqrt{\frac{3}{8}}$ or $\sqrt[3]{\frac{11}{14}}$?

PLANE GEOMETRY.

1. In what manner do the two diagonals of a parallelogram divide each other? Give proof. What angle do the two diagonals of a rhombus make with each other? Prove.

2. Prove that if all the sides of any convex polygon be produced, the sum of the exterior angles will be equal to four right angles.

3. Prove that, if the three angles of one triangle are equal, each to each, to the three angles of another triangle, the homologous sides are proportional.

4. Find the ratio of the side of a square to its diagonal.

5. If the vertical angle of a triangle be bisected by a line which cuts the base of the triangle, to what are the two segments of the base proportional? Prove.

If the line bisects an exterior angle of the triangle, what follows? State without proving.

6. Show how to inscribe a circle in a triangle; and then deduce a rule for finding the area of a triangle when the three sides of the triangle and the area of the inscribed circle are known.

MODERN AND PHYSICAL GEOGRAPHY.

1. DEFINE the following terms: *meridian, promontory, archipelago, cañon, delta.*

2. Describe the Atlantic coast of North America, beginning at a point in the same latitude as Cape Farewell. Give the names of important streams and inlets, the countries or states bordering upon the ocean, and the principal seaports, in their order, reckoning from north to south. State also, when possible, the principal articles of export.

3. Bound Italy. Name its principal mountains and streams. What is its largest city? What are its principal seaports, and in what part of the country is each situated?

4. What mountains lie between France and Spain? What name is given to the prolongation of this chain to the west? What bay lies to the north of this prolongation? What rivers empty into this bay?

5. Through what large islands does the equator pass? What large island lies opposite the mouth of the Amoor River? Where are the Shetland Islands? Orkney Islands? Hebrides? Corsica? Ceylon?

6. Into what waters do the following rivers empty: Euphrates, Ganges, Gaudiana, Indus, Saskatchewan, Tigris? State also where each rises, its course, and the countries through which it flows.

7. Describe a water route from Toronto to Suez.

8. In what part of South America are the principal highland regions? What effect does this disposition of highland have upon the drainage of the country?

ENGLISH COMPOSITION.

EACH candidate is required to write a short English composition, correct in spelling, punctuation, grammar, and expression. This composition must be at least fifty lines long, and be properly divided into paragraphs. One of the following subjects must be taken:—

The Character of Dr. Primrose.
An Account of the Tent-scene between Brutus and Cassius.
The Argument of Marmion.

FRENCH.

1. TRANSLATE INTO ENGLISH:—

Je *sortis*, et me promenai toute la matinée dans la ville, en songeant sans cesse à la réception que mon oncle me ferait. Je *crois*, disais-je en moi-même, qu'il sera ravi de me voir. Je jugeais de ses sentiments pas les miens, et je me préparais à une reconnaissance fort touchante. Je retournai chez lui en diligence à l'heure qu'on m'avait marquée. Vous arrivez à propos, me dit son valet, mon maître va bientôt sortir. Attendez ici un instant, je *vais* vous annoncer. A ces mots, il me *laissa* dans l'antichambre. Il y revint un moment après, et me fit entrer dans la chambre de son maître, dont le visage me frappa d'abord par un air de famille. Il me sembla que c'était mon oncle Thomas, tant ils se ressemblaient tous deux. Je le saluai avec un profond respect, et lui dis que j'étais fils de maître Nicolas: je lui appris aussi que j'exerçais à Madrid, depuis trois semaines, le métier de mon père en qualité de garçon, et que j'avais dessein de faire le tour de l'Espagne pour me profectionner. Tandis que je parlais, je m'aperçus que mon oncle rêvait. Il doutait apparemment s'il me désavouerait pour son neveu, ou s'il se déferait adroitement de moi : il choisit ce dernier parti. Il affecta de prendre un air riant et me dit: Eh bien! mon ami, comment se portent ton père et tes oncles ? dans quel état sont leurs affaires ? — LE SAGE.

2. State the tense of the italicized verbs in the above and give it in full.

3. Give the principal tenses of *savoir, acquérir, prendre, envoyer* (thus, INF., *être ;* PRES. PART., *étant ;* PAST. PART., *été ;* PRES. IND., *je suis ;* PRET., *je fus*).

4. Using mostly the words of 1, translate into French: (*a*) Do you think that they are delighted to see him? (*b*) They left me waiting more than an hour. (*c*) I fear that he has gone out.

PLANE TRIGONOMETRY.

1. FIND the angles of the plane right triangle in which the hypothenuse is $\frac{4}{3}$ of one of the sides.

2. Obtain, without using the tables, the natural trigonometric functions of 60°.

3. Obtain, from fundamental formulas, the sine and cosine of 270°, 270° — x, 270° + x.

4. Obtain, from fundamental formulas, $\sin x - \sin y = \ldots$.

5. In the plane oblique triangle $A\ B\ C$, B is 40°, b is 100. What values of a will give two solutions; one solution; no solution? Give the reason for each answer.

6. Obtain, from fundamental formulas,
$$\tan^2 \tfrac{1}{2}x = \frac{1 - \cos x}{1 + \cos x}.$$

7. Solve the triangle whose sides are 0.1498, 0.1596, 0.1943.

8. Prove the formula
$$\cos(x+y)\cos(x-y) = \cos^2 y - \sin^2 x.$$

ANALYTIC GEOMETRY.

1. WHAT is the equation of a line parallel to the axis of x, 3 units below it? At what point does this line intersect the line $3y + 4x + 1 = 0$? What is the *acute* angle between these two lines?

2. What are the axes and the parameter of the curve $4y^2 + 3x^2 = 36$? What is the equation of the circle whose diameter coincides with the transverse axis of this curve?

3. State and prove the relation between any ordinate of an ellipse and the corresponding ordinate of the inscribed circle.

4. Deduce formulas for passing from a rectangular to a polar system.

5. The equation of the tangent to the parabola $y^2 = 2px$ is $yy' = p(x + x')$. Find the equations of the tangent and the normal to $y^2 = 8x$, at the extremity of the positive ordinate through the focus.

6. Is the point $(-2, 1)$ situated on the hyperbola $4y^2 - 7x^2 = -24$? Why?

7. Of what is $xy + 4 = 0$ the equation? Illustrate by a figure.

8. Find the points in which the curve $y^2 = 4x$ intersects the curve $3y^2 + 2x^2 = 14$.

SOLID GEOMETRY.

1. Prove that, if a solid angle is formed by three plane angles, the sum of either two of them is greater than the third. The sum of the three angles taken together cannot exceed a certain quantity: what is it?

2. A pyramid is cut by two planes parallel to the base: prove that the two sections are similar polygons. State in the form of a proportion the relation which holds between the areas of these sections and their respective distances from the vertex of the pyramid.

3. Prove that, if from the vertices of a given spherical triangle as poles arcs of great circles are described, another triangle is formed, the vertices of which are the poles of the sides of the given triangle.

4. A ball of lead is three inches in diameter: what is its weight? A cubic foot of lead weighs 712 pounds.

5. A certain cylindrical vessel is twelve inches in diameter and eight inches deep. What are the dimensions of a vessel, *similar in form*, which will hold only one sixty-fourth as much?

6. What is a *degree of spherical surface?* How is the area of a spherical triangle measured? State without proving.

ADVANCED ALGEBRA.

1. Divide $a^2 - b^2 - c^2 - 2bc$ by $\dfrac{a+b+c}{a+b-c}$.

2. What is the equation whose roots are 1, $\dfrac{-1 \pm \sqrt{-3}}{2}$?

3. Obtain the formulas for the last term and the sum of the series in a geometrical progression. Obtain also an expression for the sum of the series, in terms of the first term, the last term, and the common factor or ratio.

4. Solve the equation $\sqrt{x} + \sqrt{(x - \sqrt{1-x})} = 1$.

5. How many words can be formed from seven letters taken all together, provided that 3 given letters are never separated?

6. Find the sum of n terms of the series 1, 3, 5, 7 ...

7. Solve the equations $x^3 - y^3 = 26$, $x^2 + xy + y^2 = 13$.

8. What is the sixth term of $(1-x)^{-2}$?

9. A courier travels from P to Q in 14 hours: a second courier starts at the same time from a place ten miles behind P, and arrives at Q at the same time as the first courier. The times in which the couriers travel 20 miles differ by half an hour. Find the distance from P to Q.

HARVARD EXAMINATION PAPERS.

OCTOBER, 1875.

ANCIENT HISTORY AND GEOGRAPHY.

1. MENTION in order (1) the ancient divisions of Greece which lie upon the eastern coast; and (2) the important islands near that coast. In both cases proceed from north to south in your enumeration.

2. Give a brief account (with the important dates) either of the great Persian wars, or of the career of Alexander the Great, at your option.

3. Mention any reasons that occur to you why Sparta should have been (1) victorious in the Peloponnesian War; and (2) vanquished by Thebes.

4. Mention in order (1) the western divisions of ancient Italy from the Alps to the Straits of Messina; and (2) any six places of historic note in these divisions. In both cases proceed from north to south in your enumeration.

5. Where is Epirus? How came a king of Epirus (Pyrrhus) to make war in Italy upon the Romans, and what grounds had he to hope for success?

6. Mention any reasons that occur to you (1) why Hannibal should have hoped to overcome Rome; and (2) why he failed in his attempt.

7. Give a brief account (with the important dates) of the

political career either of Cicero or of Augustus, at your option.

8. Mention the names of the emperors between Domitian and Commodus, in the order of their reigns.

MODERN AND PHYSICAL GEOGRAPHY.

1. WHAT range of mountains separates Russia from Siberia? What is the direction of this chain? What mountains separate Russia from Asia on the south? Give the name and altitude of any prominent peak of this range?

2. Which of the following groups of islands lies farthest north, which farthest south, and which farthest east?— Azores, Canary Islands, Cape Verd Islands. To what country does each group belong?

3. Describe the position of the highest mountain system in each of the continents. Give the name, position, and altitude of at least one prominent peak in each system.

4. Where is the Volga River? Into what does it empty? What other river approaches at one point very near the Volga, but empties into a different body of water?

5. Bound Switzerland. Name its principal lakes and rivers. What are its chief cities? What is its form of government?

6. Define the following terms: *peninsula, isthmus, sound, plateau, watershed.* Give an example of each.

7. Write as fully as you can about the following places, mentioning any fact relating to geographical position, size, commercial or political importance, and the like: (*a*) Cincinnati, (*b*) Cologne, (*c*) Frankfort-on-the-Main, (*d*) Liver-

pool, (e) Melbourne, (f) Prague, (g) Singapore, (h) Sitka, (i) Trieste, (j) Valparaiso.

8. Describe the course of the Mississippi River. What tributaries does it receive from the west? what from the east? Across or by what States does the principal stream flow? Mention any important fact relating to its outlet.

GREEK COMPOSITION.

[*Do either A or B, but not both. B consists of sentences from Jones' "Exercises in Greek Prose Composition."*]

A.

AND after this Xenophon arose and said, "O fellow soldiers, it is evident (δῆλος) that our march (πορεία) must be made on-foot; for there are no boats. But it is necessary (ἀνάγκη) to proceed at-once; for we have no supplies (ἐπιτήδεια). We therefore," said he, "will offer-sacrifice." After this the generals offered-sacrifice, and there was-present a soothsayer from-Arcadia (Ἀρκάς). But the sacrifices were not favorable. In-consequence they rested this day.

B.

1. In the time of (ἐπί) Darius many men fought on horseback.

2. He announced that the general had sent his army into the city.

3. He said that they would not have fled if no one had conquered them.

4. It was evident that he desired to cross before the rest replied.

5. The king hindered the greater part of the army from crossing.

6. Whenever he hurled (βάλλω) his javelin at any one he (always) hit (ἀκοντίζω) him.

GREEK GRAMMAR.

[*Candidates for advanced standing will omit 4 and 8, and take 9 and 10.]

1. WRITE the correct form of χαριεντσι, ισταανσι, πεφανμαι, δεδεχται.

2. Decline θρίξ, ἄστυ, in the singular, and γυνή and ναῦς in the plural.

3. Decline εὐγενής in the plural; εἷς and τίς throughout.

4.* Compare ὀλίγος, πολύς, ῥᾴδιος. Form and compare an adverb from ταχύς.

5. Give the synopsis of the Second Aorist Passive of στέλλω. Give the Imperfect Active of εἰμί in full.

6. Principal parts of τρέφω, τρέπω, πείθω, πάσχω, πίπτω. Where are the following verbs made: ἴδοιμι, ᾖσμεν, βῆναι, φασίν, and ἧκα?

7. What are the uses of ἄν with the tenses of the Indicative? What tenses of the Infinitive could be used after ἐλπίζω to denote Future *Time?*

8.* Compare the use of the article by Homer with its use in Attic Greek.

9. (a) Translate Ἔφη τοῦτο ποιῆσαι. Ἐκέλευει αὐτὸν ἐλθεῖν.

(β) Translate Ἐπειδὰν τοῦτο ἴ δ ω, ἀπέρχομαι. Translate Οὐκ ἂν ἀ π ῆ λ θ ε ν, εἰ μὴ βασιλεὺς τοῦτο ε ἶ π ε ν, ἵνα τὸν ἀδελφὸν ε ἴ δ ε ν. Explain the Mood and Tense of these verbs.

10. Give the scheme of the Tragic Iambic Trimeter. Describe the Anapaestic System.

GREEK PROSE.

[*Those offering the Greek Reader will take* 2, 3, 4. *Those offering four books of the Anabasis and the seventh book of Herodotus will take* 1, 2, 5. *Those offering the whole Anabasis will take* 1, 2, 6.]

1. TRANSLATE: —

ποταμὸς δ' εἰ μέν τις καὶ ἄλλος ἄρα ἡμῖν ἐστι διαβατέος οὐκ οἶδα· τὸν δ' οὖν Εὐφράτην οἴδαμεν ὅτι ἀδύνατον διαβῆναι κωλυόντων πολεμίων. οὐ μὲν δή, ἂν μάχεσθαί γε δέῃ, ἱππεῖς εἰσιν ἡμῖν ξύμμαχοι, τῶν δὲ πολεμίων ἱππεῖς εἰσιν οἱ πλεῖστοι καὶ πλείστου ἄξιοι· ὥστε νικῶντες μὲν τίνα ἂν ἀποκτείναιμεν; ἡττωμένων δὲ οὐδένα οἷόν τε σωθῆναι. — ANAB., II. iv. 6.

How is the stem of the verbal adjective in -τέος formed? State how many constructions (syntactical) the verbal in -τέος has, and explain them. Explain the case of πολεμίων (following κωλυόντων) and the mood of ἀποκτείναιμεν. What constitutes the protasis to ἂν ἀποκτείναιμεν?

2. Τῇ δὲ ὑστεραίᾳ ἄνευ ἡγεμόνος ἐπορεύοντο· μαχόμενοι δ' οἱ πολέμιοι, καὶ ὅπῃ εἴη στενὸν χωρίον προκαταλαμβάνοντες, ἐκώλυον τὰς παρόδους. ὁπότε μὲν οὖν τοὺς πρώτους κωλύοιεν, Ξενοφῶν ὄπισθεν ἐκβαίνων πρὸς τὰ ὄρη ἔλυε τὴν ἀπόφραξιν τῆς παρόδου τοῖς πρώτοις, ἀνωτέρω πειρώμενος γίγνεσθαι τῶν κωλυόντων· ὁπότε δὲ τοῖς ὄπισθεν ἐπιθοῖντο, Χειρίσοφος ἐκβαίνων, καὶ πειρώμενος ἀνω-

τέρω γίγνεσθαι τῶν κωλυόντων, ἔλυε τὴν ἀπόφραξιν τῆς παρόδου τοῖς ὄπισθεν· καὶ ἀεὶ οὕτως ἐβοήθουν ἀλλήλοις, καὶ ἰσχυρῶς ἀλλήλων ἐπεμέλοντο.—GOODWIN'S READER, p. 36 (ANAB., IV. ii. 24 - 26).

What relation does the participle μαχόμενοι express? Give the principal parts of λαμβάνω and βαίνω. Explain the mood of κωλύοιεν. What is the composition of ἀπόφραξιν and ἐβοήθουν? From what and how are adverbs regularly formed?

3. Ὑμῖν, ἔφη, ὦ ἐκ τοῦ ἄστεος ἄνδρες, συμβουλεύω ἐγὼ γνῶναι ὑμᾶς αὐτούς. μάλιστα δ' ἂν γνοίητε, εἰ ἀναλογίσαισθε, ἐπὶ τίνι ὑμῖν μέγα φρονητέον ἐστὶν, ὥστε ἡμῶν ἄρχειν ἐπιχειρεῖν. πότερον δικαιότεροί ἐστε; ἀλλ' ὁ μὲν δῆμος, πενέστερος ὑμῶν ὢν, οὐδὲν πώποτε ἕνεκα χρημάτων ὑμᾶς ἠδίκησεν· ὑμεῖς δὲ, πλουσιώτεροι πάντων ὄντες, πολλὰ καὶ αἰσχρὰ ἕνεκα κερδέων πεποιήκατε. ἐπεὶ δὲ δικαιοσύνης οὐδὲν ὑμῖν προσήκει, σκέψασθε εἰ ἄρα ἐπ' ἀνδρείᾳ ὑμῖν μέγα φρονητέον.—GOODWIN'S READER, p. 85 (HEL., II. iv. 40).

Analyze the forms γνοίητε and σκέψασθε, showing how they are built up from the simple stems. Explain the mood of γνοίητε, and the case of ὑμῖν (following ἐπὶ τίνι), οὐδέν, and πάντων. Explain the construction of ἄρχειν, and that of ἐπιχειρεῖν.

4. τότε μὲν οὕτω ἠγωνίσαντο, τῇ δ' ὑστεραίῃ οἱ βάρβαροι οὐδὲν ἄμεινον ἀέθλεον· ἅτε γὰρ ὀλίγων ἐόντων, ἐλπίσαντές σφεας κατατετρωματίσθαι τε καὶ οὐκ οἵους τε ἔσεσθαι ἔτι χεῖρας ἀνταείρασθαι, συνέβαλλον. οἱ δὲ Ἕλληνες κατὰ τάξις τε καὶ κατὰ ἔθνεα κεκοσμημένοι ἦσαν, καὶ ἐν μέρεϊ ἕκαστοι ἐμάχοντο, πλὴν Φωκέων· οὗτοι δὲ ἐς τὸ οὖρος ἐτάχθησαν φυλάξοντες τὴν ἀτραπόν. ὡς δὲ οὐδὲν εὕρισκον ἀλλοιότερον οἱ Πέρσαι ἢ τῇ προτεραίῃ ἐνώρων, ἀπήλαυνον.—GOODWIN'S REALER, p. 146 (HEROD., VII. 212).

5. ἐκ ταύτης ἂν τῆς νήσου ὁρμεώμενοι, φοβεόντων τοὺς Λακεδαιμονίους· παροίκου δὲ πολέμου σφι ἐόντος οἰκηίου, οὐδὲν δεινοὶ ἔσον-

ταί τοι, μὴ τῆς ἄλλης Ἑλλάδος, ἁλισκομένης ὑπὸ τοῦ πεζοῦ βοηθέ-
ωσι ταύτῃ· καταδουλωθείσης δὲ τῆς ἄλλης Ἑλλάδος, ἀσθενὲς ἤδη
τὸ Λακωνικὸν μοῦνον λείπεται. ἢν δὲ ταῦτα μὴ ποιῇς, τάδε τοι
προσδόκα ἔσεσθαι. ἔστι τῆς Πελοποννήσου ἰσθμὸς στεινός· ἐν τούτῳ τῷ χώρῳ πάντων Πελοποννησίων συνομοσάντων ἐπὶ σοὶ, μάχας
ἰσχυροτέρας ἄλλας τῶν γενομένων προσδέκεο ἔσεσθαί τοι· ἐκεῖνο
δὲ ποιήσαντι ἀμαχητὶ ὅ τε ἰσθμὸς οὗτος καὶ αἱ πόλις προσχωρή-
σουσι. —HEROD., VII. 235.

6. Χειρίσοφος δ' ἐπεὶ ᾑρέθη, παρελθὼν εἶπεν, Ἀλλ', ὦ ἄνδρες,
τοῦτο μὲν ἴστε, ὅτι οὐδ' ἂν ἔγωγε ἐστασίαζον, εἰ ἄλλον εἵλεσθε·
Ξενοφῶντα μέντοι, ἔφη, ὠνήσατε οὐχ ἑλόμενοι· ὡς καὶ νῦν Δέξιππος
ἤδη διέβαλλεν αὐτὸν πρὸς Ἀναξίβιον ὅ τι ἐδύνατο καὶ μάλα ἐμοῦ
αὐτὸν σιγάζοντος. ὁ δ' ἔφη νομίζειν αὐτὸν Τιμασίωνι μᾶλλον συνάρ-
χειν ἐθελῆσαι Δαρδανεῖ ὄντι τοῦ Κλεάρχου στρατεύματος ἢ ἑαυτῷ
Λάκωνι ὄντι. ἐπεὶ μέντοι ἐμὲ εἵλεσθε, ἔφη, καὶ ἐγὼ πειράσομαι ὅ
τι ἂν δύνωμαι ὑμᾶς ἀγαθὸν ποιεῖν. καὶ ὑμεῖς οὕτω παρασκευάζεσθε
ὡς αὔριον, ἐὰν πλοῦς ᾖ, ἀναξόμενοι. —ANAB., VI. i. 32, 33.

GREEK POETRY.

1. TRANSLATE: —

"δαιμονίη, αἰεὶ μὲν ὀΐεαι, οὐδέ σε λήθω,
πρῆξαι δ' ἔμπης οὔ τι δυνήσεαι, ἀλλ' ἀπὸ θυμοῦ
μᾶλλον ἐμοὶ ἔσεαι· τὸ δέ τοι καὶ ῥίγιον ἔσται.
εἰ δ' οὕτω τοῦτ' ἐστίν, ἐμοὶ μέλλει φίλον εἶναι.
ἀλλ' ἀκέουσα κάθησο, ἐμῷ δ' ἐπιπείθεο μύθῳ,
μή νύ τοι οὐ χραίσμωσιν, ὅσοι θεοί εἰσ' ἐν Ὀλύμπῳ,
ἆσσον ἰόνθ', ὅτε κέν τοι ἀάπτους χεῖρας ἐφείω."

ILIAD, I. 561 – 567.

(a) In what tense, mood, and voice and from what present indicatives are κάθησο and ἐφείω formed?

(β) Divide into feet the last two verses (566, 567).

(γ) Write the Attic forms corresponding to δίεαι, πρῆξαι, ἐπιπείθεο.

2. TRANSLATE:—

<blockquote>
Ὣς φάτο νεικείων Ἀγαμέμνονα, ποιμένα λαῶν,
Θερσίτης. τῷ δ᾽ ὦκα παρίστατο δῖος Ὀδυσσεύς,
καί μιν ὑπόδρα ἰδὼν χαλεπῷ ἠνίπαπε μύθῳ·
"Θερσῖτ᾽ ἀκριτόμυθε, λιγύς περ ἐὼν ἀγορητὴς
ἴσχεο, μηδ᾽ ἔθελ᾽ οἶος ἐριζέμεναι βασιλεῦσιν.
οὐ γὰρ ἐγὼ σέο φημὶ χερειότερον βροτὸν ἄλλον
ἔμμεναι, ὅσσοι ἅμ᾽ Ἀτρείδης ὑπὸ Ἴλιον ἦλθον·
τῷ οὐκ ἂν βασιλῆας ἀνὰ στόμ᾽ ἔχων ἀγορεύοις,
καί σφιν ὀνείδεά τε προφέροις, νόστον τε φυλάσσοις.
</blockquote>

ILIAD, II. 243–251.

(δ) Give the derivation of ἀκριτόμυθε and ἐριζέμεναι.

3. TRANSLATE:—

<blockquote>
Οἱ δ᾽ ἐπεὶ οὖν ἑκάτερθεν ὁμίλου θωρήχθησαν,
ἐς μέσσον Τρώων καὶ Ἀχαιῶν ἐστιχόωντο
δεινὸν δερκόμενοι· θάμβος δ᾽ ἔχεν εἰσορόωντας
Τρῶάς θ᾽ ἱπποδάμους καὶ ἐϋκνήμιδας Ἀχαιούς.
καί ῥ᾽ ἐγγὺς στήτην διαμετρητῷ ἐνὶ χώρῳ
σείοντ᾽ ἐγχείας, ἀλλήλοισιν κοτέοντε.
πρόσθε δ᾽ Ἀλέξανδρος προΐει δολιχόσκιον ἔγχος,
καὶ βάλεν Ἀτρείδαο κατ᾽ ἀσπίδα πάντοσ᾽ ἐΐσην·
οὐδ᾽ ἔρρηξεν χαλκός, ἀνεγνάμφθη δέ οἱ αἰχμὴ
ἀσπίδ᾽ ἐνὶ κρατερῇ.
</blockquote>

ILIAD, III. 340–349.

(ε) What Homeric peculiarities do you notice in the formation of εἰσορόωντας, στήτην, ἀλλήλοισιν?

(ζ) Translate the following epithets of ships: ποντόποροι, ὠκύποροι, πολυκλήϊδες, ἀμφιέλισσαι, κορωνίδες.

LATIN COMPOSITION.

TRANSLATE INTO LATIN: —

In midsummer,[1] at which time all-other[2] praetors are wont[3] to go round[4] the province and run-to-and-fro[5] or to cruise[6] in-person,[7] at that time he was not content with his own house royal,[8] which belonged to King Hiero, which the praetors are wont[9] to use. He ordered tents[10] to be pitched[11] on the beach,[12] which beach is on[13] the island at Syracuse,[14] hard-by[15] the very entrance[16] of the harbor.[17]

[1] *aestas summa.* [2] *ceteri.* [3] *consuescere:* what tense? [4] *obire,* with accusative. [5] *concursare.* [6] *navigare.* [7] *ipse.* [8] *regius.* [9] *solere.* [10] *tabernaculum.* [11] *collocare.* [12] *litus.* [13] *in.* [14] *Syracusae,* plural. [15] *prope.* [16] *introitus.* [17] *portus.*

TRANSLATE INTO ENGLISH: —

Sed quid ego plura de Gavio? quasi tu Gavio tum fueris infestus, ac non nomini, generi, iuri civium hostis. Non illi, inquam, homini, sed causae communi libertatis inimicus fuisti. Quid enim attinuit, cum Mamertini more atque instituto suo crucem fixissent post urbem in via Pompeia, te iubere in ea parte figere, quae ad fretum spectaret, et hoc addere, quod negare nullo modo potes, te idcirco illum locum deligere, ut ille, quoniam se civem Romanum esse diceret, ex cruce Italiam cernere ac domum suam prospicere posset?

LATIN GRAMMAR.

Do not translate, but answer the questions below.

Tum vero ancipiti mentem formidine pressus
Obstipui, steteruntque comae et vox faucibus haesit.
Hunc Polydorum auri quondam cum pondere magno

Infelix Priamus furtim mandarat alendum
Threicio regi, cum iam diffideret armis
Dardaniae cingique urbem obsidione videret.
Ille, ut opes fractae Teucrum, et Fortuna recessit,
Res Agamemnonias victriciaque arma secutus,
Fas omne abrumpit; Polydorum obtruncat, et auro
Vi potitur. Quid non mortalia pectora cogis,
Auri sacra fames?

Give the principal parts of *haesit, alendum, diffideret, cingi, fractae.*

Give the Subjunctive active, 1st person, of all the tenses of *pressus, haesit, mandarat.*

Give the subjunctive, 1st person in all tenses, and the Imperative throughout, of *secutus.* Give all the participles of *recessit,* and all the Infinitives of *fractae.*

Decline *infelix* and *pondere.* Decline also *ecquis.* Compare *sacra.* Compare *proximus, humilis,* and *saepe.*

Give the meaning of the derivative terminations of *victricia* and *mortalia.* Form adjectives from *auri* and *urbem.* Form a noun of agency (denoting the doer of the action) from *alendum.* How are the present and the perfect of *fractae* formed respectively? What is the construction of *armis, obsidione, auro?*

CAESAR, SALLUST, AND OVID.

[*N. B. Translate one piece of Caesar, the piece of Sallust, and two pieces of Ovid. The order in which they are done is unimportant. The second piece of Caesar is a substitute for Sallust, and the pieces of Virgil for Ovid, by those who have not read those authors.*]

I. CAESAR: BELL. GALL., Book II. § xi. Ea re constituta secunda vigilia magno cum strepitu ac tumultu castris egressi nullo certo ordine neque imperio, cum sibi quisque primum itineris locum peteret et domum pervenire properaret, fecerunt ut consimilis fugae profectio videretur. Hac re statim Caesar per speculatores cognita insidias veritus, quod qua de causa discederent nondum perspexerat, exercitum castris continuit. Prima luce omnem equitatum qui novissimum agmen moraretur praemisit.

Explain the subjunctives *discederent* and *moraretur*.

II. [*Only for those who do not offer Sallust.*] CAESAR: B. G., V. § 34. At barbaris consilium non defuit, nam duces eorum tota acie pronunciare jusserunt ne quis ab loco discederet; illorum esse praedam, atque illis reservari quaecumque Romani reliquissent; proinde omnia in victoria posita existimarent. Erant et virtute et numero pugnando pares nostri. Tamen etsi ab duce et a fortuna deserebantur, tamen omnem spem salutis in virtute ponebant, et quoties quaeque cohors procurreret, ab ea parte magnus numerus hostium cadebat.

How far north did Caesar's campaigns extend?

III. SALLUST: CAT. § 22. Fuere ea tempestate qui dicerent Catilinam oratione habita, cum ad jusjurandum popularis sceleris sui adigeret, humani corporis sanguinem vino permixtum in pateris circumtulisse; inde cum post exsecrationem omnes degustavissent, sicut in sollemnibus sacris fieri consuevit, aperuisse consilium suum, atque eo [dictitare] fecisse, quo inter se magis fidi forent, alius alii tanti facinoris conscii. Nonnulli ficta et haec et multa praeterea existumabant ab iis, qui Ciceronis invidiam, quae postea orta est, leniri credebant atrocitate sceleris eorum qui poenas dederant.

What offices had Catiline held, and which one was he aiming at in the year of his conspiracy?

(1) OVID: METAM., III. 51-60.

quae mora sit sociis, miratur Agenore natus,
vestigatque viros: tegumen direpta leonis
pellis erat, telum splendenti lancea ferro
et jaculum, teloque animus praestantior omni.
ut nemus intravit, letataque copora vidit,
victoremque supra spatiosi corporis hostem
tristia sanguinea lambentem vulnera lingua,
"Aut ultor vestrae, fidissima corpora, mortis,
aut comes" inquit "ero." Dixit dextraque molarem
sustulit, et magnum magno conamine misit.

(2) OVID: FASTI, IV. 811-818.

Contrahere agrestes et moenia ponere utrique
  Convenit: ambigitur, moenia ponat uter.
"Nil opus est" dixit "certamine" Romulus "ullo:
  Magna fides avium est; experiamur aves."
Res placet: alter adit nemorosi saxa Palati;
  Alter Aventinum mane cacumen init.
Sex Remus, hic volucres bis sex videt ordine; pacto
  Statur, et arbitrium Romulus urbis habet.

(3)   [*Only for such as do not offer Ovid.*]
VIRGIL: AEN., VIII. 671-677.

Haec inter tumidi late maris ibat imago,
Aurea, sed fluctu spumabant caerula cano;
Et circum argento clari delphines in orbem
Aequora verrebant caudis aestumque secabant.
In medio classis aeratas, Actia bella,
Cernere erat; totumque instructo Marte videres
Fervere Leucaten, auroque effulgere fluctus.

SAME: X. 96-103.

Talibus orabat Iuno, cunctique fremebant
Caelicolae adsensu vario; ceu flamina prima

Cum deprensa fremunt silvis et caeca volutant
Murmura, venturos nautis prodentia ventos.
Tum Pater omnipotens, rerum cui prima potestas,
Infit; eo dicente deum domus alta silescit,
Et tremefacta solo tellus; silet arduus aether;
Tum Zephyri posuere; premit placida aequora pontus.

(*a*) Write out, dividing into feet, and marking the quantity of every syllable, and the ictus or verse accent of every foot, the second line of each piece of verse which you translate.

(*b*) Point out any word in the first piece translated where the rules of metre will help you in its translation.

CICERO AND VIRGIL.

(*Latin Authors for Course II.*)

CICERO (*for all Candidates*).

[*If you have read Cato Major, do 1 and one other; if not, do 2 and one other. State clearly the principles of syntax that determine the forms in 1 or 2, printed at the end of each passage.*]

1. Quo in genere est in primis senectus, *quam* ut adipiscantur omnes optant, eandem accusant adepti: tanta est stultitiae inconstantia atque perversitas. Obrepere aiunt eam citius quam putavissent. Primum, quis coëgit eos falsum putare? Qui enim citius *adulescentiae* senectus quam pueritiae adulescentia obrepit? Deinde, qui minus gravis *esset* iis senectus, si octingentesimum annum agerent quam si octogesimum? Praeterita enim aetas quamvis longa cum *effluxisset,* nulla consolatio permulcere posset stultam senectutem. — CATO MAJOR, II.

Quam, adulescentiae, esset, effluxisset.

Cato says, *Quattuor reperio causas cur senectus misera videatur;* what are they?

2. Quamquam isti, qui Catilinam *Massiliam* ire dictitant, non tam hoc queruntur quam verentur. Nemo est istorum tam misericors, qui illum non ad Manlium quam ad Massiliensis ire *malit*. Ille autem, si (me hercule) hoc quod agit numquam antea cogitasset, tamen latrocinantem se interfici *mallet* quam exsulem vivere. Nunc vero, cum ei nihil adhuc praeter ipsius voluntatem cogitationemque acciderit, nisi quod vivis *nobis* Roma profectus est, optemus potius ut eat in exsilium quam *queramur*. — CATIL., II. vii.

Massiliam, mallet, nobis, queramur.

State briefly the occasion and subject of each of the orations against Catiline.

3. Ut enim cetera paria Tuberoni cum Varo fuissent, — honos, nobilitas, splendor, ingenium, quae nequaquam fuerunt, — hoc certe praecipuum Tuberonis, quod justo cum imperio ex senatus consulto in provinciam suam venerat. Hinc prohibitus non ad Caesarem, ne iratus, non domum, ne iners, non in aliquam regionem, ne condemnare causam illam quam secutus erat, videretur: in Macedoniam ad Cn. Pompei castra venit, in eam ipsam causam a qua erat rejectus injuria. — PRO LIGARIO, IX.

4. Quod enim praemium satis magnum est tam benevolis, tam bonis, tam fidelibus servis, propter quos vivit? Etsi id quidem non tanti est, quam quod propter eosdem non sanguine et volneribus suis crudelissimi inimici mentem oculosque satiavit. Quos nisi manu misisset, tormentis etiam dedendi fuerunt conservatores domini, ultores sceleris, defensores necis. Hic vero nihil habet in his malis quod minus moleste ferat, quam, etiam si quid ipsi accidat, esse tamen illis meritum praemium persolutum. — PRO MILONE, XXII.

CAESAR.

(*For Course II. only.*)

Eodem tempore equites nostri levisque armaturae pedites, qui cum iis una fuerant, quos primo hostium impetu pulsos dixeram, cum se in castra reciperent, adversis hostibus occurrebant ac rursus aliam in partem fugam petebant; et calones, qui ab decumana porta ac summo jugo collis nostris victores flumen transire conspexerant, praedandi causa egressi, cum respexissent et hostes in nostris castris versari vidissent, praecipites fugae sese mandabant.—B. G., II.

VIRGIL (*for All*).

[*Course I. omit either 2 or 3. Course II. omit both.*]

1. "Nate dea, vosque haec" inquit "cognoscite, Teucri,
Et mihi quae fuerint juvenali in corpore vires,
Et qua servetis revocatum a morte Dareta."
Dixit, et adversi contra stetit ora juvenci,
Qui donum adstabat pugnae, durosque reducta
Libravit dextra media inter cornua caestus,
Arduus, effractoque inlisit in ossa cerebro.
Sternitur exanimisque tremens procumbit humi bos.
Ille super tales effundit pectore voces:
"Hanc tibi, Eryx, meliorem animam pro morte Daretis
Persolvo; hic victor caestus artemque repono."
Aen., V.

Write out the first three lines, mark all the quantities, divide into feet, and mark the *ictus* of every foot.

2. Despectus tibi sum, nec qui sim quaeris, Alexi,
Quam dives pecoris, nivei quam lactis abundans.
Mille meae Siculis errant in montibus agnae;

Lac mihi non aestate novum, non frigore defit;
Canto quae solitus, si quando armenta vocabat,
Amphion Dircaeus in Actaeo Aracintho.
Nec sum adeo informis: nuper me in litore vidi,
Cum placidam ventis staret mare; non ego Daphnim
Judice te metuam, si numquam fallit imago. — Ecl., II.

3. Consumptis hic forte aliis, ut vertere morsus
Exiguam in Cererem penuria adegit edendi
Et violare manu malisque audacibus orbem
Fatalis crusti patulis nec parcere quadris,
"Heus, etiam mensas consumimus!" inquit Iulus;
Nec plura adludens. Ea vox audita laborum
Prima tulit finem, primamque loquentis ab ore
Eripuit pater, ac stupefactus numine pressit.
Aen., VII.

ARITHMETIC.

[*Give all the work. Reduce each answer to its simplest form.*]

1. What part of $\dfrac{12\frac{1}{2}}{\frac{1}{4}}$ is $\dfrac{\frac{2}{3} \times \frac{3}{4}}{\frac{1}{2}}$?

2. What is the cost of a pile of wood whose dimensions are 2, 1.9, and 42.5 metres, at $2 per stere?

3. Find, by logarithms, the third power of the fourth root of $\dfrac{12\frac{1}{2} \times .01}{317.4}$.

4. A and B gain in business $5,040, of which A is to have ten per cent more than B. What is the share of each?

5. If 2 cubic inches of iron weigh as much as 15 cubic

inches of water, and a cubic foot of water weigh 1000 ounces, find the weight, in tons, of a cubic yard of iron.

6. If 12 pipes, each delivering 12 gallons a minute, fill a cistern in 3 hours 24 minutes, how many pipes, each delivering 16 gallons a minute, will fill a cistern 6 times as large in 6 hours 48 minutes?

7. How many kilometres make a mile?

8. How many bags, each containing 2 bu. 1 pk. 3 qt., will be required to hold 111 bu. 2 pk. 4 qt. of grain?

9. What is the compound interest of $1 for 143 years, allowing it to double once in 11 yr. 11 mo.?

ALGEBRA.

[Give the whole work.]

1. FIND the greatest common divisor and the least common multiple of $(243a^{10}b^5 + 1)$ and $(81a^8b^4 - 1)$, by resolving each expression into factors.

2. Solve the equation
$$\frac{x+a}{x-a} - \frac{x-a}{x+a} = \frac{1}{x-a} - \frac{1}{x^2-a^2} + \frac{1}{x+a}.$$
What is the value of x, if $6a + 7 = 0$?

3. Divide $\dfrac{6\sqrt{b}}{25\sqrt[5]{a^3}}$ by $\dfrac{20c\sqrt[4]{b^3}}{21ab\sqrt[3]{a^2}}$; and express the result without fractional or negative exponents.

4. Solve the equations $2x - y = 21$, $2x^2 + y^2 = 153$.

5. A person buys some cloth for $90. If he had got two yards more for the same sum, the price per yard would have been fifty cents less. How much did he buy, and at what price per yard?

6. Find $(a-b)^{12}$ by the Binomial Theorem.

ADVANCED ALGEBRA.

[*Give the whole work.*]

1. Solve the equations $x^3 - y^3 = 215$, $x^2 + xy + y^2 = 43$.

2. A certain number consists of three digits, in arithmetical progression. If it be divided by the sum of the digits, the quotient is 48; but if 198 be subtracted from it, the digits are inverted. Find the number.

3. Prove the formula for the sum of a geometric progression, in terms of a, r, and n.

4. The first term of a geometric progression is 512, the last term is 162, and the sum is 1562. Find the whole series. Find also what the sum of this series would be, if continued to infinity.

5. Solve the equation $\sqrt{(x+4)} - \sqrt{x} = \sqrt{(x+\frac{3}{2})}$.

6. Simplify $\left(\dfrac{a+b}{a-b} + \dfrac{a-b}{a+b} \right) \div \left(\dfrac{a^2+b^2}{a^2-b^2} - \dfrac{a^2-b^2}{a^2+b^2} \right)$.

7. Find the greatest common divisor of
$2x^5 - 11x^2 - 9$ and $4x^5 + 11x^4 + 81$.

PLANE GEOMETRY.

1. When are two polygons said to be *similar?* What are similar *arcs?* similar *segments?*

2. If a triangle has two sides equal, what is it called? Prove what is true of the angles opposite the equal sides.

3. If, in any triangle, a line be drawn parallel to the base, it will divide the other two sides proportionally. Prove.

4. At a given point in the circumference of a circle a tangent to the circle is drawn. What is the measure of the angle between the tangent and a chord drawn from the point of contact? Prove. What will this angle be if the chord passes through the centre of the circle?

5. Prove that the perimeters of regular polygons, of the same number of sides, are to each other as the radii of the circumscribed circles. State, without proving, what the ratio of the *areas* of the polygons is.

6. Find the area of the circle in which a square, each side of which is $\sqrt{8}$ inches long, can be inscribed; and then find the radius of a second circle which shall be nine times as large as the first.

SOLID GEOMETRY.

1. DEFINE the following terms: *prism; right prism; pentagonal prism; altitude of a zone; spherical sector; lunary surface.*

2. Given two planes perpendicular to each other, and a line in one of them perpendicular to their common intersection; prove that the line is perpendicular to the other plane.

3. How may the frustum of a right cone be generated? How is its convex surface found? Give proof.

4. The altitude of a given right cone is ten inches: how far from the vertex of the cone must two planes be passed, parallel to the base of the cone, in order to divide the lateral surface into three equal parts.

5. Prove that, if two spherical triangles on the same sphere, or on equal spheres, are *equiangular* with respect to each other, they are also *equilateral* with respect to each other. If the radius of one sphere is three times as great as

that of another, what will be the ratio of the sides of two mutually equiangular spherical triangles, one on one sphere and the other on the other?

ANALYTIC GEOMETRY.

[*Give the whole work.*]

1. What angle does the line $y + 4x + 2 = 0$ make with $2y + 8x = 0$? with $4y = x$? with $5y + 3x = 1$?

2. Which of the four lines in the previous question pass through the origin, and which do not? Prove.

3. The general equation of a circle referred to rectangular axes is $(y - n)^2 + (x - m)^2 = r^2$. At what points is the circle whose radius is $\sqrt{\frac{31}{4}}$, and whose centre is at $(-3, -\frac{1}{2})$, cut by the line $y + 1 = 0$?

4. Deduce formulas for passing from a rectangular to a polar system. [Denote the polar coördinates by ρ, φ; the coördinates of the pole with reference to the rectangular system by m, n; the angle which the initial line makes with x by a.]

5. The equation of the tangent to a circle is $xx' + yy' = r^2$. Lines are drawn through (7, 1) tangent to the circle $x^2 + y^2 = 25$. Find the points of tangency.

6. What is meant by the *parameter* of a curve? What is the parameter of $y^2 = 2px$? Prove. Of $a^2y^2 + b^2x^2 = a^2b^2$? Prove.

7. Explain in full *one* method of drawing a tangent to a parabola at a given point of the parabola.

8. Find whether the line $4y - 3x = 0$ intersects the hyperbola $5y^2 - 2x^2 + 15 = 0$, or its conjugate. What is the tangent of the angle which the asymptotes of this curve make with the axis of x?

PLANE TRIGONOMETRY.

[*Give the whole work.*]

1. Tan $x = m$. What is the cotangent of $(180° - x)$? of $(270° + x)$? What angle in the third quadrant has a cotangent equal to m?

2. Explain by a figure the changes, both numerical and algebraic, through which the sine of an arc passes, as the arc increases from 0° to 360°.

3. Obtain, from fundamental formulas, all the trigonometric functions of *the negative of an angle.*

4. Solve the plane right triangle whose sides are 13, 12, 5.

5. Obtain, from fundamental formulas, an expression for $\cot(x + y)$, in terms of $\cot x$ and $\cot y$.

6. Two angles of a plane oblique triangle are 13° 17′ 48″ and 114° 47′ 9″, and the included side is 0.1493. Solve the triangle.

7. Given, $\quad b^2 = c^2 + d^2 - 2cd \cos B,$
$1 - \cos x = 2 \sin^2 \tfrac{1}{2}x,$
$s = \tfrac{1}{2}(b + c + d).$

Find an expression for $\sin^2 \tfrac{1}{2}B$.

9. In the plane oblique triangle BCD, explain in full the method of solution when the sides c, d, and the angle B are given.

ENGLISH COMPOSITION.

You are required to write a short English composition of not less than fifty lines, correct in spelling, punctuation, grammar, and expression. Make several paragraphs. Take one of these subjects: —
An account of the Trial in the Merchant of Venice.
The story of Fergus Mac Ivor, in Waverley.
Prospero's Life on the Island, and how he came there.

FRENCH.

1. TRANSLATE INTO ENGLISH : —

Pendant le règne du fameux Crésus, il y avait en Lydie un jeune homme bien fait, plein d'esprit, très-vertueux, de la race des anciens rois, et devenu si pauvre qu'il fut réduit à se faire berger. Se promenant un jour sur des montagnes escarpées où il *rêvait* sur ses malheurs en menant son troupeau, il s'*assit* au pied d'un arbre pour se délasser. Il *aperçut* auprès de lui une ouverture étroite dans un rocher. La curiosité l'engage à y entrer. Il trouve une caverne large et profonde. D'abord il n'y *voit* goutte; enfin ses yeux s'accoutument à l'obscurité. Il entrevoit dans une lueur sombre une urne d'or sur laquelle ces mots étaient gravés: "Ici tu trouveras l'anneau de Gygès. O mortel, qui que tu sois, à qui les dieux destinent un si grand bien, montre-leur que tu n'es pas ingrat, et garde-toi d'envier jamais le bonheur d'aucun autre homme." Il ouvre l'urne, trouve l'anneau, le prend, et, dans le transport de sa joie il laissa l'urne. quoiqu'il fût très-pauvre et qu'elle fût d'un grand prix. Il sort de la caverne et se hâte d'éprouver l'anneau enchanté dont il avait si souvent entendu parler depuis son enfance. — FÉNELON.

2. State the tense of the italicized verbs in the above, and give it in full.

3. Give the principal tenses of *venir, mourir, valoir, prendre.* (Thus, INF., *être;* PRES PART., *étant;* PAST PART., *été;* IND. PRES., *je suis;* PRET., *je fus.*)

4. Translate into French: (*a*) He is the best boy in the school. (*b*) We read[1] French better than you think.[2] (*c*) I do not doubt[3] that you will come. (*d*) We have no more money[4]; have the kindness[5] to send[6] us some.

[1] lire. [2] croire. [3] douter. [4] argent, *m.* [5] bonté, *f.* [6] envoyer.

ELEMENTARY BOTANY.

1. DESCRIBE the germination of a grain of corn, a pine-seed, and an acorn.
2. What are buds?
3. Describe equitant leaves.
4. What are stipules?
5. What is a leaf-tendril?
6. How does it differ from a branch-tendril?
7. In how many ways may stamens be united together?
8. What is a two-lipped flower?
9. Describe the fruit of a rose, strawberry, blackberry, apple, pea, and squash.
10. Describe the leaves of this plant.

HARVARD EXAMINATION PAPERS.

JUNE, 1876.

ANCIENT HISTORY AND GEOGRAPHY.

[*Answer five questions, including the first.*]

1. DRAW a map of Italy, showing the position of the divisions and of the following places: Vercellae, Beneventum, Brundusium, Nola, Roma, Capua, Antium, Ostia, Asculum Apulum, Neapolis, Veii, Tibur, Tusculum, Tarentum, Pisa, Florentia, Placentia, Cannae, Cumae, Baiae.

2. The alliance of Athens and Sparta, 464–461 B. C., and its connection with Athenian politics.

3. The Theban supremacy.

4. Philip of Macedon.

5. The expedition of Pyrrhus to Italy.

6. State the extent of the Roman Empire at the time of Augustus. When and how were the various provinces subjected to Rome?

7. Give some account of the writers of the Augustan age.

MODERN AND PHYSICAL GEOGRAPHY.

1. DEFINE *latitude* and *longitude*. What is the latitude of the Tropic of Capricorn? what that of the Arctic Circle? What countries of Europe are crossed by the 40th parallel of latitude? Between what parallels does Australia lie?

2. In what zone do the three northern continents chiefly lie? in what the three southern? How do the southern continents compare with the northern in respect to coast indentations and projections? What continent has its coast relatively most indented? Name its chief projections.

3. Describe the three great river systems of South America, naming the principal affluents of each. Name the chief rivers of Siberia, Chinese Empire, India, and Burmah, and state their courses and where they empty.

4. What States and Territories of the United States lie wholly west of the Mississippi River? Through what States does that river run?

5. Name the states of Europe and their capitals.

6. Name the chief cities of the Prussian and Austrian Empires, and give as complete an account of one of them as time will allow.

7. Name the larger East India Islands. Give some account of the physical character of Java, and of its productions. What large island lies near the southern extremity of Hindostan?

8. What mountains are included in the Appalachian system, and in what course, or courses, do its chains run? How do the Rocky and Appalachian Mountains compare in height and extent? Name the principal mountain chains

in or adjacent to Asia. Where do they lie, and what are their directions? Name some of the highest peaks, and state their altitudes.

GREEK COMPOSITION.

[*Do A if you can; if not, do B; but do not do both.*]

A.

1. When Clearchus saw the messengers[1] he asked[2] (them) what they wished:
2. and they said that they came for the purpose of making a truce,[3] being empowered[4] to announce the King's (terms)[5] to the Greeks;
3. and that they would lead them (to a place) whence[6] they could obtain [*have*] supplies,[7] if there should be a truce.
4. And after having the King's (terms), Clearchus asked if the truce would extend [*be*] to all;
5. and they said, "To all, until your (terms) shall be announced to the King."

[1] ἄγγελος, ἀγγέλλω. [2] ἐρωτάω. [3] σπονδή, σπένδειν. [4] ἱκανός.
[5] τὰ παρὰ βασιλέως. [6] ὅθεν. [7] ἐπιτήδεια.

B.

1. If we should send arms to the general, we think he would be grateful to us.
2. He seized the soldier, declaring he would inflict punishment on him for his cowardice.

3. He said most of the soldiers would have crossed, if the enemy had not hindered.

4. I fear that it will be necessary for me to go with the generals, in order to see Menon.

5. He used to delay in each city until I arrived.

GREEK GRAMMAR.

1. Give Accusative and Vocative singular, and Genitive and Dative plural of δικαστής, γίγας, λέων, βασιλεύς, παῖς, and γλυκύς.

2. Decline the Greek words for *one, three*, and *both*. Decline τίς in the singular, and οὗτος in the plural.

3. Explain the formation of these words, giving the meaning of root and ending: ἡδέως, παιδίον, χρυσέος, ἀληθεύω.

4. Give a synopsis of the Aorist Active of λείπω, and all the participles of τίθημι.

5. Where are these verbs made, and from what Present Indicative: ἡρώτων, τιθεῖσι, μαθών, ἐξόν, ἀφεῖναι?

6. What case or cases regularly follow, κατηγορέω (*I accuse*); χράομαι (*I use*); ἀποδείκνυμι (*I appoint*)?

7. Translate ὁ Σωκράτης (ἐστὶ) σοφός — ὁ αὐτὸς Σωκράτης (ἐστὶ) σοφός — αὐτὸς ὁ Σωκράτης (ἐστὶ) σοφός.

Translate, *This man; every man; another man; most men.*

8. What time, relatively to the main verb, does the Infinitive express in the following phrases: φησὶν ἐλθεῖν; βούλεται ἐλθεῖν; δεῖ τοῦτο ποιῆσαις.

GREEK PROSE.

[*N. B. Those who offer the Greek Reader will take* 2, 3, 4. *Those who offer four books of the Anabasis and the Seventh Book of Herodotus will take* 1, 2, 5. *Candidates in Course II. will take* 1 *and* 2, *or* 2 *and* 3.]

1. TRANSLATE : —

Πρὸς ταῦτα μεταστάντες οἱ ῞Ελληνες ἐβουλεύοντο καὶ ἀπεκρίναντο· Κλέαρχος δ᾽ ἔλεγεν. ῾Ημεῖς οὔτε συνήλθομεν ὡς βασιλεῖ πολεμήσοντες, οὔτε ἐπορευόμεθα ἐπὶ βασιλέα· ἀλλὰ πολλὰς προφάσεις Κῦρος εὕρισκεν, ὡς καὶ σὺ εὖ οἶσθα, ἵνα ὑμᾶς τε ἀπαρασκευάστους λάβοι καὶ ἡμᾶς ἐνθάδε ἀναγάγοι. ᾽Επεὶ μέντοι ἤδη ἑωρῶμεν αὐτὸν ἐν δεινῷ ὄντα, ᾐσχύνθημεν καὶ θεοὺς καὶ ἀνθρώπους προδοῦναι αὐτὸν, ἐν τῳ πρόσθεν χρόνῳ παρέχοντες ἡμᾶς αὐτοὺς εὖ ποιεῖν. ἐπεὶ δὲ Κῦρος τέθνηκεν, οὔτε βασιλεῖ ἀντιποιούμεθα τῆς ἀρχῆς, οὔτ᾽ ἔστιν ὅτου ἕνεκα βουλοίμεθ᾽ ἂν τὴν βασιλέως χώραν κακῶς ποιεῖν. — ANAB., II. iii. 21 – 23.

Explain the tense of *πολεμήσοντες* and the case of *βασιλεῖ* (line 2); construction of *πρόσθεν* (line 7), of *ὄντα* (line 6), of *ποιεῖν* (line 7), and of *ὅτου* (line 9).

2. TRANSLATE : —

᾽Ακούσας δὲ Ξενοφῶν ἔλεγεν ὅτι ὀρθῶς ᾐτιῶντο, καὶ αὐτὸ τὸ ἔργον αὐτοῖς μαρτυροίη. Ἀλλ᾽ ἐγὼ, ἔφη, ἠναγκάσθην διώκειν, ἐπειδὴ ἑώρων ἡμᾶς ἐν τῷ μένειν κακῶς μὲν πάσχοντας, ἀντιποιεῖν δ᾽ οὐδὲν δυναμένους. ᾽Επειδὴ δὲ ἐδιώκομεν, ἀληθῆ, ἔφη, ὑμεῖς λέγετε. κακῶς μὲν γὰρ ποιεῖν οὐδὲν μᾶλλον ἐδυνάμεθα τοὺς πολεμίους, ἀνεχωροῦμεν δὲ πάνυ χαλεπῶς. Τοῖς οὖν θεοῖς χάρις ὅτι οὐ σὺν πολλῇ ῥώμῃ ἀλλὰ σὺν ὀλίγοις ἦλθον· ὥστε βλάψαι μὲν μὴ μεγάλα, δηλῶσαι δὲ ὧν δεόμεθα. — ANAB., III. iii. 12 – 14.

In what voice, mood, and tense, and from what verbs, are *ᾐτιῶντο, ἑώρων, ἐδυνάμεθα,* and *ἀνεχωροῦμεν* ? Explain the construction of μαρτυροίη, μένειν, πάσχοντας, and δηλῶσαι, and the case of πολεμίους and of ὧν (last line).

3. TRANSLATE: —

Καὶ πρῶτον μὲν αὐτῶν ἐσκόπει, πότερά ποτε νομίσαντες ἱκανᾶς ἤδη τἀνθρώπινα εἰδέναι, ἔρχονται ἐπὶ τὸ περὶ τῶν τοιούτων φροντίζειν, ἢ τὰ μὲν ἀνθρώπινα παρέντες, τὰ δαιμόνια δὲ σκοποῦντες, ἡγοῦνται τὰ προσήκοντα πράττειν. ἐθαύμαζε δὲ εἰ μὴ φανερὸν αὐτοῖς ἐστιν, ὅτι ταῦτα οὐ δυνατόν ἐστιν ἀνθρώποις εὑρεῖν· ἐπεὶ καὶ τοὺς μέγιστον φρονοῦντας ἐπὶ τῷ περὶ τούτων λέγειν οὐ ταὐτὰ δοξάζειν ἀλλήλοις, ἀλλὰ τοῖς μαινομένοις ὁμοίως διαχεῖσθαι πρὸς ἀλλήλους. — MEMOR., I. i. 12, 13.

Explain the case of αὐτῶν (first line) and of μαινομένοις, and the construction of φροντίζειν. In what voice and tense, and from what verb, is παρέντες? How is its meaning here related to that of the simple verb?

4. TRANSLATE: —

Ὡς δέ σφι διετέτακτο καὶ τὰ σφάγια ἐγίνετο καλά, ἐνθαῦτα ὡς ἀπείθησαν οἱ Ἀθηναῖοι, δρόμῳ ἵεντο ἐς τοὺς βαρβάρους. ἦσαν δὲ στάδιοι οὐκ ἐλάσσονες τὸ μεταίχμιον αὐτῶν ἢ ὀκτώ. οἱ δὲ Πέρσαι ὁρέοντες δρόμῳ ἐπιόντας, παρεσκευάζοντο ὡς δεξόμενοι· μανίην τε τοῖσι Ἀθηναίοισι ἐπέφερον καὶ πάγχυ ὀλεθρίην, ὁρέοντες αὐτοὺς ὀλίγους, καὶ τούτους δρόμῳ ἐπειγομένους, οὔτε ἵππου ὑπαρχούσης σφι οὔτε τοξευμάτων. ταῦτα μέν νυν οἱ βάρβαροι κατείχαζον· Ἀθηναῖοι δέ, ἐπεί τε ἀθρόοι προσέμιξαν τοῖσι βαρβάροισι, ἐμάχοντο ἀξίως λόγου. πρῶτοι μὲν γὰρ Ἑλλήνων πάντων τῶν ἡμεῖς ἴδμεν δρόμῳ ἐς πολεμίους ἐχρήσαντο, πρῶτοι δὲ ἀνέσχοντο ἐσθῆτά τε Μηδικὴν ὁρέοντες, καὶ τοὺς ἄνδρας ταύτην ἐσθημένους· τέως δὲ ἦν τοῖσι Ἕλλησι καὶ τὸ οὔνομα τὸ Μήδων φόβος ἀκοῦσαι. — HEROD., VI. 112.

Give the Attic forms for ἀπείθησαν (from what verb?), ὁρέοντες, μανίην, and οὔνομα.

5. TRANSLATE: —

Λακεδαιμονίων δὲ καὶ Θεσπιέων τοιούτων γενομένων, ὅμως λέγεται ἄριστος ἀνὴρ γενέσθαι Σπαρτιήτης, Διηνέκης, τὸν τόδε φασὶ εἰπεῖν τὸ

GREEK POETRY.

ἔπος πρὶν ἢ συμμῖξαί σφεας τοῖσι Μήδοισι· πυθόμενον πρός τευ τῶν Τρηχινίων ὡς, ἐπεὰν οἱ βάρβαροι ἀπιέωσι τὰ τοξεύματα, τὸν ἥλιον ὑπὸ τοῦ πλήθεος τῶν ὀϊστῶν ἀποκρύπτουσι, — τοσοῦτό τι πλῆθος αὐτῶν εἶναι. τὸν δὲ οὐκ ἐκπλαγέντα τούτοισι εἰπεῖν, ἐν ἀλογίῃ ποιεύμενον τὸ τῶν Μήδων πλῆθος, ὡς πάντα σφι ἀγαθὰ ὁ Τρηχίνιος ξεῖνος ἀγγέλλοι, εἰ, ἀποκρυπτόντων τῶν Μήδων τὸν ἥλιον, ὑπὸ σκιῇ ἔσοιτο πρὸς αὐτοὺς ἡ μάχη, καὶ οὐκ ἐν ἡλίῳ. — HEROD., VII. 226.

Give the Attic forms for τόν (line 2), τευ, ἐπεάν, ἀπιέωσι (from what verb?), and ἀλογίῃ.

GREEK POETRY.

[*You are advised to do the translation first, and answer the questions (a—f) afterward. Candidates in Course II. will do the translation in 1 and 2, and answer the questions (a), (b), (c), and (e).*]

1. TRANSLATE: —

εἷος ὁ ταῦθ' ὥρμαινε κατὰ φρένα καὶ κατὰ θυμόν,
ἕλκετο δ' ἐκ κολεοῖο μέγα ξίφος, ἦλθε δ' Ἀθήνη
οὐρανόθεν· πρὸ γὰρ ἧκε θεά, λευκώλενος Ἥρη,
ἄμφω ὁμῶς θυμῷ φιλέουσά τε κηδομένη τε.
στῆ δ' ὄπιθεν, ξανθῆς δὲ κόμης ἕλε Πηλείωνα,
οἴῳ φαινομένη· τῶν δ' ἄλλων οὔ τις ὁρᾶτο.
θάμβησεν δ' Ἀχιλεύς, μετὰ δ' ἐτράπετ', αὐτίκα δ' ἔγνω
Παλλάδ' Ἀθηναίην· δεινὼ δέ οἱ ὄσσε φάανθεν.

IL., I. 193 – 200.

(a) Divide the last two verses into feet. Designate any one of these eight verses that has the *feminine* caesura.

(b) Who first collected the poems of Homer in their present form? What is the theme of the *Iliad?*

2. Translate:—

τῶν δ', ὥς τ' ὀρνίθων πετεηνῶν ἔθνεα πολλά,
χηνῶν ἢ γεράνων ἢ κύκνων δουλιχοδείρων,
Ἀσίῳ ἐν λειμῶνι, Καϋστρίου ἀμφὶ ῥέεθρα,
ἔνθα καὶ ἔνθα ποτῶνται ἀγαλλόμενα πτερύγεσσιν,
κλαγγηδὸν προκαθιζόντων, σμαραγεῖ τε λειμών,
ὣς τῶν ἔθνεα πολλὰ νεῶν ἄπο καὶ κλισιάων
ἐς πεδίον προχέοντο Σκαμάνδριον· αὐτὰρ ὑπὸ χθὼν
σμερδαλέον κονάβιζε ποδῶν αὐτῶν τε καὶ ἵππων.

Iliad, II. 459–466.

(c) Write the Attic forms of οὐρανόθεν and φάανθεν in the first passage, and give the derivation of ῥέεθρα and κλαγγηδόν in the second.

(d) Attic for οἱ in the last verse of the third passage?

3. Translate:—

" Ζεῦ πάτερ, οὔ τις σεῖο θεῶν ὀλοώτερος ἄλλος.
ἦ τ' ἐφάμην τίσασθαι Ἀλέξανδρον κακότητος·
νῦν δέ μοι ἐν χείρεσσιν ἄγη ξίφος, ἐκ δέ μοι ἔγχος
ἠίχθη παλάμηφιν ἐτώσιον, οὐδὲ δάμασσα."
ἦ καὶ ἐπαΐξας κόρυθος λάβεν ἱπποδασείης,
ἕλκε δ' ἐπιστρέψας μετ' ἐϋκνήμιδας Ἀχαιούς·
ἄγχε δέ μιν πολύκεστος ἱμὰς ἁπαλὴν ὑπὸ δειρήν,
ὅς οἱ ὑπ' ἀνθερεῶνος ὀχεὺς τέτατο τρυφαλείης.

Iliad, III. 365–372.

(e) State in the order of their occurrence the details of a sacrifice as described by Homer in Books both I. and II. of the Iliad.

(f) What is the meaning of the phrase, μηρούς τ' ἐξέταμον? What were the σπλάγχνα?

LATIN COMPOSITION.

TRANSLATE INTO LATIN: —

1. THE ninth year after[1] the expulsion of the kings,[2] when Tarquin's son-in-law[3] had collected[4] a mighty[5] army for avenging[6] his father-in-law's[7] wrong,[8] a new dignity[9] was created[10] at Rome, which is called[11] the dictatorship,[12] — greater than the consulship.[13] The same year a master of the horse, too, was appointed,[14] to be-under-the-orders-of[15] the dictator.

2. To the Sabines begging[16] the dictator and the senate to give pardon[17] for their[18] mistake[19] to men who[18] were young,[20] answer was made[21] that the young could[22] be pardoned,[23] the old[24] could not be pardoned.

[1] *post* (with participle of *exigo*). [2] *rex*. [3] *gener*. [4] *colligo*. [5] *ingens*. [6] *vindico*. [7] *socer*. [8] *iniuria*. [9] *dignitas*. [10] *creo*. [11] *appello*. [12] *dictatura*. [13] *consulatus*. [14] *fio*. [15] *obsequor*. [16] *oro*. [17] *venia*. [18] omit. [19] *error*. [20] *adulescens*. [21] *respondeo*. [22] *possum*. [23] *ignosco*. [24] *senex*.

TRANSLATE INTO ENGLISH: —

A. Postumius dictator, T. Aebutius magister equitum magnis copiis peditum equitumque profecti ad lacum Regillum in agro Tusculano agmini hostium occurrerunt; et quia Tarquinios esse in exercitu Latinorum auditum est, sustineri ira non potuit, quin extemplo confligerent.

LATIN GRAMMAR.

1. MARK the quantity of the penult and ultima of each of the following words: *custodis, radices, decorus, fidei, veni* (Imperat.), *fieri, circumdare, liceret*.

2. Indicate, by English spelling, the proper pronunciation of each syllable of the following sentence: *Gaius Iulius Caesar multas nationes vicit.*

3. Give rule for the gender of *palus, decus, Tenedos.*

4. Decline *ego; filia quaedam; vetus miles; alia manus.*

5. Compare *malus, dives, liber.* Form and compare adverbs from *audax, durus, libens.*

6. State where each of the following forms is made, and give principal parts of the verb to which it belongs: *iacĕret, pactus, dedidisses, oderit, fugem, arcessit, severas, peperit, gaudet, sanximus.*

7. Give a synopsis (*i. e.* one form for every tense in each mood, besides Participles, etc.) of *eo;* of the tenses formed on the Perfect stem of *pello.* Inflect the Fut. Indic. of *doceo* and *fero*, and the Pres. Subj. of *facio* and *sto*, in both voices. Give all the participles of *labor.*

8. What is the root of *amo?* of *frango?* of *paciscor?* Show how the three stems of each verb are formed from the root.

9. Separate each of the following words into its component parts, — stem, suffix, — and give the meaning of each: *similitudo, documentum, particula, deditio, flebilis, nosco.*

10. What case or cases follow *ob, sub; rogare, imperare, donare, potiri, paenitere; peritus, similis, dignus?* Write in Latin: *at Athens; he came to Rome by the Appian way, he is not believed in this by his friends.*

CAESAR, SALLUST, AND OVID.

[*N. B. Each candidate is expected to translate the first and one other piece of verse; also the first and one other of prose; and to answer all the questions. Any candidate who has* read no Ovid *can obtain a substitute paper in verse by application at once to the Examiner.*]

TRANSLATE: —

I. Nec tibi quadrupedes animosos ignibus illis,
   Quos in pectore habent, quos ore et naribus efflant,
   In promptu regere est. Vix me patiuntur, ubi acres
   Incaluere animi, cervixque repugnat habenis.
   At tu, funesti ne sim tibi muneris auctor,
   Nate, cave, dum resque sinit, *tua* corrige vota.
   Scilicet ut nostro genitum te sanguine credas,
   Pignora certa petis. Do pignora certa timendo,
   Et patrio pater esse metu probor. Aspice vultus
   Ecce meos. Utinamque oculos in pectora posses
   Inserere, et patrias intus deprendere curas!

II. Hunc ego, me Cyclops nulla cum fine petebat:
    Nec, si quaesieris, odium Cyclopis, amorne
    Acidis in nobis fuerit praesentior, edam:
    Par utrumque fuit. Pro quanta potentia regni
    Est, Venus alma, tui! nempe ille inmitis et ipsis
    Horrendus silvis, et visus ab hospite nullo
    Inpune, et magni cum dis contemptor Olympi,
    Quid sit amor, sentit, nostrique cupidine captus
    Uritur, *oblitus* pecorum antrorumque suorum.

III. Victor abes. Nec scire mihi, quae causa morandi,
     Aut in quo lateas ferreus orbe, licet.
     Quisquis ad haec vertit peregrinam littora puppim,
     Ille mihi de te multa rogatus abit:

Quamque tibi reddat, si te *modo* viderit usquam,
　Traditur huic digitis charta novata meis.
Nos Pylon, antiqui Neleïa Nestoris arva,
　Misimus. Incerta est fama remissa Pylo.

IV. Aut semel in nostras quoniam nova puppis harenas
　Venerat, audaces attuleratque viros,
Isset anhelatos non praemedicatus in ignes
　Immemor Aesonides oraque adunca boum,
Semina sevisset, totidem sevisset et hostes,
　Et caderet cultu cultor ab ipse suo.
Quantum perfidiae tecum, scelerate, perisset,
　Dempta forent capiti quam *mala* multa meo!

(*a*) Divide into feet the second line in each piece translated, marking the quantity of every syllable and ictus of every foot. (*b*) Show how the metre helps to determine the meaning of any two of the words in italics.

TRANSLATE:—

1. Veneti reliquaeque item civitates cognito Caesaris adventu certiores facti, simul quod quantum in se facinus *admisissent* intelligebant, legatos, quod nomen ad omnes nationes sanctum inviolatumque semper fuisset, retentos a se et in vincula conjectos, pro magnitudine periculi bellum parare et maxime ea quae ad usum navium pertinent providere instituunt, *hoc* majore spe quod multum natura loci confidebant.

2. Vix agmen novissimum extra munitiones processerat, quum Galli cohortati inter se ne speratam praedam ex manibus dimitterent, longum esse perterritis Romanis Germanorum auxilium exspectare, neque suam parti dignitatem ut tantis copiis tam exiguam manum, praesertim fugientem atque impeditam, adoriri non *audeant*, flumen transire et iniquo loco committere non dubitant. Quae fore suspicatus

Labienus, ut omnes citra flumen eliceret eadem usus *simulatione* itineris, placide progrediebatur.

3. Ceterum juventus pleraque, sed maxime nobilium, Catilinae *inceptis* favebat; quibus in otio vel magnifice vel molliter vivere copia erat, incerta pro certis, bellum quam pacem malebant. Fuere item ea tempestate qui *crederent* M. Licinium Crassum non ignarum ejus consili fuisse; quia Cn. Pompeius invisus ipsi magnum exercitum ductabat, cujusvis opes voluisse contra illius potentiam crescere; simul confisum, si conjuratio valuisset, facile apud illos principem se fore.

4. Ita compositis rebus in loca quam maxime occulta discedit, ac post paulo cognoscit Marium ex itinere frumentatum cum paucis cohortibus *Siccam* missum, quod oppidum primum omnium post malam pugnam ab rege defecerat. Eo cum delectis equitibus noctu pergit, et jam egredientibus Romanis in porta pugnam facit; simul magna voce Siccenses hortatur uti cohortes ab tergo circumveniant; fortunam illis praeclari facinoris casum dare; si id *fecerint*, postea sese in regno, illos in libertate sine metu aetatem acturos.

(*a*) What was the extent of Caesar's province? What was the place to which his attention was directed on first crossing the Alps? (*b*) Give the syntax of the words in italics in (1) and in any one of the other prose pieces.

CICERO AND VIRGIL.

[*Each candidate will do two selections of prose and two of poetry, with the questions attached to each.*
Candidates in Course II. will substitute the passage from Caesar for their second selection in poetry.
Those who do not select 1 will be presumed not to have read Cato Major.
Explain the construction of the words printed below each passage.]

1. QUIN etiam memoriae proditum est, cum Athenis ludis quidam in theatrum grandis natu venisset, magno consessu locum nusquam ei datum a suis civibus; cum autem ad Lacedaemonios accessisset, qui, legati cum essent, certo in loco consederant, consurrexisse omnes illi dicuntur et senem sessum recepisse. — CAT. MAJ., 18.

Athenis, ludis, essent, sessum.

2. Quem enim imperatorem possumus ullo in numero putare, cujus in exercitu centuriatus veneant atque venierint? Quid hunc hominem magnum aut amplum de re publica cogitare, qui pecuniam, ex aerario depromptam ad bellum administrandum, aut propter cupiditatem provinciae magistratibus diviserit, aut propter avaritiam Romae in quaestu reliquerit? — LEG. MAN., 37.

Cogitare, administrandum, diviserit.

3. Vidimus tuam victoriam proeliorum exitu terminatam: gladium vagina vacuum in urbe non vidimus. Quos amisimus civis, eos Martis vis perculit, non ira victoriae; ut dubitare debeat nemo quin multos, si fieri posset, C. Caesar ab inferis excitaret, quoniam ex eadem acie conservat quos potest. Alterius vero partis nihil amplius dicam

quam (id quod omnes verebamur) nimis iracundam futuram fuisse victoriam.— Pro Marc., 17.

Vagina, excitaret, futuram fuisse.

4. Volturcius vero subito litteras proferri atque aperiri jubet, quas sibi a Lentulo ad Catilinam datas esse dicebat. Atque ibi vehementissime perturbatas Lentulus tamen et signum et manum suam cognovit. Erant autem [scriptae] sine nomine sed ita: *Qui sim scies ex eo quem ad te misi. Cura ut vir sis, et cogita quem in locum sis progressus; vide ecquid tibi jam sit necesse, et cura ut omnium tibi auxilia adjungas, etiam infimorum.*— Cat., III. 12.

Sim, adjungas.

5. Itaque, credo, si civis Romanus Archias legibus non esset, ut ab aliquo imperatore civitate donaretur perficere non potuit. Sulla cum Hispanos donaret et Gallos, credo hunc petentem repudiasset: quem nos in contione vidimus, cum ei libellum malus poëta de populo subjecisset, quod epigramma in eum fecisset, tantummodo alternis versibus longiusculis, statim ex eis rebus quas tunc vendebat jubere ei praemium tribui, sed ea condicione, ne quid postea scriberet.— Pro Arch., 25.

Repudiasset, scriberet, donaret. Explain the circumstances of this oration.

6. Jamque adeo donati omnes opibusque superbi
   Puniceis ibant evincti tempora taenis,
Cum saevo e scopulo multa vix arte revolsus,
Amissis remis atque ordine debilis uno,
Inrisam sine honore ratem Sergestus agebat.
Qualis saepe viae deprensus in aggere serpens,
   Aerea quem obliquum rota transiit, aut gravis ictu
Seminecem liquit saxo lacerumque viator,
Nequiquam longos fugiens dat corpore tortus,
Parte ferox, ardensque oculis, et sibila colla

Arduus attollens; pars volnere clauda retentat
Nexantem nodis seque in sua membra plicantem:
Tali remigio navis se tarda movebat.—AEN., V. 268.

Write out the first and fifth lines, marking feet and caesura. Give the argument of this book.

7. Tu mihi seu magni superas jam saxa Timavi,
Sive oram Illyrici legis aequoris, en erit umquam
Ille dies, mihi cum liceat tua dicere facta?
En erit, ut liceat totum mihi ferre per orbem
Sola Sophocleo tua carmina digna cothurno?
A te principium, tibi desinam: accipe jussis
Carmina coepta tuis, atque hanc sine tempora circum
Inter victrices hederam tibi serpere laurus.
        Ec., VIII. 6.

Explain allusions in *cothurno, hederam*. Write out the first line, marking feet and caesura.

8. Parte alia ventis et dis Agrippa secundis
Arduus agmen agens; cui, belli insigne superbum,
Tempora navali fulgent rostrata corona.
Hinc ope barbarica variisque Antonius armis,
Victor ab Aurorae populis et litore rubro,
Aegyptum virisque Orientis et ultima secum
Bactra vehit; sequiturque, nefas! Aegyptia coniunx.
Una omnes ruere, ac totum spumare reductis
Convolsum remis rostrisque tridentibus aequor.
        AE., VIII. 682.

Explain allusion in the whole passage; in *conjunx*.

9. Dum in his locis Caesar navium parandarum causa moratur, ex magna parte Morinorum ad eum legati venerunt, qui se de superioris temporis consilio excusarent, quod homines barbari et nostrae consuetudinis imperiti bellum populo Romano fecissent, seque ea, quae imperasset, facturos pollicerentur. Hoc sibi satis opportune Caesar acci-

disse arbitratus, quod neque post tergum hostem relinquere volebat, neque belli gerendi propter anni tempus facultatem habebat, neque has tantularum rerum occupationes sibi Britanniae anteponendas judicabat, magnum his obsidum numerum imperat. — B. G., IV. 22.

Excusarent, imperasset.

ARITHMETIC.

[*Give all the work. Give each answer in its simplest form.*]

1. How many hectares make a square mile? Use logarithms (if you desire) in the computation.

2. Divide, by means of logarithms, $\dfrac{0.347\frac{2}{3}}{.439}$ by $\sqrt[3]{0.9}$.

3. What is the ratio of 15 A. 1 R. 2 P. to $2\frac{3}{4}$ times 2 A. 3 R. 4 P.?

4. Divide $460 into three parts which shall be to each other as $\frac{1}{2}$, $\frac{2}{3}$, and $\frac{3}{4}$.

5. What are the prime factors of 1716? How many integral divisors has this number, and what are they? What is the smallest integer by which this number can be multiplied, so that the product shall be a square?

6. A man paints two sides of a wall 7 feet high in 31 hours 6 minutes 40 seconds. If he can paint 4 square yards in an hour, how long is the wall?

7. A man sells flour at $6.50 a barrel, and gains 10 per cent. What per cent would he gain if he sold the flour for $8.25 a barrel?

8. In what time will $4,500, at 5 per cent, gain $181.25?

9. Find the cube root of 1027243.729.

ALGEBRA.

Course I.

[*Write legibly and without crowding; give the whole work, and reduce the answers to their simplest forms.*]

1. Divide $16x^3y - [13x^2y^2 + 11xy^3 - 6(y^4 + 2x^4)]$ by $-2x^2 - 5xy - 3y^2$.

2. A merchant who had two brands of flour sold a barrels of the first and b barrels of the second at an average price of c dollars per barrel; and, at the same rates, he sold m barrels of the first and n barrels of the second at an average price of p dollars per barrel. Find the price of each brand.

3. Solve the equation $\dfrac{x}{m^2p(x+a)} = \dfrac{x+a}{n^2px}$.

4. Two men, A and B, set out at the same time on the same walking journey, in opposite directions; A to go from M to N, and B to go from N to M. When they meet, the distance that A has already gone exceeds that which B has gone by 100 miles, and it is found that A will require 8 days more to reach N, while B will require 18 days more to reach M. Required, the distance MN, and the rate of each traveller.

5. Divide $\dfrac{2}{x} - \dfrac{3}{2x-1} - \dfrac{2x-3}{4x^2-1}$ by $\dfrac{16}{2x+1} - \dfrac{6x-1}{x^2}$.

6. Reduce $\dfrac{x^3 + x - 10}{x^4 - 16}$ to its lowest terms.

7. Divide $\dfrac{(\sqrt[5]{a}\sqrt[4]{b})^3}{\sqrt[3]{c^2}}$ by $\dfrac{\sqrt[6]{c^2b^5}}{\sqrt[5]{a^7}}$.

8. Write out $(x-y)^9$.

ADVANCED ALGEBRA.

[*Give the whole work.*]

1. WHAT is the meaning of $x^{1/3}$? Of $x^{3/4}$? Of $x^{-3/4}$? Show that such meaning may properly be given to such expressions. What is the continued product of these three quantities?

2. Find four values of x in the equation
$$x^2 + \frac{1}{x^2} = a^2 + \frac{1}{a^2}.$$

3. In the quadratic equation $ax^2 + bx + c = 0$, prove to what the sum and the product of the roots are respectively equal. If a is 8 and c is 2, what value of b will make the two roots equal to each other?

4. The sum of two numbers is nine times their difference, and if their product is diminished by the greater number, the result is twelve times the greater number divided by the less. Find the numbers.

5. The interior angles of a rectilinear figure are in Arithmetical Progression: the smallest angle is 120°, and the common difference is 5°. Find the number of sides. If you obtain two results, see if both are possible.

6. What is the sum of n terms of the series $3, 2, \frac{4}{3} \ldots$? What is the sum, if n is infinity?

7. What is the middle term of $(x + y)^{40}$?

8. Having 4 single books, and 3 sets containing respectively 8, 5, and 3 volumes, in how many ways can I arrange them on a shelf, provided the volumes of each set are kept together?

PLANE GEOMETRY.

1. Prove that the angle formed by two secants of a circle, and which has its vertex without the circumference, has for its measure half the concave arc intercepted between its sides, minus half the convex arc.

2. If, in a right triangle, a perpendicular is drawn from the vertex of the right angle to the hypothenuse, what relations exist between the three triangles thus formed? Prove.

How is this proposition useful in proving the Pythagorean proposition?

3. Find an expression for the length of any chord EF of a circle in terms of the segments AD and BD, into which it divides the diameter AB perpendicular to it.

4. If from a point, without a circle, a tangent and a secant are drawn, the tangent is a mean proportional between the entire secant and the part without the circle.

Prove without using the corresponding theorem for two secants.

5. How can the area of a trapezoid be found? The area of any regular polygon? Give the proof in each case.

6. Draw in your book any pentagon. Find a triangle equivalent to it. Explain and prove the method of your solution.

SOLID GEOMETRY.

1. Define a *straight line perpendicular to a plane*, and prove that, when a straight line is perpendicular to two straight lines drawn through its foot in a plane, it is perpendicular to the plane.

2. Prove that, if two solids have equal bases and heights, and if their sections, made by any plane parallel to the common plane of their bases, are equal, they are equivalent.

3. How is the area of the convex surface of a regular pyramid of any number of sides measured? Prove.

4. The altitude of a certain solid is 2 in., its surface 15 sq. in., and solid contents 4 cu. in. What is the altitude and surface of a similar solid whose solidity is 256 cu. in.?

5. Prove that the sum of the angles of a spherical triangle is greater than two right angles.

6. What is the measure of the area of a lunary surface? *State without proving.*

ANALYTIC GEOMETRY.

1. What are the slope and intercept of the line $2x - 5y - 10 = 0$? What is the equation of the perpendicular let fall upon this line from the point $(-1, 2)$?

2. Find the equation of a circle referred to its principal vertex, either from the equation referred to the centre or directly from a figure.

3. What curve is represented by the equation $4x^2 - 9y^2 + 25 = 0$? What is here the origin? what the coördinate axes? Find the parameter and excentricity of the curve.

4. Find the equation of the circle passing through the vertex of the parabola $y^2 = 10x$ and the extremities of the double ordinate through the focus.

5. Deduce the equation of the normal to any point (x', y') of an ellipse, and prove that this normal bisects the angle between the focal lines to the point. The lengths of the focal lines are $r = a - ex'$, $r' = a + ex'$, where a is the semi-transverse axis, and e the excentricity

TRIGONOMETRY.

1. OBTAIN the formulae
$$\sin^2 x + \cos^2 x = \ldots$$
$$\sin(x - y) = \ldots$$
$$2\cos^2 x = 1 + \ldots$$

2. Compare the tangent and cotangent of $(90° + y)$ with the same functions of $(y - 90°)$.

3. Give the formulae for solving a plane oblique triangle ABC, when a b and C are given; and explain fully the method of solution.

4. Find all the parts of the plane oblique triangle for which $B = 39° \, 43'$, $C = 62° \, 9'$, $a = 143.7$.

5. At a distance of 100 feet from a tree, the angle of elevation of its top is observed to be $23° \, 3'$. If the height of the instrument above the ground is 5 feet, how high is the tree?

ENGLISH COMPOSITION.

You are required to write a short English composition, correct in spelling, punctuation, grammar, and expression. This composition must be at least fifty lines long, and be properly divided into paragraphs. One of the following subjects must be taken: —

The story of the Tempest.
The story and character of Portia.

FRENCH.

1. TRANSLATE INTO ENGLISH:—

Frémyn arriva ; je le regardai fixement, et je lui trouvai une physionomie dure qui ne promettait rien de bon. Il n'avait pas mis plus de trente heures à faire ses soixante lieues. Je commençai par lui montrer les misérables dont j'avais à plaider la cause. Ils étaient tous debout devant lui ; les femmes pleuraient ; les hommes, appuyés sur leur bâton, la tête nue, avaient la main dans leurs bonnets. F., assis, les yeux fermés, la tête penchée, et le menton appuyé sur sa poitrine, ne les regardait pas. Je parlai en leur faveur ; je ne *sais* où l'on prend ce qu'on dit en pareil cas. Je lui fis toucher au doigt combien il était incertain que cet héritage lui fût légitimement acquis ; je le conjurai par son opulence, par la misère qu'il avait sous les yeux ; je *crois* même que je me jetai à ses pieds : je n'en *pus* tirer un sou. .... Je lui jetai les clefs au nez ; il les ramassa, s'empara de tout ; et je m'en revins si troublé, si peiné, si changé, que votre mère, qui vivait encore, crut qu'il m'était arrivé quelque grand malheur... Ah ! mes enfants, quel homme que ce F. !— DIDEROT.

2. State the tense of the italicized verbs in the above, and give it in full.

3. Give the principal tenses of *connaître, tenir, vouloir, peindre* (thus, INF., *être ;* PRES. PART., *étant ;* PAST PART., *été ;* IND. PRES., *je suis ;* PRET., *je fus*).

4. TRANSLATE INTO FRENCH :—

(*a*) I had no time[1] to speak[2] to him, but I will write[3] him a long letter.[4] (*b*) Bitter[5] fruits[6] are often the most wholesome.[7] (*c*) Have you taken a walk[8] this morning?

(*a*) To which of those pupils⁹ have you promised¹⁰ a reward¹¹?

¹ temps, *m.* ² parler. ³ écrire. ⁴ lettre, *f.* ⁵ amer. ⁶ fruit, *m.* ⁷ sain. ⁸ se promener. ⁹ écolier, *m.* ¹⁰ promettre. ¹¹ récompense, *f.*

GERMAN.

1. (*a*) EXPLAIN the sounds of vowels and diphthongs by English equivalents wherever it is possible.

(*b*) What are modified vowels?

(*c*) Which words are written with capital letters in German?

(*d*) Explain gutturals and such consonants as may differ from the English in their application.

2. (*a*) How many parts of speech are there in German? What are they?

(*b*) State your observations on the affinity of German and English words.

3. (*a*) Decline, —

guter Mann, liebe Mutter, kleines Haus;
Der freundliche Knabe, die wundervolle Ruine, das prächtige Schloß;
Ein heftiger Sturm, eine schöne Gabe, ein artiges Kind.

(*b*) Decline the personal pronouns ich, du, er, singular and plural.

4. Give the general rules for the gender of German nouns, with examples demonstrating the same.

5. (*a*) Give the synopsis (first person singular of all tenses in the indicative and subjunctive) of the auxiliaries haben, sein, and werden.

(*b*) Give the third person plural of all tenses (both indicative and subjunctive) of the regular verb loben. -

(*c*) Give the synopsis of the impersonal verb regnen, and also of the verb ausgehen.

6. (*a*) Conjugate the present tense of the verbs können, mögen, wollen, müssen, dürfen, sollen. Also state their meaning.

(*b*) Give the principal parts (infinitive, imperfect, and past participle) of the verbs sehen, gehen, stehlen, waschen, preisen, wissen, denken, halten, helfen, sitzen, finden, bringen.

7. Write out some story in German that you have read, and if you cannot give it in German, write its translation in English.

8. (*a*) Write in letters the following numbers: 6789; 704,532; 1876; 778,899.

(*b*) Give the German for the days of the week, and the names of the months.

9. Translate the following story into English: —

„Ein junger Student der Harvard Universität reiste nach Europa, um seine Studien auf einer deutschen Universität fortzusetzen. Auf seiner Reise durch Italien fand er einen Freund in Rom, der mit ihm nach Athen reiste, um die wundervollen Ruinen des alten Griechenlands zu sehen. Sie fanden einen amerikanischen Consul in Athen, der sie auf die Akropolis begleitete. Voll Enthusiasmus über die Erinnerungen ihrer klaßischen Studien riefen sie aus: Wären wir keine Amerikaner, so möchten wir Griechen sein.

CHEMISTRY AND PHYSICS.

CHEMISTRY.

1. What happens when a candle burns? Describe experiments which illustrate the subject, and state clearly what each experiment proves.

2. What goes on when we breathe the air? Illustrate the subject by familiar facts and experiments.

3. What sort of action do plants exert on the air? Illustrate the subject as before.

4. What is water made up of? Illustrate by experiments, and state the law of chemical combination which may be deduced from them.

5. Describe the process represented by the following symbols, and state fully what the symbols express: —

$$Zn + H_2SO_4 = H_2 + ZnSO_4.$$

PHYSICS.

6. Define the terms velocity and force, and name the chief forces of nature.

7. Define the term specific gravity, and state the principle of Archimedes by which the specific gravity of solids is most easily found.

8. How is a barometer made, and what does it measure?

9. How is a thermometer made, and what does it measure?

10. What is meant by *the latent heat of water*, and how is this quantity measured?

PHYSICS AND ASTRONOMY.

1. Why is the height of the barometer less at the summit of a mountain than at its foot?
2. Describe the common pump and its action.
3. What is latent heat?
4. Why is a spectrum formed when sunlight is passed through a prism?
5. On what principle does the use of lightning-rods depend?
6. What is the theory of Copernicus?
7. State Kepler's three laws of planetary motion.
8. What is a sidereal day, and how is its length determined?
9. How is the sun's period of rotation ascertained?
10. What is the cause of solar eclipses? When will an eclipse be annular?

BOTANY.

1. What are the organs of vegetation?
2. Of what parts does an embryo consist?
3. Describe the germination of a maple-seed, and a grain of corn.
4. What is a biennial plant?
5. Draw an outline sketch of a twice-pinnate leaf.
6. In what ways are leaves arranged upon the stem?
7. How does a cyme differ from a corymb?
8. What is the difference between an imperfect and an incomplete flower?

9. Explain the structure of the "fruit" of the strawberry.

10. Describe upon the schedule the plant given for analysis.

SCHEDULE FOR PLANT-ANALYSIS.

1. STATE whether this plant is *exogenous* or *endogenous*, and give reasons for your answer.

2. Describe the *arrangement, venation, shape, margin, apex,* and *base* of the leaves.

3. What kind of flower-clusters does this plant have?

4. THE FLOWER. — State whether it is or is not *complete, regular,* and *symmetrical*. Give your reasons for each answer.

CALYX. — State whether free from, or coherent with, the ovary.

SEPALS. — Give their number.

COROLLA. — State whether *polypetalous* or *monopetalous*.

STAMENS. — (1) Give number. (2) State whether distinct or united together. (3) To what are they attached?

PISTIL. — (1) State whether *simple* or *compound*. (2) If possible, give the number of cells in the ovary. (3) Is the ovary *superior* or *inferior?*

HARVARD EXAMINATION PAPERS.

SEPTEMBER, 1876.

ANCIENT HISTORY AND GEOGRAPHY.

1. TELL the situation of the following places, and name (with date) some important event connected with each: Saguntum, Mantinea, Zama, Cynocephalae, Cunaxa, Philippi, Beneventum, Actium, Leuctra, Pharsalia, L. Trasimenus, Furculae Caudinae.
2. Give a brief account either of Aristides, of Cimon, or of Alcibiades.
3. The expedition of Alexander the Great against Persia.
4. What magistrates at Rome, and what were their respective duties?
5. Brief outline of Roman history from the death of Julius Caesar to the year 27 B. C.

MODERN AND PHYSICAL GEOGRAPHY.

1. WHAT is meant by *relief?* State some common features of continental relief. Represent in profile the relief of the United States along an east and west line.

2. Define, with precision, *latitude* and *longitude*. State definitely what portions of the earth's surface are crossed by the Tropic of Cancer,— what by the Tropic of Capricorn.

3. In what parts of North and South America are the great plains? By what names are they popularly known in the regions where they lie? In what portions of the old world are the plains most extensive and unbroken? By what names are those plains severally designated?

4. British America.— Name the larger bays, lakes, and rivers, and state the courses of the latter, and where they empty. Name the provinces into which it is now divided. Give some account of its physical character.

5. Name all the larger West India Islands. Give as complete an account of one of them as time will allow. Name the political divisions of Central America.

6. Europe.— Name its seas, bays, and gulfs, and state where they lie. Name the larger rivers, and state their courses and where they empty. Name the principal mountain ranges, and give their positions and directions. What country extends farthest north, and what one farthest south?

7. Hindostan.— Name its three principal cities. What cape forms its southern extremity, and what is its latitude? State the leading physical characters of Hindostan.

8. Where (definitely) are the cities here named, viz.: (*a*) Manilla, (*b*) Acapulco, (*c*) Melbourne, (*d*) Valparaiso, (*e*) Matanzas, (*f*) Rangoon, (*g*) Yeddo? Where (definitely) are the capes here named, viz.: (*a*) North Cape, (*b*) St. Lucas, (*c*) Guardafui, (*d*) Tarifa, (*e*) Mendocino, (*f*) Finisterre, (*g*) St. Roque, (*h*) Blanco?

GREEK COMPOSITION.

1. I AM glad,[1] Clearchus, to hear these words from you; for while you think thus, if you should plot[2] any evil against me it seems to me that you would be ill-disposed[3] to your own self as well as to me.

2. And in order that you may learn that you would not justly distrust[4] either the King or myself, listen: If we wished to destroy[5] you, do you think that we have not plenty[6] of both cavalry and foot? or do you think we would have no place suitable[7] to attack[8] you?

3. And if we should be defeated[9] in battle, surely by burning the crops[10] we could oppose[11] a famine[12] to you against which you would not be able to fight if you were ever so[13] brave.

[1] ἥδομαι. [2] βουλεύω. [3] κακόνους. [4] ἀπιστέω. [5] ἀπόλλυμι. [6] πλῆθος. [7] ἐπιτήδειος. [8] ἐπιτίθεμαι. [9] ἡττάομαι. [10] καρπός. [11] ἀντιτάσσω. [12] λιμός. [13] πάνυ. ἱππεύς, πεζός, χωρίον, μάχη, ἀγαθός.

GREEK GRAMMAR.

[* *Candidates for advanced standing will omit 2 and 4, and do 7 and 8.*]

1. DECLINE throughout, with the article, γλῶσσα and γένος. Give the nominative and vocative singular, and dative plural of ἐλπίδος, ποιμένος, νεώς, and πατρός.

2.* Decline in the singular the pres. part. act. of τιμάω, giving both the contracted and uncontracted forms. Give the rule for the accent of contracted forms of words.

3. Decline τὶς. Translate the following pronouns: ἐκείνου, τούτου, ταὐτοῦ, αὐτοῦ, and ὅτου.

4.* Inflect the imperf. ind. act. of τίθημι, pres. opt. mid. of δηλόω, aorist ind. pass. of στέλλω.

5. State where these verbs are made, and give the principal parts of each: βούλει, ἴην, εἰδῶ, μεῖναι, μάθε, and ἔσχε.

6. Describe particular and general suppositions, giving examples. Define the term Indirect Discourse. What time do the tenses of the infinitive mode denote in Indirect Discourse?

7. Give the different forms in which a wish is expressed. What is the implication of each? Explain the origin of these constructions.

8. Give a scheme of the Tragic Iambic Trimeter line showing what substitutions for the Iambus are admissible.

GREEK PROSE.

[*N. B. Those who offer the Greek Reader will take* 2, 3, 4. *Those who offer four books of the Anabasis and the seventh book of Herodotus will take* 1, 2, 5. *Candidates in Course II. will take* 1 *and* 2, *or* 2 *and* 3.]

1. TRANSLATE: —

'Ἀλλ' ἥδομαι μέν, ὦ Κλέαρχε, καὶ ἀκούων σου φρονίμους λόγους· ταῦτα γὰρ γιγνώσκων εἴ τι ἐμοὶ κακὸν βουλεύοις, ἅμα ἄν μοι δοκεῖς καὶ σαυτῷ κακόνους εἶναι. Ὡς δ' ἂν μάθῃς ὅτι οὐδ' ἂν ὑμεῖς δικαίως οὔτε βασιλεῖ οὔτ' ἐμοὶ ἀπιστοίητε, ἀντάκουσον. Εἰ γὰρ ὑμᾶς ἐβουλόμεθα ἀπολέσαι, πότερά σοι δοκοῦμεν ἱππέων πλήθους ἀπορεῖν ἢ πεζῶν ἢ ὁπλίσεως, ἐν ᾗ ὑμᾶς μὲν βλάπτειν ἱκανοὶ εἴημεν ἄν, ἀντιπάσχειν δὲ οὐδεὶς κίνδυνος; Ἀλλὰ χωρίων ἐπιτηδείων ὑμῖν ἐπιτίθεσθαι ἀπορεῖν ἅ σοι δοκοῦμεν; Οὐ τοσαῦτα μὲν πεδία ἡμῖν φίλια ὄντα σὺν πολλῷ πόνῳ διαπορεύεσθε, τοσαῦτα δὲ ὄρη ὑμῖν ὁρᾶτε ὄντα πορευτέα ἃ ἡμῖν ἔξεστι προκαταλαβοῦσιν ἄπορα ὑμῖν παρέχειν, τοσοῦτοι δ᾽ εἰσὶ ποταμοὶ ἐφ᾽ ὧν ἔξεστιν ἡμῖν ταμιεύεσθαι, ὁπόσοις ἂν ὑμῶν βουλώμεθα μάχεσθαι; Εἰσὶ δ᾽ αὐτῶν οὓς οὐδ᾽ ἂν παντάπασι διαβαίητε, εἰ μὴ ἡμεῖς ὑμᾶς διαπορεύοιμεν. — ANAB., II. v. 16-18.

Explain the case of σαυτῷ (line 3); the construction of εἶναι (line 3); the construction of ἐβουλόμεθα (lines 4, 5); the case of κίνδυνος (line 7); the case of ὑμῶν, and construction of βουλώμεθα (line 12).

2. TRANSLATE:—

Ἐπειδὴ δὲ ἑώρα ὁ Χειρίσοφος προκατειλημμένην τὴν ἀκρωνυχίαν, καλεῖ Ξενοφῶντα ἀπὸ τῆς οὐρᾶς, καὶ κελεύει λαβόντα τοὺς πελταστὰς παραγενέσθαι εἰς τὸ πρόσθεν. Ὁ δὲ Ξενοφῶν τοὺς μὲν πελταστὰς οὐκ ἦγεν· ἐπιφαινόμενον γὰρ ἑώρα Τισσαφέρνην καὶ ἅπαν τὸ στράτευμα· αὐτὸς δὲ προσελάσας ἠρώτα· Τί καλεῖς; Ὁ δὲ λέγει αὐτῷ· Ἔξεστιν ὁρᾶν· προκατείληπται γὰρ ἡμῖν ὁ ὑπὲρ τῆς καταβάσεως λόφος, καὶ οὐκ ἔστι παρελθεῖν, εἰ μὴ τούτους ἀποκόψομεν. Ἀλλά τί οὐκ ἦγες τοὺς πελταστάς; Ὁ δὲ λέγει ὅτι οὐκ ἐδόκει αὐτῷ ἔρημα καταλιπεῖν τὰ ὄπισθεν, πολεμίων ἐπιφαινομένων. Ἀλλὰ μὴν ὥρα γ᾽, ἔφη, βουλεύεσθαι πῶς τις τοὺς ἄνδρας ἀπελᾷ ἀπὸ τοῦ λόφου. — ANAB., III. iv. 38-40.

Explain the construction of προκατειλημμένην (line 1), and of ἐπιφαινομένων (line 9). In what voice, mood, and tense, and from what verbs are ἑώρα (line 4), ἠρώτα (line 5), and προκατείληπται (line 6), and ἀπελᾷ (line 10)? What would be the form of προκατείληπται (line 6), if it were quoted indirectly with change of mood?

3. TRANSLATE:—

Ὁ δ᾽ Ἐπαμεινώνδας αὖ καὶ τοῦ ἱππικοῦ ἔμβολον ἰσχυρὸν ἐποιήσατο, καὶ ἀμίππους πεζοὺς συνέταξεν αὐτοῖς, νομίζων τὸ ἱππικὸν ἐπεὶ διακόψειεν, ὅλον τὸ ἀντίπαλον νενικηκὼς ἔσεσθαι· μάλα γὰρ χαλεπὸν εὑρεῖν τοὺς ἐθελήσοντας μένειν, ἐπειδάν τινας φεύγοντας τῶν ἑαυτῶν ὁρῶσι· καί ὅπως μὴ ἐπιβοηθῶσιν οἱ Ἀθηναῖοι ἀπὸ τοῦ εὐωνύμου κέρατος ἐπὶ τὸ ἐχόμενον, κατέστησεν ἐπὶ γηλόφων τινῶν ἐναντίους αὐτοῖς καὶ ἱππέας καὶ ὁπλίτας, φόβον βουλόμενος καὶ τούτοις παρέχειν ὡς, εἰ βοηθήσαιεν, ὄπισθεν οὗτοι ἐπικείσοιντο αὐτοῖς. τὴν μὲν δὴ συμβρλὴν οὕτως ἐποιήσατο, καὶ οὐκ ἐψεύσθη τῆς ἐλπίδος· κρατήσας γὰρ ᾗ προσέβαλεν ὅλον ἐποίησε φεύγειν τὸ τῶν ἐναντίων. — HELL., VII. v. 24.

From what stems are ἔμβολον (line 1) and ἀμίππους (line 2) formed? Explain the construction of ὁρῶσι (line 5) and of ἐπιβοηθῶσιν (line 5). State briefly the principal events in the life of Epaminondas, with dates.

4. TRANSLATE: —

Ἀθηναίοισι δὲ τεταγμένοισι ἐν τεμένεϊ Ἡρακλέος ἐπῆλθον βοηθέοντες Πλαταιέες πανδημεί· καὶ γὰρ καὶ ἐδεδώκεσαν σφέας αὐτοὺς τοῖσι Ἀθηναίοισι οἱ Πλαταιέες, καὶ πόνους ὑπὲρ αὐτῶν οἱ Ἀθηναῖοι συχνοὺς ἤδη ἀναραιρέατο· ἔδοσαν δὲ ὧδε. πιεζόμενοι ὑπὸ Θηβαίων οἱ Πλαταιέες ἐδίδοσαν πρῶτα παρατυχοῦσι Κλεομένεΐ τε τῷ Ἀναξανδρίδεω καὶ Λακεδαιμονίοισι σφέας αὐτούς, οἱ δὲ οὐ δεχόμενοι ἔλεγόν σφι τάδε· Ἡμεῖς μὲν ἑκαστέρω τε οἰκέομεν, καὶ ὑμῖν τοιήδε τις γίνοιτ' ἂν ἐπικουρίη ψυχρή· φθαίητε γὰρ ἂν πολλάκις ἐξανδραποδισθέντες ἤ τινα πυθέσθαι ἡμέων. συμβουλεύομεν δὲ ὑμῖν δοῦναι ὑμέας αὐτοὺς Ἀθηναίοισι, πλησιοχώροισί τε ἀνδράσι καὶ τιμωρέειν ἐοῦσι οὐ κακοῖσι.
— HEROD., VI. 108.

Give the Attic forms of ἀναραιρέατο (line 4) and of ἐοῦσι (line 10). What terminations does Herodotus use in the genitive and dative plural of the first declension?

5. TRANSLATE: —

Ξέρξης δὲ ἐπεί τε διέβη ἐς τὴν Εὐρώπην, ἐθηεῖτο τὸν στρατὸν ὑπὸ μαστίγων διαβαίνοντα. διέβη δὲ ὁ στρατὸς αὐτοῦ ἐν ἑπτὰ ἡμέρῃσι καὶ ἐν ἑπτὰ εὐφρόνῃσι, ἐλινύσας οὐδένα χρόνον. ἐνθαῦτα λέγεται, Ξέρξεω ἤδη διαβεβηκότος τὸν Ἑλλήσποντον, ἄνδρα εἰπεῖν Ἑλλησπόντιον· Ὦ Ζεῦ, τί δὴ ἀνδρὶ εἰδόμενος Πέρσῃ καὶ οὔνομα ἀντὶ Διὸς

GREEK POETRY.

Ξέρξεα θέμενος, ἀνάστατον τὴν Ἑλλάδα ἐθέλεις ποιῆσαι, ἄγων πάντας ἀνθρώπους; καὶ γὰρ ἄνευ τούτων ἐξῆν τοι ποιέειν ταῦτα.
Ὁ δὲ ναυτικὸς ἔξω τὸν Ἑλλήσποντον πλώων παρὰ γῆν ἐκομίζετο, τὰ ἔμπαλιν πρήσσων τοῦ πεζοῦ. Ὁ δὲ Δορίσκος ἐστὶ τῆς Θρηίκης αἰγιαλός τε καὶ πεδίον μέγα, διὰ δὲ αὐτοῦ ῥέει ποταμὸς μέγας Ἕβρος. ἔδοξε ὦν τῷ Ξέρξῃ ὁ χῶρος εἶναι ἐπιτήδεος ἐνδιατάξαι τε καὶ ἐξαριθμῆσαι τὸν στρατόν, καὶ ἐποίεε ταῦτα. — HEROD., VII. 56 – 59.

Give the Attic forms of ἐνθαῦτα (line 3), οὔνομα (line 5), and of πλώων (line 8). What is the chief peculiarity in syntax in Herodotus?

GREEK POETRY.

[*You are advised to do the translation before answering the questions. Candidates in Course II. will translate 1 and 2 and answer the questions under (a) — (δ) inclusive.*]

1. Τὸν δ᾽ ἠμείβετ᾽ ἔπειτα Θέτις, κατὰ δάκρυ χέουσα·
ὤμοι, τέκνον ἐμόν, τί νύ σ᾽ ἔτρεφον, αἰνὰ τεκοῦσα;
αἴθ᾽ ὄφελες παρὰ νηυσὶν ἀδάκρυτος καὶ ἀπήμων
ἧσθαι· ἐπεὶ νύ τοι αἶσα μίνυνθά περ, οὔτι μάλα δήν.
νῦν δ᾽ ἅμα τ᾽ ὠκύμορος καὶ ὀϊζυρὸς περὶ πάντων
ἔπλεο· τῷ σε κακῇ αἴσῃ τέκον ἐν μεγάροισιν.
τοῦτο δέ τοι ἐρέουσα ἔπος Διὶ τερπικεραύνῳ
εἶμ᾽ αὐτὴ πρὸς Ὄλυμπον ἀγάννιφον, αἴ κε πίθηται.
IL., I. 413 – 420.

(a) From what stems are the following words formed: ὠκύμορος (417), μεγάροισιν (418), and ἀγάννιφον (420)?

(β) Translate the following epithets of Zeus: μητίετα, αἰγίοχος, ὑψιβρεμέτης, τερπικέραυνος, and νεφεληγερέτα.

2. νῦν δ᾽ ἔρχεσθ᾽ ἐπὶ δεῖπνον, ἵνα ξυνάγωμεν Ἄρηα.
εὖ μέν τις δόρυ θηξάσθω, εὖ δ᾽ ἀσπίδα θέσθω,

εὖ δέ τις ἵπποισιν δεῖπνον δότω ὠκυπόδεσσιν,
εὖ δέ τις ἅρματος ἀμφὶς ἰδὼν, πολέμοιο μεδέσθω·
ὥς κε πανημέριοι στυγερῷ κρινώμεθ᾽ Ἄρηϊ.
οὐ γὰρ παυσωλή γε μετέσσεται, οὐδ᾽ ἠβαιὸν,
εἰ μὴ νὺξ ἐλθοῦσα διακρινέει μένος ἀνδρῶν.
ἱδρώσει μέν τευ τελαμὼν ἀμφὶ στήθεσσιν
ἀσπίδος ἀμφιβρότης, περὶ δ᾽ ἔγχεϊ χεῖρα καμεῖται·
ἱδρώσει δέ τευ ἵππος, ἐΰξοον ἅρμα τιταίνων.

Il., II. 381–390.

(γ) Give the Attic forms of ὠκυπόδεσσιν (383), μετέσσεται (386), and τευ (390).

(δ) Define a spondaic verse, and specify an example in the above passage.

3. Ἦ, καὶ ἀπὸ στομάχους ἀρνῶν τάμε νηλέϊ χαλκῷ·
καὶ τοὺς μὲν κατέθηκεν ἐπὶ χθονὸς ἀσπαίροντας,
θυμοῦ δευομένους· ἀπὸ γὰρ μένος εἵλετο χαλκός.
οἶνον δ᾽ ἐκ κρητῆρος ἀφυσσάμενοι δεπάεσσιν
ἔκχεον, ἠδ᾽ εὔχοντο θεοῖς αἰειγενέτῃσιν.
ὧδε δέ τις εἴπεσκεν Ἀχαιῶν τε Τρώων τε·
Ζεῦ κύδιστε, μέγιστε, καὶ ἀθάνατοι θεοὶ ἄλλοι,
ὁππότεροι πρότεροι ὑπὲρ ὅρκια πημήνειαν,
ὧδέ σφ᾽ ἐγκέφαλος χαμάδις ῥέοι, ὡς ὅδε οἶνος,
αὐτῶν, καὶ τεκέων, ἄλοχοι δ᾽ ἄλλοισι μιγεῖεν.

Il., III. 292–301.

(ε) Give the Attic forms of δεπάεσσιν (295), αἰειγενέτῃσιν (296), and τεκέων (301).

(ζ) Comment on the form εἴπεσκεν (297).

LATIN COMPOSITION.

TRANSLATE INTO ENGLISH: —

Nam reges Syriae, regis Antiochi filios, scitis Romae nuper fuisse; qui venerant non propter Syriae regnum, — nam id sine controversia obtinebant ut a patre et a maioribus acceperant, — sed regnum Egypti ad se et ad Selenen matrem suam pertinere arbitrabantur. Hi ipsi, posteaquam per senatum agere quae voluerant non potuerunt, in Syriam, in regnum patrium profecti sunt. Eorum alter, qui Antiochus vocatur, iter per Siciliam facere voluit; itaque isto praetore, venit Syracusas.

TRANSLATE INTO LATIN: —

These kings that I speak of[1] had brought[2] to Rome a candelabrum, made,[3] with wonderful[4] workmanship[5], of[6] most brilliant[7] gems, in order to set it up[8] in the Capitol; but since[9] they had found[10] the temple not yet finished, they determined[11] to take it back[12] to Syria. The matter came, I know not how[13], to the ears of this man; for the king had wished it kept-secret[14], — not because he feared or suspected[15] anything, but in order that not many persons should see it before the Roman people. This man begs[16] the king to send it to him, saying he wishes to examine[17] it, and will not let[18] others see it.

[1] dico. [2] adfero. [3] perficere. [4] mirabilis. [5] opus. [6] e. [7] clarus. [8] ponere. [9] quod. [10] offendere. [11] statuere. [12] reportare. [13] quomodo. [14] celare. [15] suspicor. [16] petere ab. [17] inspicere. [18] potestatem facere.

LATIN GRAMMAR.

[*Do not crowd your work.*]

1. How do you pronounce *pars;* in- in *ingens;* op- in *optimus;* ia- and ci- in *iacio;* ti in *ratio?*

2. What is the root and what is the stem of *rex?* of *fama?* of *acus?* Give any other words that you can remember, from the same roots. Give the meanings of the derivative suffixes of *cautio, lumen, stabulum.* Form from the stem *aspero-* a word meaning *rough-ness;* one from *crimin-* meaning *reproach-ful;* one from *lauda-* meaning *praise-worthy;* one from *favere,* meaning *patron;* one from *tristi-* meaning *somewhat sad.*

[*Mark the quantities of the penult and ultima of every Latin word you write in answering the 3d and 5th questions.*]

3. Decline *ille vir; locus celeber.* Compare *tenax, asper, frugi.* Form Comparative and Superlative Adjectives from *infra.* Form and compare Adverbs from *bonus, atrox.*

4. What is the root and what are the stems of *rumpere?* of *canere?* of *regere?* of *nancisci?*

5. State where each of the following forms is made, and give the principal parts of the verb to which it belongs: *usserit, vetaberis, texitis, sprevissent, vinxeram, vīdĕris, vĭdēris.* Form the III. S. Fut. Pf. Ind. Act. of *parĕre;* II. P. Perf. Ind. Act. of *gignere*; II. P. Fut. Ind. Pass. of *capere;* I. S. Pres. Subj. Pass. of *vocare;* III. P. Imp. Subj. Pass. of *oblivisci;* III. P. Fut. Imperat. Pass. of *haurire;* the Fut. Inf. Pass. of *premere;* the Fut. Part. Pass. of *mittere.* Give a synopsis of *patior.* Inflect the active voice of *fero* throughout.

6. Write in Latin: They made him king; he is made king; he spares (*parcere*) the city; the city is spared; he hides (*celare*) this from me; the letter (*epistola*) is written; he is loved; for how much did he buy (*emere*) this? for a shilling (*denarius*); I fear he will not come to Athens for many days yet.

CAESAR, SALLUST, AND OVID.

[*You are expected to translate two pieces of verse, and two of prose,— by preference I., II., 1, 2 — and to answer all the questions.*]

Bacchus and the Sailors.

I. Forte petens Delon Chiae telluris ad oras
    Applicor, et dextris adducor littora remis,
    Doque *leves sultus*, udaeque immittor arenae.
    Nox ubi consumpta est (Aurora rubescere primum
    Coeperat;) exsurgo, laticesque inferre recentes
    Admoneo, monstroque viam, quae ducit ad undas.
    Ipse, quid aura mihi tumulo promittat ab alto,
    Prospicio, comitesque voco, repetoque carinam.
    "Adsumus en!" inquit sociorum primus Opheltes;
    Utque putat, praedam deserto nactus in agro,
    Virginea puerum ducit per littora forma.

Deianira to Hercules.

II. I nunc, tolle animos, et fortia facta recense:
    Quod tu non esses jure, vir illa fuit;
    Illi procedit rerum mensura tuarum:
    Cede *bonis*; heres laudis amica tuae.
    Pro pudor! hirsuti costis exuta leonis

Aspera texerunt vellera molle latus.
Falleris, et nescis: non sunt spolia ista leonis,
Sed tua; tuque ferae victor es, illa tui.

Baucis and Philemon.

III. Accubuere dei. Mensam succincta tremensque
Ponit anus; mensae sed erat pes tertius impar:
Testa parem fecit. Quae postquam subdita clivum
Sustulit, aequatam mentae tersere virentes.
Ponitur hic bicolor sincerae bacca Minervae,
Conditaque in liquida corna autumnalia faece,
Intubaque et radix et lactis massa coacti,
Omnia fictilibus. Post haec caelatus eodem
Sistitur argento crater, fabricataque fago
Pocula, qua cava sunt, flaventibus illita ceris.

(*a*) Divide into *feet*, marking the quantity of every syllable, and ictus of every foot, the fourth line of each piece translated. [The caesural pause need *not* be indicated.]

(*b*) How do the quantities of the words in italics help to define the meaning?

(1) His difficultatibus duae res erant *subsidio*, scientia atque usus militum, quod superioribus proeliis exercitati quid fieri oporteret non minus commode ipsi sibi praescribere, quam ab aliis doceri poterant; et quod ab opere singulisque legionibus singulos legatos Caesar discedere, nisi munitis castris, vetuerat. Hi propter propinquitatem et celeritatem hostium, nihil jam Caesaris imperium exspectabant, sed per se, quae videbantur, administrabant.

(2) Saepenumero, patres conscripti, multa verba in hoc ordine feci, saepe de luxuria atque avaritia nostrorum civium questus sum, multosque mortalis ea causa adversos habeo: qui mihi atque animo meo nullius umquam delicti

gratiam *fecissem,* haud facile alterius lubidini male facta condonabam. Sed ea tametsi vos *parvi* pendebatis, tamen res publica firma erat: opulentia neglegentiam tolerabat.

(3) Equestris autem proelii ratio et cedentibus et insequentibus par atque idem periculum inferebat. Accedebat huc, ut nunquam *conferti,* sed rari magnisque intervallis *proeliarentur,* stationesque dispositas haberent, atque alios alii deinceps exciperent, integrique et recentes defatigatis succederent. Postero die procul a castris hostes in collibus constiterunt, rarique se ostendere et lenius, quam pridie, nostros equites proelio lacessere coeperunt.

(*a*) Give the syntax of *subsidio* (1), *parvi* (2); and either *proeliarentur* (3), or *fecissem* (2); and the parts of *conferti* (3).

CICERO AND VIRGIL.

[*Translate two pieces of prose and two of poetry, and explain the construction of the words in italics under the extracts you choose.*]

1. Quid tam inusitatum quam ut, cum duo consules clarissimi fortissimique essent, eques Romanus ad bellum maximum formidolosissimumque pro consule mitteretur? Missus est: quo quidem tempore cum esset non nemo in senatu qui diceret, non oportere mitti hominem privatum pro consule, L. Philippus dixisse dicitur, non se illum sua sententia pro consule, sed pro consulibus mittere.

Mitteretur, tempore, oportere, mitti.

2. Neque enim quisquam est tam aversus a Musis qui non mandari versibus aeternum suorum laborum facile praeconium patiatur. Themistoclem illum, summum Athenis virum, dixisse aiunt, cum ex eo quareretur, quod acro-

ama aut cujus vocem libentissime audiret, ejus a quo sua virtus optime praedicaretur. Itaque ille Marius item eximie L. Plotium dilexit, cujus ingenio putabat ea quae gesserat, posse celebrari.

Praedicaretur, audiret, patiatur.

3. Sed tamen cum in animis hominum tantae latebrae sint et tanti recessus, augeamus sane suspicionem tuam: simul enim augebimus diligentiam. Nam quis est omnium tam ignarus rerum, tam rudis in re publica, tam nihil umquam nec de sua nec de communi salute cogitans, qui non intelligat tua salute contineri suam et ex unius tua vita pendere omnium?

Sint, augeamus, intelligat.

4. Etenim quaero, si quis pater familias, liberis suis a servo interfectis, uxore occisa, incensa domo, supplicium de servo non quam acerbissimum sumpserit, utrum is clemens ac misericors an inhumanissimus et crudelissimus esse videatur? Mihi vero importunus ac ferreus, qui non dolore et cruciatu nocentis suum dolorem cruciatumque lenierit.

Videatur, lenierit.

5. Non ulli pastos illis egere diebus
   Frigida, Daphni, boves ad flumina; nulla nec amnem
   Libavit quadrupes, nec graminis attigit herbam.
   Daphni, tuum Poenos etiam ingemuisse leones
   Interitum montesque feri silvaeque loquuntur.
   Daphnis et Armenias curru subiungere tigris
   Instituit, Daphnis thiasos inducere Bacchi
   Et foliis lentas intexere mollibus hastas.

6. Heu vatum ignarae mentes! quid vota furentem,
   Quid delubra iuvant? Est mollis flamma medullas
   Interea, et tacitum vivit sub pectore volnus.
   Uritur infelix Dido totaque vagatur
   Urbe furens, qualis coniecta cerva sagitta,

Quam procul incautam nemora inter Cresia fixit
Pastor agens telis, liquitque volatile ferrum
Nescius; illa fuga silvas saltusque peragrat
Dictaeos; haeret lateri letalis arundo.

How is the translation indicated by the quantity in 5, line 1? What is the quantity of *u* in *saltus* in 6, line 8? Why?

Write out metrically in 6, lines 6 and 7.

ARITHMETIC.

[*Give the whole work.*]

1. The sum of $\dfrac{\frac{3}{5} \text{ of } \frac{5}{6}}{0.5}$ and $\dfrac{\frac{4}{7} \text{ of } \frac{7}{12}}{\frac{4}{5} \times 2.25}$ is how many times their difference?

2. A owns $\frac{5}{13}$ of a field, and B owns the remainder; $\frac{3}{4}$ of the difference between their shares is 5 A. 3 R. 16$\frac{1}{2}$ P. What is B's share in acres?

3. A man earns $325 in 2$\frac{1}{4}$ months, and spends in 6 months what he earns in 4$\frac{1}{2}$ months. What does he save in a year?

4. Find, by logarithms, $\frac{3}{4}$ of $\dfrac{11.846 \times .004}{\sqrt[3]{.0777}}$.

5. One decagramme is 0.3527 oz. Avdp. How many pounds Avdp. are there in a quintal?

6. What per cent is gained in buying oil at 80 cents a gallon, and selling it at 12 cents a pint?

7. If 12 pipes, each delivering 12 gallons a minute, fill a cistern in 3 h. 24 min., how many pipes, each delivering 16 gallons a minute, will fill a cistern 6 times as large in 6 h. 48 min.?

8. Find the cube root of 0.001295029.

ALGEBRA.

Course I.

[*Write legibly and without crowding; give the whole work; and reduce the answers to their simplest forms.*]

1. SUBSTITUTE $y+3$ for x in $x^4-x^3+2x^2-3$, and simplify and arrange the result.

2. Divide $\frac{3x}{4y}$ by the product of $\frac{a^2-x^2}{c^2-x^2}$, $\frac{bc+bx}{a^2+ax}$, and $\frac{c-x}{a-x}$.

3. Solve the equation $\frac{3}{8-x} - \frac{8-x}{3} = \frac{x-11}{12}$.

4. Add

$\sqrt{20\,a^2m - 20acm + 5c^2m}$ to $\sqrt{20c^2m - 60acm + 45a^2m}$.

5. Solve the equations

$\frac{1}{x} + \frac{1}{y} = 2,\qquad \frac{1}{x} + \frac{1}{z} = 3,\qquad \frac{1}{y} + \frac{1}{z} = 3.$

6. Find the least common multiple and greatest common divisor of $x^2+4x-21$ and x^2-x-56.

7. It takes A 10 days longer to do a piece of work than it takes B: and both together can do it in 12 days. In how many days can each do it alone?

ADVANCED ALGEBRA.

[*Give the whole work.*]

1. SOLVE one of the following equations: —

    (a) $\dfrac{x^3 - 4x}{x - 2} + \dfrac{x^2 - 1}{x + 1} = 39;$

    (b) $2x^2 - 2x + 6\sqrt{2x^2 - 3x + 2} = x + 14;$

    (c) $x^{-1} + x^{-2} = 6.$

2. One root of the equation $x^3 - 37x = 84$ is -3. What are the other two roots?

3. The sum of a certain number of terms of the series 21, 19, 17 ... is 120. Find the number of terms, and the last term.

4. The sum of three numbers in Arithmetical Progression is 15; if 1, 4, and 19 be added to them respectively, the results are in Geometrical Progression. Find the numbers.

5. With the digits 1, 2, 4, 5, 7, 0 how many even numbers between 100 and 1000 can be formed?

6. Find the middle terms of $\left(x - \dfrac{1}{x} \right)^{2p+1}$.

7. A sets off from London to York, and B at the same time sets off from York to London, and each travels uniformly: A reaches York 16 hours, and B reaches London 36 hours, after they have met on the road. Find in what time each has performed the journey.

PLANE GEOMETRY.

1. Define a plane, a parallelogram, a trapezoid, a tangent to a circle.

2. Prove that when two triangles have two sides of the one respectively equal to two sides of the other and the included angle of the first greater than the included angle of the second, the third side of the first is greater than the third side of the second.

3. Show how to draw a tangent to a circle from a point without the circle, and prove your method correct.

4. Draw from one of the vertices of a triangle a line cutting the opposite side into parts proportional to the other two sides. Give proof.

5. Prove that the square described on the hypothenuse of a right triangle is equivalent to the sum of the squares described upon the other two sides.

6. Given two similar polygons, to construct one similar to them both and equivalent to their sum.

7. Given π (the ratio of circumference to diameter) and r (radius). Find expressions in terms of π and r for the circumference and area of a circle.

SOLID GEOMETRY.

1. Prove that, if two planes are perpendicular to each other, the straight line, drawn through any point of the common intersection perpendicular to one of the planes, must be in the other plane.

2. Prove that the solidity of *any* parallelopiped is the product of its base by its altitude.

3· The area of the surface described by a straight line revolving about another straight line in the same plane with it as an axis, is the product of the revolving line by the circumference described by its middle point. Give proof in each of the cases to which this theorem applies.

4. The cubic contents of two similar polyedrons are respectively 3 cubic inches and 24 cubic inches, and one side of the first is 5 inches; what is the homologous side of the second?

5. Prove that the angles of a spherical triangle are respectively supplements of the sides of the corresponding polar triangle.

6. Prove that, of two sides of a spherical triangle, that is the greater which is opposite the greater angle, and the converse.

ANALYTIC GEOMETRY.

1. What is the locus of each of the following equations:—

(1) $3x^2 + y^2 - 7 = 0$, (2) $2y^2 + 3x = 0$,
(3) $y^2 - x^2 + 1 = 0$, (4) $y = 0$,
(5) $2x^2 - x + 2y^2 - 3y - 2 = 0$,

the system of coördinates being rectangular?

How is each of these loci situated with respect to the coördinate axes?

2. The vertices of a triangle are $A = (-1, 2)$, $B = (2, -3)$, $C = (-3, -1)$; find the equations of the sides AB and BC, and some trigonometric function of the angle B.

3. Deduce the equation of the parabola referred to its principal vertex.

4. Deduce the equation of the tangent to an ellipse at the point (x', y') of the curve.

5. The equation of a curve referred to a certain rectangular system is $x^2 - y^2 = 1$; what is the equation of the same curve referred to a second rectangular system, having the same origin as the first, and in which the axis of x makes an angle of 45° with the old axis of x? Sin 45° = cos 45° = $\sqrt{\tfrac{1}{2}}$.

PLANE TRIGONOMETRY.

1. TRACE the changes in the value and sign of the cosine, tangent, and cosecant of φ when φ increases from 0° to 360°.

2. Deduce the fundamental formula

$$\cos(x+y) = \underline{\hspace{5cm}}.$$

3. Deduce the formulae

$$\operatorname{ctn}(x-y) = \frac{1 + \underline{\hspace{2cm}}}{\underline{\hspace{3cm}}},$$

$$\cos^2 \tfrac{1}{2}x = \underline{\hspace{6cm}},$$

$$\sin x + \sin y = 2 \sin \tfrac{1}{2}(x+y) \cos \tfrac{1}{2}(x-y).$$

4. Given the three sides of a plane oblique triangle. Show how to find the three angles giving the necessary formulae.

5. A ladder 51.42 ft. long, placed with its foot 10 ft. from a house, just reaches the top of the house. How high is the house? What angle does the ladder make with the vertical?

ENGLISH COMPOSITION.

You are required to write a short English composition, correct in spelling, punctuation, grammar, and expression. This composition must be at least fifty lines long, and be properly divided into paragraphs. One of the following subjects must be taken:—

Moses at the Fair.
The Story of Ariel.
The Character of Flora MacIvor.

FRENCH.

1. TRANSLATE INTO ENGLISH:—

De ma position présente, il ne faut pas conclure que j'ai eu la Fortune pour marraine. Mes ancêtres, si le mot n'est pas bien ambitieux, *étaient* des pêcheurs; mon père était le dernier de onze enfants, et mon grand-père avait eu bien du mal à élever sa famille, car dans ce métier-la plus encore que dans les autres le gain n'est pas en proportion du travail; compter sur de la fatigue, du danger, c'est le certain, sur un peu d'argent, le hasard.

A dix-huit ans, mon père fut pris par l'inscription maritime; c'est une espèce de conscription, au moyen de laquelle l'État *peut* se faire servir par tous les marins pendant trente-deux ans, — de dix-huit à cinquante. Il partit ne sachant ni lire ni écrire. Il revint premier maitre, ce qui est le plus beau grade auquel *parviennent* ceux qui n'ont point passé par les écoles du gouvernement.

Le Port-Dieu, notre pays, étant voisin des îles anglaises, l'Etat y fait stationner un cutter de guerre, qui a pour mission d'empêcher les gens de Jersey de venir nous prendre notre poisson, en même temps qu'il force nos marins à observer les règlements sur la pêche: ce fut sur ce cutter que mon père fut envoyé pour continuer son service. — MALOT, *Romain Kalbris.*

2. State the tense of the italicized verbs in the above, and give it in full.

3. Give the principal tenses of *connaître, faire, recevoir, sentir* (thus, INF., *être ;* PRES. PART., *étant ;* PAST PART., *été ;* IND. PRES., *je suis ;* PRET., *je fus*).

4. TRANSLATE INTO FRENCH : —

(*a*) Where are you going to-morrow ? (*b*) I do not know yet, I think I shall go to my uncle's. (*c*) How do you wish me to do this?

GERMAN.

1. TRANSLATE INTO ENGLISH : —

Ariosto baute sich ein kleines Haus. Ein Freund fragte ihn, wie er sich mit einem so kleinen Hause begnügen könne, da er so schöne Paläste in seinem Orlando beschreibe. Der Dichter antwortete: Worte sind billiger als Steine.

2. Parse or explain the grammatical forms and relations of words in the first sentence of the above.

3. TRANSLATE INTO GERMAN : —

Diogenes saw a youth blushing, and said to him: Well done, my son, that is the color of virtue.

4. Write out a short extract of some story that you have read in German

5. **Translate into English:**—

Verschwunden ist die finstre Nacht
Die Lerche schlägt,[1] der Tag erwacht
Die Sonne kommt mit Prangen[2]
Am Himmel aufgegangen
Sie scheint in König's Prunkgemach[3]
Sie scheinet durch des Bettlers Dach
Und was in Nacht verborgen war
Das macht sie kund[4] und offenbar.

[1] warbles.   [2] splendor.   [3] palace.   [4] known.

CHEMISTRY AND PHYSICS.

Chemistry.

1. Describe the preparation of hydrogen, and give its two most striking properties.

2. Define the terms, acid, base, salt.

3. What is coal? Describe the method of preparing gas from it.

4. How can pure water be obtained? What parallel process occurs in nature?

Physics.

5. Describe the Leyden jar and the Grove battery.

6. In what three ways can heat be distributed?

7. What is the action of a prism on light?

8. What is the acoustic distinction between noise and music?

PHYSICS AND ASTRONOMY.

1. DESCRIBE the barometer.
2. How is sound propagated? What is the difference between a noise and a musical tone?
3. Describe the different methods by which heat is distributed?
4. Illustrate electrical induction by means of the gold-leaf electroscope.
5. What causes the change of the seasons?
6. How is the moon's distance determined?
7. How is the velocity of light ascertained by the eclipses of Jupiter's satellites?
8. What is known of the sun's chemical constitution, and by what means?

BOTANY.

1. WHAT is a biennial plant?
2. Describe underground stems and branches.
3. How do "endogens" differ from "exogens"?
4. Make a sketch of a ternately decompound leaf.
5. How are leaves arranged on the stem?
6. Describe the different kinds of flower-clusters.
7. What is meant by "calyx superior"?
8. What is the function of green leaves?
9. Describe upon the annexed schedule the plant given for examination.

SCHEDULE FOR PLANT-ANALYSIS.

1. STATE whether this plant is *exogenous* or *endogenous*, and give reasons for your answer.

2. Describe the *arrangement, venation, shape, margin, apex,* and *base* of the leaves.

3. What kind of flower-clusters does this plant have?

4. THE FLOWER. — State whether it is or is not *complete, regular,* and *symmetrical.* Give your reasons for each answer.

CALYX. — State whether free from, or coherent with, the ovary.

SEPALS. — Give their number.

COROLLA. — State whether *polypetalous* or *monopetalous.*

STAMENS. — (1) Give number. (2) State whether distinct or united together. (3) To what are they attached?

PISTIL. — (1) State whether *simple* or *compound.* (2) If possible, give the number of cells in the ovary. (3) Is the ovary *superior* or *inferior?*

HARVARD EXAMINATION PAPERS.

JUNE, 1877.

ANCIENT HISTORY AND GEOGRAPHY.

[*A number marked with an asterisk may be substituted for the same number without it, but for no other.*]

1. DRAW a map of Greece, indicating its chief physical characteristics (rivers, lakes, mountains, capes, bays, etc.) with their ancient names. Can you state any points in which the physical character of the country affected the history of the people?

1.* Name and describe the situation of the various branches of the *Mare Internum*, and of the principal rivers flowing into it. Can you show any points in which the position of this sea influenced the history of the nations living on its borders?

2. Name and describe one or two events in Greek and one or two events in Roman history which exhibit the peculiar traits of Greek and Roman character.

2.* State what you know of the life of any two of the following: Socrates, Demosthenes, Cato the elder, Cicero.

3. The Thirty Tyrants.

4. Give an account either of the expedition of Alexander or of that of Hannibal, describing by map or otherwise the position of the places you mention.

5. Say as much as your time will allow about the life, character, and policy of Augustus.

MODERN AND PHYSICAL GEOGRAPHY.

1. What is the ratio of the land and water surfaces upon the globe, and how many square miles does each comprise? State the position of the land masses, with their northern and southern limits of latitude.

2. What are isothermal lines? Illustrate by examples. Explain contrasts of climate in the same latitude, taking as examples the climates of Labrador and Britain, New York and Naples, San Francisco and Washington.

3. Describe the position of the highest mountain system in each of the continents. Give the name, position and altitude of at least one principal peak in each system.

4. Through which of the United States do the parallels of 40° and 35° run? Name the nine largest lakes of North America, and the five largest in Europe.

5. Name and give the location of the three principal cities of Hindostan, and give as full an account as possible of one of them. Name the four largest cities of England, France, and Italy, respectively.

6. Where are the Balearic islands, and what are their names? What large islands lie west of Italy? What large

island lies east of southern Africa? Where are the Galapagos, Aleutian, Kurile, and Philippine islands?

7. Where are the following straits, and what bodies of water do they connect: Behring, Messina, Sunda, Davis, Otranto, Magellan?

8. Where are the following rivers, what are their courses, and into what do they empty: the Volga, Orinoco, Amoor, Indus, Ganges, Zambesi, Mackenzie, Churchill?

GREEK COMPOSITION.

[*Do either A or B, but not both.*]

A.

AND Lysander[1] commanded those who were following to raise-on-high [2] a shield whenever they should see that the enemy had disembarked; and if they did as he ordered, he expected [3] to capture the whole hostile fleet.[4] But Conon [5] seeing their approach [6] signalled [7] [to his fleet] to flee at-full-speed [8] so that he himself with seven other ships got-off-to-sea [9]; but the rest of the ships Lysander captured, since most of the men were on shore to get-water.[10]

[1] Λύσανδρος. [2] αἴρω. [3] οἴομαι. [4] Plural of ναῦς. [5] Κόνων. [6] ἐπίπλους. [7] σημαίνω. [8] κατὰ κράτος. [9] ἀνάγω. [10] ὑδρεύω.

B.

1. He said he should delay till the king arrived.

2. I should fear to follow the guide whom he might give us.

3. He would not have done this, if I had not bid him.

4. If I should escape the notice of these men, I should be saved; but if I should be taken, I should suffer death.

5. He hunted on horseback whenever he wished to exercise himself.

6. I was the first to announce to him that Cyrus was making an expedition against him.

7. I said that we had many fair hopes of safety.

8. They said that they had come with guides who, if a truce should be made, would bring them [to a place] whence they would get provisions.

GREEK GRAMMAR.

1. DECLINE, in the singular, ὁ σοφιστής. Write the Accusative and Vocative singular, and the Genitive and Dative plural, of ὁ ποιμήν, τὸ ὄρος, ἡ χάρις, ὁ θής, and τὸ κέρας.

2. Decline εὐγενής in the plural. Decline the comparative degree of ἡδύς. Decline τίς. Decline ἐγώ. Give the cardinal and ordinal numerals, from one to five. Which are declinable?

3. Inflect the Perfect Indicative Passive of πράσσω; the Imperfect Active of βοάω; the Aorist Indicative Active of τίθημι; the Present Optative Active of δίδωμι.
Give a synopsis of the Aorist Middle of λείπω, and Aorist Passive of στέλλω.

4. Where are these forms made, and from what Present Indicative Active: πεσεῖν, ἰέναι, ἀναγνούς, κατεβήτην, εἰδείην, ᾔεσαν, ἑστᾶσι?

5. What is the force of the derivative ending in each of these words: πρᾶγμα, Μεγαρεύς, ποιητής, δουλόω, δικαστήριον?

GREEK PROSE:

[*Those who offer the Greek Reader will take* 1, 2, 3. *Those who offer four books of the Anabasis and the seventh book of Herodotus will take* 1, 4, 5. *Candidates in Course II. will take* 1 *and* 6, *or* 1 *and* 4.]

1. TRANSLATE:—

'Ηνίκα δ' ἦν ἤδη δείλη, ὥρα ἦν ἀπιέναι τοῖς πολεμίοις· οὔποτε γὰρ μεῖον ἀπεστρατοπεδεύοντο οἱ βάρβαροι τοῦ Ἑλληνικοῦ ἑξήκοντα σταδίων, φοβούμενοι μὴ τῆς νυκτὸς οἱ Ἕλληνες ἐπιθῶνται αὐτοῖς. πονηρὸν γὰρ νυκτός ἐστι στράτευμα Περσικόν. οἵ τε γὰρ ἵπποι αὐτοῖς δέδενται, καὶ ὡς ἐπὶ τὸ πολὺ πεποδισμένοι εἰσί, τοῦ μὴ φεύγειν ἕνεκα εἰ λυθείησαν· ἐάν τέ τις θόρυβος γίγνηται, δεῖ ἐπισάξαι τὸν ἵππον Πέρσῃ ἀνδρί, καὶ χαλινῶσαι δεῖ, καὶ θωρακισθέντα ἀναβῆναι ἐπὶ τὸν ἵππον.—ANAB., III. iv. 34, 35.

Explain the use of the subjunctives ἐπιθῶνται (line 3) and γίγνηται (line 6). Explain the case of αὐτοῖς (line 5) and the construction of φεύγειν (line 5).

2. Translate:—

Ταῦτα λέγοντος Θεμιστοκλέος, αὖτις ὁ Κορίνθιος Ἀδείμαντος ἐπεφέρετο, σιγᾶν τε κελεύων τῷ μή ἐστι πατρὶς, καὶ Εὐρυβιάδεα οὐκ ἐῶν ἐπιψηφίζειν ἀπόλι ἀνδρί· πόλιν γὰρ τὸν Θεμιστοκλέα παρεχόμενον οὕτω ἐκέλευε γνώμας συμβάλλεσθαι. ταῦτα δέ οἱ προέφερε, ὅτι ἡλώκεσάν τε καὶ κατείχοντο αἱ Ἀθῆναι. τότε δὴ ὁ Θεμιστοκλέης ἐκεῖνόν τε καὶ τοὺς Κορινθίους πολλά τε καὶ κακὰ ἔλεγε, ἑωυτοῖσί τε ἐδήλου λόγῳ ὡς εἴη καὶ πόλις καὶ γῆ μέζων ἤπερ ἐκείνοισι, ἔστ' ἂν διηκόσιαι νέες σφι ἔωσι πεπληρωμέναι· οὐδαμοὺς γὰρ Ἑλλήνων αὐτοὺς ἐπιόντας ἀποκρούσεσθαι.—HEROD., VIII. 61.

Why μή and not οὐ (line 2)? Explain the use of the Optative εἴη (line 7) and of the Subjunctive ἔωσι (line 8), and the construction and tense of ἀποκρούσεσθαι (line 9), Give the Attic equivalents of τῷ (line 2), οἱ (line 4), and ἑωυτοῖσι (line 6).

3. Translate:—

Οἱ μὲν οὖν Λακεδαιμόνιοι τοσαῦτα εἶπον, νομίζοντες τοὺς Ἀθηναίους ἐν τῷ πρὶν χρόνῳ σπονδῶν μὲν ἐπιθυμεῖν, σφῶν δὲ ἐναντιουμένων κωλύεσθαι, διδομένης δὲ εἰρήνης ἀσμένως δέξεσθαί τε καὶ τοὺς ἄνδρας ἀποδώσειν. οἱ δὲ τὰς μὲν σπονδὰς, ἔχοντες τοὺς ἄνδρας ἐν τῇ νήσῳ, ἤδη σφίσιν ἐνόμιζον ἑτοίμους εἶναι, ὁπόταν βούλωνται, ποιεῖσθαι πρὸς αὐτοὺς, τοῦ δὲ πλέονος ὠρέγοντο.—THUCYD., IV. 21.

Why τοσαῦτα and not τοσάδε (line 1)? Explain the tense of the Infinitives ἐπιθυμεῖν (line 2), δέξεσθαι (line 3), and εἶναι (line 5).

4. Translate:—

καὶ γὰρ ἔργῳ ἐπεδείκνυτο καὶ ἔλεγεν ὅτι οὐκ ἄν ποτε προοῖτο, ἐπεὶ ἅπαξ φίλος αὐτοῖς ἐγένετο, οὐδ' εἰ ἔτι μὲν μείους γένοιντο ἔτι δὲ

GREEK PROSE. 341

κάκιον πράξειαν. φανερὸς δ᾽ ἦν καὶ, εἴ τίς τι ἀγαθὸν ἢ κακὸν ποιήσειεν αὐτὸν, νικᾶν πειρώμενος· καὶ εὐχὴν δέ τινες αὐτοῦ ἐξέφερον ὡς εὔχοιτο τοσοῦτον χρόνον ζῆν ἔστε νικῴη καὶ τοὺς εὖ καὶ τοὺς κακῶς ποιοῦντας ἀλεξόμενος. — ANAB. I. ix. 10, 11.

Explain the use of the Optatives ποιήσειεν (line 4), εὔχοιτο (line 5), and νικῴη (line 5), and the construction and tense of πειρώμενος (line 4).

5. TRANSLATE: —

Λακεδαιμόνιοι δὲ ἐμάχοντο ἀξίως λόγου, ἄλλα τε ἀποδεικνύμενοι ἐν οὐκ ἐπισταμένοισι μάχεσθαι ἐξεπιστάμενοι καὶ ὅκως ἐντρέψειαν τὰ νῶτα, ἁλέες φεύγεσκον δῆθεν· οἱ δὲ βάρβαροι ὁρέοντες φεύγοντας βοῇ τε καὶ παταγῷ ἐπήϊσαν, οἱ δ᾽ ἂν καταλαμβανόμενοι ὑπέστρεφον ἀντίοι εἶναι τοῖσι βαρβάροισι, μεταστρεφόμενοι δὲ κατέβαλλον πλήθεϊ ἀναριθμήτους τῶν Περσέων· ἔπιπτον δὲ καὶ αὐτῶν τῶν Σπαρτιητέων ἐνθαῦτα ὀλίγοι. ἐπεὶ δὲ οὐδὲν ἐδυνέατο παραλαβεῖν οἱ Πέρσαι τῆς ἐσόδου, πειρεόμενοι καὶ κατὰ τέλεα καὶ παντοίως προσβάλλοντες, ἀπήλαυνον ὀπίσω. — HEROD., VII. 211.

To what does ἄν (line 4) belong, and with what effect? Give the Attic equivalents of ὁρέοντες (line 3), πλήθεϊ (line 5), ἐνθαῦτα (line 7), and ἐδυνέατο (line 7).

6. TRANSLATE: —

Ἐννοήσωμεν δὲ καὶ τῇδε, ὡς πολλὴ ἐλπίς ἐστιν ἀγαθὸν αὐτὸ εἶναι. δυοῖν γὰρ θάτερόν ἐστι τὸ τεθνάναι· ἢ γὰρ οἷον μηδὲν εἶναι, μηδ᾽ αἴσθησιν μηδεμίαν μηδενὸς ἔχειν τὸν τεθνεῶτα, ἢ κατὰ τὰ λεγόμενα μεταβολή τις τυγχάνει οὖσα, καὶ μετοίκησις τῇ ψυχῇ τοῦ τόπου τοῦ ἐνθένδε εἰς ἄλλον τόπον. Καὶ εἴτε μηδεμία αἴσθησίς ἐστιν, ἀλλ᾽ οἷον ὕπνος ἐπειδάν τις καθεύδων μηδ᾽ ὄναρ μηδὲν ὁρᾷ, θαυμάσιον κέρδος ἂν εἴη ὁ θάνατος. — PLATO, 40 C.

Explain the construction of the pronoun οἶον (line 2), the use of the negatives μηδ'... μηδεμίαν μηδενός (line 3), and the mood of ὁρᾷ (line 6).

GREEK POETRY.

[*You are advised to do the translation first, and answer the questions* (a—f) *afterward. Candidates in Course II. will do the translation in 1 and 2, and answer the first four questions* (a—d).]

1. TRANSLATE:—

ἦ γάρ κεν δειλός τε καὶ οὐτιδανὸς καλεοίμην,
εἰ δὴ σοὶ πᾶν ἔργον ὑπείξομαι, ὅττι κεν εἴπῃς·
295 ἄλλοισιν δὴ ταῦτ' ἐπιτέλλεο, μὴ γὰρ ἔμοιγε
σήμαιν'· οὐ γὰρ ἔγωγ' ἔτι σοὶ πείσεσθαι ὀίω.
ἄλλο δέ τοι ἐρέω, σὺ δ' ἐνὶ φρεσὶ βάλλεο σῇσιν·
χερσὶ μὲν οὔτοι ἔγωγε μαχήσομαι εἵνεκα κούρης
οὔτε σοὶ οὔτε τῳ ἄλλῳ, ἐπεί μ' ἀφέλεσθέ γε δόντες·
300 τῶν δ' ἄλλων ἅ μοί ἐστι θοῇ παρὰ νηὶ μελαίνῃ,
τῶν οὐκ ἄν τι φέροις ἀνελὼν ἀέκοντος ἐμεῖο.

ILIAD, I. 293–301.

(*a*) Select from this passage six Homeric words or forms, and give their Attic equivalents.

(*b*) Translate the following names either of parts or of the rigging of the Homeric ship: πρύμνη, ἱστίον, ἱστοδόκη, ἱστός, πρότονοι, πρυμνήσια.

GREEK POETRY.

2. TRANSLATE: —

360 ἀλλά, ἄναξ, αὐτός τ' εὖ μήδεο πείθεό τ' ἄλλῳ.
οὔτοι ἀπόβλητον ἔπος ἔσσεται, ὅττι κεν εἴπω·
κρῖν' ἄνδρας κατὰ φῦλα, κατὰ φρήτρας, Ἀγάμεμνον,
ὡς φρήτρη φρήτρηφιν ἀρήγῃ, φῦλα δὲ φύλοις.
εἰ δέ κεν ὣς ἔρξῃς καί τοι πείθωνται Ἀχαιοί,
365 γνώσῃ ἔπειθ' ὅς θ' ἡγεμόνων κακὸς ὅς τέ νυ λαῶν
ἠδ' ὅς κ' ἐσθλὸς ἔῃσι· κατὰ σφέας γὰρ μαχέονται·
γνώσεαι δ' εἰ καὶ θεσπεσίῃ πόλιν οὐκ ἀλαπάξεις,
ἦ ἀνδρῶν κακότητι καὶ ἀφραδίῃ πολέμοιο.

ILIAD, II. 360–368.

(c) Explain the difference between φῦλον and φρήτρη.

(d) Explain the use of the Subjunctive ἔρξῃς (verse 364), and the force of the suffix of κακότητι (verse 368).

3. TRANSLATE: —

310 Ἦ ῥα καὶ ἐς δίφρον ἄρνας θέτο ἰσόθεος φώς,
ἂν δ' ἄρ' ἔβαιν' αὐτός, κατὰ δ' ἡνία τεῖνεν ὀπίσσω·
πὰρ δέ οἱ Ἀντήνωρ περικαλλέα βήσετο δίφρον.
τὼ μὲν ἄρ' ἄψορροι προτὶ Ἴλιον ἀπονέοντο·
Ἕκτωρ δὲ Πριάμοιο πάϊς καὶ δῖος Ὀδυσσεύς
315 χῶρον μὲν πρῶτον διεμέτρεον, αὐτὰρ ἔπειτα
κλήρους ἐν κυνέῃ χαλκήρεϊ πάλλον ἑλόντες,
ὁππότερος δὴ πρόσθεν ἀφείη χάλκεον ἔγχος.
λαοὶ δ' ἠρήσαντο, θεοῖσι δὲ χεῖρας ἀνέσχον.

ILIAD, III. 310–318.

(e) Give the stems from which the following words are formed: φώς (verse 310), ἄψορροι (verse 313), χαλκήρεϊ (verse 316).

(f) Explain the use of the Optative ἀφείη (verse 317).

LATIN COMPOSITION.

TRANSLATE INTO ENGLISH:—

Non videtur esse praetermittendum de virtute militis veterani quintae legionis. Nam cum in sinistro cornu elephans, vulnere ictus et dolore concitatus, in lixam[1] inermem impetum fecisset, eumque sub pede subditum dein genu innixus pondere suo proboscide erecta vibrantique, stridore maximo premeret atque enecaret, miles hic non potuit pati, quin se armatus bestiae offerret.

[1] servant.

TRANSLATE INTO LATIN:—

When[1] the elephant saw[2] that he was coming at him with a hostile weapon, leaving the dead[3] body, he encircles[4] the soldier with his trunk and hoists[5] him in-the-air,[6] armed [as he was].[7] The soldier,[8] seeing[9] that he must act with-pluck,[10] did not stop[11] hacking[12] the trunk with his sword. Overcome[13] by pain the elephant dropped[14] the soldier, and turning-round[15] with a tremendous-roar,[16] went-back-to-join[17] the rest of the beasts.

[1] postquam. [2] animadvertere. [3] cadaver. [4] circumdare. [5] extollere. [6] in sublime. [7] omit. [8] the soldier, qui. [9] with, cum. [10] constanter. [11] desistere [12] caedere. [13] adductus. [14] abicere. [15] conversus. [16] maximus stridor. [17] se recipere ad.

LATIN GRAMMAR.

[*Write Latin words very distinctly, particularly the endings. Do not crowd your work.*]

1. How do you pronounce *mon-* and *-ti-* in *mŏnitio; pul-* in *pulsus; ci-* and *-ves* in *cives?*

2. What is the root and what is the stem of *gens;* of *profugus;* of *nomen?* Give any other words that you can remember from the same roots. Give the meanings of the derivative suffixes, of *loquax, libertas, flebilis.* Form from the stem *simili-* a noun meaning *like-ness;* one from *digno-* meaning *worth;* one from *venari* meaning *a huntress.*

[*Mark the quantities of the penult and ultima of every Latin word you write in answering the 3d and 5th questions.*]

3. Give the Gen. Pl. of *ignis, cohors, custos;* the Abl. Sing. of *animal.* Decline *Gaius ipse; ingens portus; eadem vis.* Compare *humilis, magnus, celeber.* Form and compare adverbs from *sapiens, malus.*

4. What is the root and what are the stems of *fundere;* of *ducere;* of *cantare?*

5. State where each of the following forms is made, and give the principal parts of the verb to which it belongs: *pendas, scribēris, didicimus, ēmĕris, speret, parueram.* Form (α) the 3d S. Impf. Subj. Act. of *cupere;* (β) 2d Pl. Pf. Ind. Act. of *cedere;* (γ) 2d S. Pres. Imperat. Pass. of *ducere;* (δ) 3d Pl. Plupf. Subj. Act. of *fingere;* (ε) 2d

S. Pf. Ind. of *fidere;* (ς) 1st S. Imp. Subj. Pass. of *gerere;* (η) 1st Pl. Plupf. Ind. Act. of *parcere.* Inflect (1) the Pres. Ind. Pass. of *ferere;* (2) the Pres. Imperat. of *nolle;* (3) the Fut. Ind. of *fieri;* (4) the Fut. Pf. Ind. Act. of *dare.*

6. Write in Latin: I am ashamed of my brother; this concerns (*refert*) me; skilled (*peritus*) in war; they obey (*parēre*) the leader; the leader is obeyed; he came to Fidenae three years before; having taken the city, he departed.

CAESAR, SALLUST, AND OVID.

[*Translate two pieces of prose and two of poetry; and write out the first two lines of IV. or VI., marking the feet, quantity, ictus, and caesura of each line.*]

I. Mercatoribus est ad eos aditus magis eo, ut, quae bello ceperint, quibus vendant, habeant, quam quo ullam rem ad se importari desiderent: quin etiam jumentis, quibus maxime Gallia delectatur, quaeque impenso parant pretio, Germani importatis non utuntur, sed qnae sunt apud eos nata, parva atque deformia, haec quotidiana exercitatione summi ut sint laboris, efficiunt. Equestribus proeliis saepe ex equis desiliunt ac pedibus proeliantur; equosque eodem remanere vestigio assuefaciunt: ad quos se celeriter, cum usus est, recipiunt. — CAESAR, *Bell. Gall.*, IV. 2.

*II. At Catilina ex itinere plerisque consularibus, praeterea optumo cuique literas mittit: se falsis criminibus circumventum, quoniam factioni inimicorum resistere nequiverit, fortunae cedere, Massiliam in exilium proficisci;

non quo sibi tanti sceleris conscius esset, sed uti res publica quieta foret, neve ex sua contentione seditio oriretur. Ab his longe divorsas literas Q. Catulus in senatu recitavit, quas sibi nomine Catilinae redditas dicebat; earum exemplum infra scriptum est.—SALLUST, *Cat.*, 34.

*III.   Clauserat Hippotades aeterno carcere ventos,
      admonitorque operum coelo clarissimus alto
      Lucifer ortus erat. Pennis ligat ille resumtis
      parte ab utraque pedes, teloque accingitur unco,
      et liquidum motis talaribus aëra findit.
      Gentibus innumeris circumque infraque relictis
      Aethiopum populus Cepheaque conspicit arva.
      Illic immeritam maternae pendere linguae
      Andromedam poenas iniustus iusserat Ammon.
                    OVID, *Metam.* IV. 662 – 670.

*IV.   Sauromatae cingunt, fera gens, Bessique Getaeque
        quam non ingenio nomina digna meo!
      Dum tamen aura tepet, medio defendimur Istro:
        ille suis liquidus bella repellit aquis.
      At quum tristis hiems squalentia protulit ora,
        terraque marmoreo candida facta gelu:
      *   *   *   *   *   *   *   *
        *   *   *   *   *   *   *
      nix iacet: et iactam nec Sol pluviaeve resolvunt;
        indurat Boreas, perpetuamque facit.
      Ergo, ubi delicuit nondum prior, altera venit:
        et solet in multis bima manere locis;
      tantaque commoti vis est Aquilonis, ut altas
        aequet humo turres, tectaque rapta ferat.
                    OVID, *Trist.*, III. 10, 5 – 13.

[*Those who have not read Sallust and Ovid may substitute the following for II., III., IV.*]

V. (for II.) Cognito ejus adventu Acco, qui princeps ejus consilii fuerat, jubet in oppida multitudinem convenire; conantibus, priusquam id effici posset, adesse Romanos nunciatur; necessario sententia desistunt legatosque deprecandi causa ad Caesarem mittunt; adeunt per Aeduos, quorum antiquitus erat in fide civitas. Libenter Caesar petentibus Aeduis dat veniam excusationemque accipit; quod aestivum tempus instantis belli, non quaestionis, esse arbitrabatur.— CAESAR, *Bell. Gall.*, VI. 4.

VI. (*for* III).

Insula Sicanium iuxta latus Aeoliamque
Erigitur Liparen, fumantibus ardua saxis,
Quam subter specus et Cyclopum exesa caminis
Antra Aetnaea tonant, validique incudibus ictus
Auditi referunt gemitum, striduntque cavernis
Stricturae Chalybum, et fornacibus ignis anhelat,
Volcani domus, et Volcania nomine tellus.
Hoc tunc Ignipotens caelo descendit ab alto.
Ferrum exercebant vasto Cyclopes in antro,
Brontesque Steropesque et nudus membra Pyracmon.
           VIRG., *Aen.*, VIII. 416-425.

VII. (*for* IV.)

Sed fugit interea, fugit inreparabile tempus,
Singula dum capti circumvectamur amore.
Hoc satis armentis: superat pars altera curae,
Lanigeros agitare greges hirtasque capellas.
Hic labor; hinc laudem fortes sperate coloni.
Nec sum animi dubius, verbis ea vincere magnum

Quam sit, et angustis hunc addere rebus honorem;
Sed me Parnasi deserta per ardua dulcis
Raptat amor; iuvat ire iugis, qua nulla priorum
Castaliam molli devertitur orbita clivo.
                    VIRG., *Georg.*, III. 284-293.

CICERO AND VIRGIL.

[*N. B.* — *Translate two pieces of prose and two of poetry, and answer the questions on those passages. If you have read Cato Major, take the first passage. Candidates in Course II. will take for the prose one passage from Cicero with the passage from Caesar.*]

Faciam ut potero, Laeli. Saepe enim interfui querelis meorum aequalium — pares autem, vetere proverbio, cum paribus facillime congregantur — quae C. Salinator, quae Sp. Albinus, homines consulares nostri fere aequales deplorare solebant, tum quod voluptatibus carerent sine quibus vitam nullam putarent, tum quod spernerentur ab iis a quibus essent coli soliti. Qui mihi non id videbantur accusare quod esset accusandum. — CATO MAJOR, III. 7.

Explain construction of *carerent, putarent*. Explain the meaning of *consulares*.

Quis nostrum tam animo agresti ac duro fuit, ut Roscii morte nuper non commoveretur? Qui cum esset senex mortuus, tamen propter excellentem artem ac venustatem videbatur omnino mori non debuisse. Ergo illi corporis motu tantum amorem sibi conciliarat a nobis omnibus:

nos animorum incredibiles motus celeritatemque ingeniorum negligemus? Quoties ego hunc Archiam vidi, judices, — utar enim vestra benignitate, quoniam me in hoc novo genere dicendi tam diligenter attenditis —, quoties ego hunc vidi, cum litteram scripsisset nullam, magnum numerum optimorum versuum de eis ipsis rebus, quae tum agerentur, dicere ex tempore ! — Pro Arch. 8.

Explain construction of *commoveretur, animo, motu.*
Why does Cicero say *novo genere?* Who was Roscius?

Jam accepta in Ponto calamitate ex eo proelio, de quo vos paulo ante invitus admonui, cum socii pertimuissent, hostium opes animique crevissent, satis firmum praesidium provincia non haberet, amisissetis Asiam, Quirites, nisi ad ipsum discrimen ejus temporis divinitus Cn. Pompeium ad eas regiones fortuna populi Romani attulisset. . . . Et quisquam dubitabit quid virtute perfecturus sit qui tantum auctoritate perfecerit? Aut quam facile imperio atque exercitu socios et vectigalia conservaturus sit, qui ipso nomine ac rumore defenderit? — Cn. Pomp. Or. 15.

Construction of *amisissetis, perfecturus sit, accepta calamitate,* and *Cn. Pompeium . . . attulisset.* What are the allusions?

Imitari, Castor, potius avi mores disciplinamque debebas quam optimo et clarissimo viro fugitivi ore male dicere. Quod si saltatorem patrem habuisses neque eum virum, unde pudoris pudicitiaeque exempla peterentur, tamen hoc maledictum minime in illam aetatem conveniret. Quibus ille studiis ab ineunte aetate se imbuerat, non saltandi, sed bene ut armis, optime ut equis uteretur, ea tamen illum cuncta jam exacta aetate defecerant. Itaque Deiotarum cum plures in equum sustulissent, quod haerere in eo senex posset, admirari solebamus. — Pro Deiot.

Explain the construction of *viro, habuisses, quibus, studiis, armis.*

Interim milites legionum duarum, quae in novissimo agmine praesidio impedimentis fuerant, praelio nunciato, cursu incitato, in summo colle ab hostibus conspiciebantur. Et Titus Labienus castris hostium potitus et ex loco superiore, quae res in nostris castris gererentur, conspicatus, decimam legionem subsidio nostris misit. Qui cum ex equitum et calonum fuga, quo in loco res esset, quantoque in periculo et castra et legiones et imperator versaretur cognovissent, nihil ad celeritatem sibi reliqui fecerunt. — CAESAR, II. 26.

Construction of *praesido, impedimentis, castris, gererentur, cognovissent.*

Hanc pro Palladio moniti pro numine laeso
Effigiem statuere, nefas quae triste piaret.
Hanc tamen inmensam Calchas attollere molem
Roboribus textis caeloque educere iussit,
Ne recipi portis, aut duci in moenia possit,
Neu populum antiqua sub religione tueri.
Nam si vestra manus violasset dona Minervae,
Tum magnum exitium — quod di prius omen in ipsum
Convertant! — Priami imperio Phrygibusque futurum.
AENEID, II. 183 – 191.

Where is *statuere* made? How do you know? Construction of *piaret, violasset.* What is the allusion in *Palladio?* Who was Calchas? Write out metrically line 2, marking the caesura.

Nec minor in terris, Xanthum Simoentaque testor,
Aeneae mihi cura tui. Cum Troia Achilles

Exanimata sequens impingeret agmina muris,
Milia multa daret leto, gemerentque repleti
Amnes, nec reperire viam atque evolvere posset
In mare se Xanthus, Pelidae tunc ego forti
Congressum Aenean nec dis nec viribus aequis
Nube cava rapui, cuperem cum vetere ab imo
Structa meis manibus periurae moenia Troiae.
                                V. 803 – 811.

Xanthum, etc.; allusion? *Periurae;* why so? Write out metrically the first line.

Fer cineres, Amarylli, foras, rivoque fluenti
Transque caput iace; nec respexeris.   His ego Daphnim
Adgrediar: nihil ille deos, nil carmina curat.
Ducite ab urbe domum, mea carmina, ducite Daphnim.
Aspice, corripuit tremulis altaria flammis
Sponte sua, dum ferre moror, cinis ipse.   Bonum sit!
Nescio quid certe est, et Hylax in limine latrat.
Credimus? an, qui amant, ipsi sibi somnia fingunt?
Parcite, ab urbe venit, iam, carmina, parcite, Daphnis.
                                ECL. VIII. 101 – 109.

Write out metrically line 6. Construction of *respexeris*.

[*Only for those who have substituted Aen. VII. for some other reading.*]

Quanta per Idaeos saevis effusa Mycenis
Tempestas ierit campos, quibus actus uterque
Europae atque Asiae fatis concurrerit orbis,
Audiit, et si quem tellus extrema refuso
Submovet Oceano, et si quem extenta plagarum

ARITHMETIC. 353

Quattuor in medio dirimit plaga Solis iniqui.
Diluvio ex illo tot vasta per aequora vecti
Dis sedem exiguam patriis litusque rogamus
Innocuum et cunctis undamque auramque patentem.
AENEID, VII. 222-230.

Explain any allusion you may see in this passage.

ARITHMETIC.

[*Give all the work. Reduce each answer to its simplest form.*]

1. Divide $\frac{5}{6} \times \dfrac{3\frac{2}{3}}{\frac{5}{6} \times 1\frac{2}{3}}$ by $\dfrac{7\frac{2}{3} - 4\frac{1}{2}}{\frac{1}{14} \times 3\frac{5}{6}} \times \dfrac{7\frac{2}{3}}{31\frac{3}{8}}$.

2. Multiply 31.49 by 0.001297 *by logarithms*, and extract the fifth root of the product.

3. What sum must I invest in six per cent bonds, selling at 2½ per cent premium, to secure an annual income of $840?

4. The metre is 39.37 inches. Find how many hectares make an acre. [Use logarithms if you wish.]

5. If 144 pounds Advoirdupois be equivalent to 175 pounds Troy, what is the ratio of the pennyweight Troy to the dram Advoirdupois?

6. A's gain is $840; B's gain is $1,125; C's gain is $1,820. A's capital was in trade 7 months; B's, 9 months; C's, 14 months. How much of the capital $13,875, did each own?

ALGEBRA.

[*Write legibly, and without crowding; give the work clearly; and reduce the answers to their simplest forms. The shortest methods are preferred.*]

1. Two pipes, which supply the same reservoir, fill it in 4 hours and 12 minutes when both run together; but the first pipe alone can fill it in one hour less than half the time in which the second pipe alone can fill it. Find the time in which each pipe alone can fill the reservoir.

2. Solve the equation
$$\frac{2b-x-2a}{bx} = \frac{x-4a}{ab-b^2} - \frac{4b-7a}{ax-bx}.$$

3. Solve the equations
$$2x + 4y + 3z + 7 = 0,$$
$$\tfrac{2}{3}x - \tfrac{1}{2}y - 5z - 4 = 0,$$
$$3x + 5y - 6z - 5 = 0.$$

4. Find the greatest common divisor and the least common multiple of
$$2x^3 + 9x^2 - 8x - 15 \text{ and } 6x^3 - 5x^2 - 8x + 3.$$

5. Reduce the following expression to its simplest form as a single fraction:—
$$\frac{9a-5b}{(a-b)^2} - \left(\frac{4b}{a^2-b^2} - \frac{a-3b}{(a+b)^2} \right) - \frac{4}{a-b};$$
and divide it by $\dfrac{a^2-ab}{(a+b)^3}$, reducing the answer to its lowest terms, and freeing it from parentheses.

6. Expand $\left(\dfrac{\sqrt{a}}{2\sqrt[3]{b^2}} - 3\sqrt{b}\right)^8$, reducing each term of the result to its simplest form.

7. Write out the first four and the last four terms of $(x-y)^{100}$. How many terms does this power contain in all?

8. What is the logarithm of 1 in any system? Why?

ADVANCED ALGEBRA.

[*The shortest methods are preferred. Work clearly.*]

1. PROVE that the sum of the antecedents of a proportion is to the sum of the consequents as either antecedent is to its consequent.

2. Solve the equation
$$(1-x)^{1/2} - (2x+7)^{1/2} = (3x+10)^{1/2}.$$

3. Solve the equations
$$x^2 + y^2 = 22 + x + y,$$
$$\frac{2x+y}{x-7} + \frac{x+7}{y} = 0.$$

4. I am one of twenty men from whom seven are to be drafted for a dangerous service. Find the number of different combinations which can occur, and also the number of combinations which will include me among the seven.

5. Divide the number 520 into four parts which shall be in geometric progression, and such that the difference of the extremes shall be to the difference of the means as 19 to 6.

6. Divide $(y-z)x^3 - 2z(y+z)x^2 - 2y(y^2+3z^2)x + (y^2-z^2)^2$ by $(y-z)x - (y+z)^2$

PLANE GEOMETRY.

[*Number your answers carefully, but do not restate any proposition stated in the question.*]

1. Two right triangles having the hypothenuse and a side of one equal to the hypothenuse and a side of the other are equal. Prove.

2. If one side of a regular decagon be extended, how many degrees will the external angle contain? State and prove the proposition on which your answer is based.

3. A line drawn perpendicular to a radius at its extremity is tangent to the circle. Prove.

4. A line drawn parallel to one side of a triangle divides the other two sides proportionally. State and prove the converse proposition.

5. Show that the lines joining the middle points of adjacent sides of any quadrilateral form a parallelogram. *Suggestion:* Draw the diagonals of the quadrilateral.

6. If two chords intersect within the circle, the product of the segments of the one is equal to the product of the segments of the other. Prove. What does this proposition become when the chords are replaced by secants intersecting without the circle? By a secant and a tangent which intersect?

7. What ratio does the letter π represent in Geometry? Give its approximate numerical value to four decimal places. Find expressions for the circumference and the area of a circle in terms of π and the radius.

Example. Compute the circumference and the area of a circle having a radius of 4 feet.

SOLID GEOMETRY.

[*N. B.* — *Give as complete proofs as you can for the propositions stated in Nos.* 2, 3, 4, *and* 5.]

1. DEFINE a prism, a parallelopiped, a regular pyramid

2. The intersection of two planes is a straight line.

3. The sections of a prism made by parallel planes are equal polygons.

4. A point on the surface of a sphere, which is at the distance of a quadrant from two other points, is a pole of the great circle which passes through these two points.

5. Symmetrical spherical triangles are equivalent.

6. The angles of a spherical triangle are 60°, 70°, and 80°. The radius of the sphere is 10 feet. Find the area of the triangle in square feet.

ANALYTIC GEOMETRY.

[*N. B.— Give all the work.*]

1. Find the points of intersection of the right line which passes through the points (4, 2) and (— 3, —5), with the circle the circumference of which passes through the point (— 6, 8), the origin being the centre. Find the inclination of the same right line to the axis of X. Give your reasons for each step.

2. Deduce the equation of a parabola referred to its principal vertex. From this equation obtain the equation of the same curve when the directrix is the axis of X, and the principal axis the axis of Y.

3. Prove that any ordinate of an ellipse is to the corresponding ordinate of the circumscribing circle as the conjugate axis of the ellipse is to its transverse axis.

4. Define an asymptote to a curve. Prove that the diagonals of the parallelogram formed on the axes of two conjugate hyperbolas are asymptotes to both.

TRIGONOMETRY.

1. Define the Trigonometric Functions as ratios, and prove the formulas, $\sin^2\varphi + \cos^2\varphi = 1$, $\dfrac{\sin\varphi}{\cos\varphi} = \tan\varphi$.

2. By the aid of a circle whose radius is unity draw lines representing the functions of an angle in the fourth quadrant, and tell which are positive and which are negative.

3. Find expressions for the functions of $180° - \varphi$ in terms of the functions of φ. Write all the functions of $180.°$

Given $\sin 35° = .574$, $\cos 35° = .819$, $\tan 35° = .700$, required sin, cos, and tan of $215°$.

4. Find formulas for $\sin 2a$ and $\sin 3a$ in terms of $\sin a$ and $\cos a$.

5. Write the formulas for solving a triangle when two angles and a side are given, and state and prove the theorem from which the formulas are derived.

6. A ladder 45 feet long leans against a house, and just reaches a window 40 feet above the ground. Required the inclination of the ladder to the horizon.

7. Two sides of a triangle are 1427 feet and 1232 feet respectively, and the included angle is $27° 15'$. Solve the triangle.

ENGLISH COMPOSITION.

You are required to write a short English composition, correct in spelling, punctuation, division by paragraphs, and expression. You are recommended to arrange what you have to say before beginning to write; to pay more attention to quality than to quantity of work, and to make a fair copy from a rough draft.

One of the following subjects must be taken:—

I. Mark Antony's Speech in Julius Caesar.
II. Christmas at Bracebridge Hall.
III. The Combat between Sir Kenneth and Conrade.

FRENCH.

[*The translation should be in good English. French idioms should be rendered by corresponding English ones, whenever it is possible. If you fear you are going too far from the literal sense, the word-for-word meaning may be enclosed in parentheses. Leave blanks for the words you do not know.*]

TRANSLATE:—

IL était une fois une Reine si vieille, qu'elle n'avait plus ni dents ni cheveux; sa tête branlait comme les feuilles que le vent remue; elle ne voyait goutte même avec ses lunettes; le bout de son nez et celui de son menton se touchaient; elle était rapetissée de la moitié, et toute en un peloton, avec le dos si courbé, qu'on *aurait* cru qu'elle avait toujours été contrefaite. Une Fée, qui avait assisté à sa naissance, l'aborda et lui dit: " *Voulez*-vous rajeunir?"— Volontiers, répondit la Reine: je donnerais tous mes joyaux pour n'avoir que vingt ans. — Il faut donc, continua la Fée, donner votre vieillesse à quelque autre dont vous prendrez la jeunesse et la santé. A qui donnerons-nous vos cent ans?" La Reine *fit* chercher partout quelqu'un qui voulût être vieux pour la rajeunir. Il vint beaucoup de gueux qui voulaient vieillir pour être riches; mais quand ils avaient vu la Reine tousser, cracher, vivre de bouillie, être sale, hideuse, puante, souffrante, et radoter un peu, ils ne voulaient plus se charger de ses années : ils aimaient mieux mendier et porter des haillons. Il venait aussi des ambitieux, à qui elle promettait de grands rangs et de grands honneurs. " Mais que faire de ces rangs ? disaient-ils après l'avoir vue, nous n'oserions nous montrer étant si dégoûtants." Mais enfin il se présenta une jeune fille de village, belle comme le jour, qui demanda la couronne pour prix de sa jeunesse. La Reine s'en fâcha d'abord : mais que faire ? à quoi sert-il de se fâcher ? elle voulait rajeunir. " Partageons, dit-elle, mon royaume ; vous en aurez la moitié et moi l'autre : c'est bien assez pour vous qui êtes une petite paysanne."— FÉNELON.

[*Any one may pass who does well the above translation even should he not answer a single question in Grammar. The following questions are set to give a better chance to those whose translation may not be quite satisfactory, and who by answering them will show that they have a fair knowledge of elementary grammar. Even should the candidate feel that his translation is good enough, it will be well for him, if he can, to answer the last question (No. 5), and thereby make sure of a high mark for this examination.*]

2. Give in full the tense of the italicized verbs in the above.

3. Give the principal tenses of *voir, toucher, croire, vivre, sentir* (thus, INF., *être;* PRES. PART., *étant;* PAST PART., *été;* IND. PRES., *je suis;* PRET., *je fus*).

4. Say what you know about the position of personal pronouns.

5. TRANSLATE INTO FRENCH: —

(a) What an old woman that is! (b) I am going there to-morrow. (c) I don't know whether he has ever spoken French.

GERMAN GRAMMAR.

TRANSLATE INTO GERMAN: —

In the year[1] 1863 there lived[2] in a certain[3] village[4] in France,[5] a very[6] charming[7] girl[8] whose name[9] was Mariette. She was loved by Colin, the richest[10] young man in the village. Had Colin not been so bashful[11] he might[12] easily[13] have won[14] Mariette's heart.[15] But[16] he did not dare[17] to tell[18] his love,[19] and in order to conceal[20] it he acted[21] very unkindly[22] towards[23] the poor[24] child[25] when he saw her. One might[26] think[27] that it would be difficult[28] to win Mariette's heart in this way,[29] but one can never[30] explain[31] the wonderful[32] workings[33] of a young girl's heart. So-then[34] Mariette began[35] to love Colin almost[36] without knowing[37] it. But being very bashful[38] herself[39] she did not let him notice[40] it, and had it not been for a lucky accident (had lucky[41] accident[42] not been) they might have been very unhappy.[43] Colin sent[44] a beautiful[45] pitcher to Mariette without saying from[46] whom it came. The whole[47] history[48] of this pitcher is too long to tell[49] now,[50] the examination[51] being only[52] one hour[53] long (since the examination is only one hour long). But it is enough[54] to say that when Colin accidentally[55] broke[56] the pitcher one day, he was obliged to confess[57] that he had sent it in order to save himself from punishment (in-order [58] not to be punished [59]). Thus [60] they found-out [61] that they had long loved each-other, [62] and they were married [63] the next day. [65]

[1] das Jahr. [2] leben. [3] gewiß. [4] das Dorf. [5] Frankreich. [6] sehr [7] reizend. [8] das Mädchen. [9] der Name. [10] reich. [11] schüchtern. [12] hätte können. [13] leicht. [14] gewinnen. [15] das Herz. [16] aber. [17] wagen. [18] erklären. [19] die Liebe. [20] verbergen. [21] sich betragen. [22] unfreundlich. [23] gegen. [24] Arm. [25] Kind. [26] sollen (use preterit). [27] glauben. [28] schwer. [29] in this way = auf diese Weise. [30] niemals. [31] erklären. [32] wunderbar. [33] Empfindung. [34] also. [35] anfangen. [36] fast. [37] wessen (use infinitive with zu). [38] scheu. [39] selbst. [40] merken. [41] glücklich. [42] der Zufall. [43] unglücklich. [44] senden. [45] schön. Pitcher = der Krug. [46] von. [47] ganz. [48] die Geschichte. [49] erzählen. [50] jetzt. [51] Prüfung. [52] nur. [53] die Stunde. [54] genug. [55] zufällig. [56] zerbrechen. [57] gestehen. [58] um (dependent order). [59] bestrafen. [60] so. [61] erfahren. [62] sich. [63] vermählen. [64] ander (use accusative). [65] der Tag. [66] müssen (use Preterit Active).

1. Decline der used as a relative (masculine only). Translate: whose house is that? which book have you? the man whose book I have is not at home.

2. What influence has the relative pronoun upon the position of the object to which it refers? Translate: I have given a book to the man who was here yesterday.

3. Decline in full: the old man; a large house; his good daughter (Tochter); good bread.

4. Explain the comparison of adjectives.

GERMAN TRANSLATION.

For the canditates who offer Translation at sight.

1. Zur Zeit Kaiser Karls des Großen, welcher 800 Jahre nach Christi Geburt lebte, gab es noch wenige oder gar keine Schulen. Um die Bildung seines Volkes zu fördern, ließ der Kaiser an seinem Hofe eine Schule errichten, welche seine eigenen Kinder, wie die seiner sämmtlichen Hofleute und seiner Dienerschaft besuchen mußten. Von Zeit zu Zeit besuchte der Kaiser die Schule selbst und ließ die Kinder von dem Lehrer prüfen, oder legte ihnen auch wol selbst einige Fragen vor. Als er einst längere Zeit mit fernen Völkern Krieg geführt hatte und endlich aus demselben zurückgekehrt war, besuchte er sogleich seine Schule und ließ eine Prüfung anstellen. Hierbei fand es sich, daß die Söhne der Vornehmen und Reichen sehr unwissend waren, während die der Geringen und Armen vortreffliche Fortschritte gemacht hatten. Da ließ der Kaiser sämmtliche Schüler vortreten und stellte die fleißigen zu seiner Rechten, die trägen und nachlässigen aber zu seiner Linken. Hierauf sprach er, indem er sich liebreich zu den erstern wandte: „Es freut mich, daß ihr euch Mühe gegeben und eurem Lehrer Freude gemacht habt. Fahret fort in eurem Bestreben, und seid versichert, daß ich einst gut für euch sorgen werde." Dann aber wandte er sich mit zornigem Angesicht zu denen, die zu seiner Linken standen, und sprach: „Schämt euch, ihr Söhne der Reichen und Vornehmen; denkt ihr, eure schönen Kleider und eure hübschen Gesichter werden bei mir etwas gelten? Ich sage euch, wenn ihr euch nicht ändert und ernstliche Besserung zeigt, so habt ihr nicht das Geringste von mir zu hoffen."

2. Thomas Morus, Kanzler, von England, zeichnete sich durch seine strenge Rechtlichkeit als Richter aus. Eines Tages schickte ihm ein angesehener Mann, für welchen er einen Prozeß zu führen hatte, zwei schöne silberne Kannen, um ihm damit ein Geschenk zu machen.

Thomas Morus, welcher sogleich die Absicht merkte, in welcher ein so werthvolles Geschenk gemacht wurde, befahl seinem Diener, die beiden Gefäße mit dem besten Weine aus seinem Keller zu füllen. Nachdem dies geschehen, ließ er sie dem Ueberbringer zurückgeben und dem Herrn desselben sagen, daß es ihm ein Vergnügen machen würde, ihm öfter einige Proben von seinem Weine zukommen zu lassen. Auf diese Weise vermied er ein Geschenk, welches augenscheinlich aus Eigennutz gemacht worden war, ohne denjenigen zu beleidigen, der es ihm zugesandt hatte.

3. Give the principal parts (Present Infinite, Præterite first person, Past Participle) of all the verbs in 2.

CHEMISTRY AND PHYSICS.

CHEMISTRY.

1. Describe in full one complete method for determining the composition of water.

2. How is oxygen prepared?

3. What are the three forms of carbon? How can they be shown to be identical.

4. Describe as fully as you can the element magnesium and its compounds.

5. How many grammes of NaCl can be made from 106 grammes of Na_2CO_3.

$$Na_2CO_3 + 2\,HCl = 2\,NaCl + H_2O + CO_2$$
$$H = 1 \quad Na = 23 \quad Cl = 35.5$$
$$C = 12 \quad O = 16$$

PHYSICS.

6. Explain the hydrostatic press, also called the water press, or Bramah's press.

7. Describe the method of finding the number of vibrations in one second corresponding to any note.

8. Describe the refraction of light.

9. Describe the electrical machine.

PHYSICS AND ASTRONOMY.

1. How is the position of the centre of gravity of a body determined?

2. What are some of the properties of liquids?

3. How is specific gravity determined.

4. How is the common thermometer graduated?

5. How can a magnet be produced by means of the electric current?

6. Describe the apparent motion of Mercury and Venus.

9. Explain the phases of the moon.

8. What are the causes of the disappearances of the satellites of Jupiter?

9. Explain the motion of the sun-spots.

10. What is known in regard to the masses of comets?

BOTANY.

1. DESCRIBE the germination of an *almond* and of an *acorn*.

2. Describe some of the forms of underground stems and branches.

3. Give an outline sketch of a thrice-pinnate leaf.

4. Describe the fruits of the *strawberry, raspberry*, and *cranberry*.

5. Is the flower given for examination regular and perfect?

6. What kind of a flower-cluster does this plant have?

7. Is the calyx of the given flower free from or adherent to the ovary?

8. Describe the arrangement, venation, and shape of the leaves; and write answers to the following questions: —

(*a*) Is the plant *exogenous* or *endogenous?* (Give your reasons for the answer.)

(*b*) Upon what are the stamens inserted?

(*c*) Is the flower *monopetalous, polypetalous,* or *apetalous?*

HARVARD ADMISSION EXAMINATION PAPERS.

SEPTEMBER, 1877.

ANCIENT HISTORY AND GEOGRAPHY.

[*Take* five *of these six numbers; three in Greek history and two in Roman, or two in Greek and three in Roman.*]

1. TRY to account (briefly) for the rise and the decline of the power of Athens. What was the most brilliant period of Athenian history?

2. Classify or group, geographically, the principal Greek colonies, and mention some important facts in their history.

3. Give some account of three men famous in Greek literature.

4. What do you understand by an Agrarian law? What did the Gracchi propose? By what class and for what reason were they opposed.

5. Give the geographical situation of the countries or districts in which were carried on the chief wars of the Romans before those against Carthage. Name the wars in the order of time.

6. The two Triumvirates; — their purpose and result.

MODERN AND PHYSICAL GEOGRAPHY.

[*Where it is in his power, the student should use diagrams in answering the questions.*]

1. DEFINE the terms *tropic, equator, latitude, longitude.*

2. Give approximately the latitude and the longitude, from Greenwich, of the following points: Cape of Good Hope, Cape Horn, Cape St. Roque, Cape Race.

3. In making a coasting voyage from Copenhagen to Odessa, name the important cities at which a vessel could touch.

4. Draw a sketch map of North America, and show on it the places of four important river systems.

5. Name the larger islands of the West Indies in their order from west to east.

6. Name in their order of entrance into the main stream the principal tributaries of the Mississippi.

7. What is the Gulf Stream? Trace its course.

8. What European states lie in whole or in part within the basin of the Rhine? what within that of the Danube?

GREEK COMPOSITION.

[*Do either A or B.*]

A.

But Tiribazus ordered that those who were willing to agree to (ὑπακούω) the peace should come together before him. When they had come, he exhibited (ἐπιδείκνυμι) the king's seal (σημεῖα) and read aloud (ἀναγιγνώσκω) the things written. They were as follows: "Artaxerxes, the king, thinks it is just that the cities in Asia be his; but the Greek cities be independent (αὐτόνομος). If any do not accept (δέχομαι) this peace, I will make war on them."

B.

1. He said that if we had not come, they would all be marching against the king.

2. They fear that the Greeks will attack them during the night.

3. I know he will do this, if it is possible.

4. They waited until the men went away from the city.

5. With you I will suffer whatever seems good to the citizens.

GREEK GRAMMAR.

1. DECLINE ἡ ναῦς, τὸ εἶδος, and, in the plural only, the adjectives γλυκύς and ἀσφαλής.

2. Compare the adjectives σαφής, σοφός, μέγας, κακός, ῥᾴδιος. Form and compare adverbs from the adjectives φίλος, ταχύς.

3. Decline οὗτος, τίς, and ἑαυτοῦ.

4. Give the Active Infinitives and Participles of λείπω and φαίνω.

5. Give the present Optative Active and Middle of δηλόω; Aorist Optative Active of λύω; Perfect Optative Middle of λέγω.

6. Principal parts of αἱρέω, βαίνω, δίδωμι, φέρω, ἔρχομαι, θνήσκω, κλίνω, μανθάνω.

7. Write and accent correctly: τιμαοι, τετριβται, ἐλειπθην, δεδεχμαι, λυουσι, συνχεω.

8. Describe the derivation of these words, giving the meaning of root and ending: λυτήρ, λύσις, τάχος, μισθόω, πολέμικος.

GREEK PROSE.

[*N. B.* — Those who offer the Greek Reader will take 1, 2, 3. Those who offer four books of the Anabasis and the seventh book of Herodotus will take 4, 1, 3. Candidates in Course II. will take 1 and 2, or 4 and 1.

1. TRANSLATE: —

ἀπορουμένοις δ' αὐτοῖς προσελθών τις ἀνὴρ ʿΡόδιος εἶπεν. Ἐγὼ θέλω, ὦ ἄνδρες, διαβιβάσαι ὑμᾶς κατὰ τετρακισχιλίους ὁπλίτας, ἂν ἐμοὶ ὧν δέομαι ὑπηρετήσητε, καὶ τάλαντον μισθὸν πορίσητε. ἐρωτώμενος δὲ ὅτου δέοιτο, Ἀσκῶν, ἔφη, δισχιλίων δεήσομαι· πολλὰ δ' ὁρῶ πρόβατα καὶ αἶγας καὶ βοῦς καὶ ὄνους, ἃ ἀποδαρέντα καὶ φυσηθέντα ῥᾳδίως ἂν παρέχοι τὴν διάβασιν. — ANAB., III. v. 8, 9.

Explain the use of the Subjunctive πορίσητε (line 3), and of the Optative δέοιτο (line 4). Give the value of the *talent* (line 3) both in Greek minae and American dollars.

2. TRANSLATE: —

Πολλάκις ἐθαύμασα, τίσι ποτὲ λόγοις Ἀθηναίους ἔπεισαν οἱ γραψάμενοι Σωκράτην, ὡς ἄξιος εἴη θανάτου τῇ πόλει. ἡ μὲν γὰρ γραφὴ κατ' αὐτοῦ τοιάδε τις ἦν.

Ἀδικεῖ Σωκράτης οὓς μὲν ἡ πόλις νομίζει θεοὺς οὐ νομίζων, ἕτερα δὲ καινὰ δαιμόνια εἰσφέρων. ἀδικεῖ δὲ καὶ τοὺς νέους διαφθείρων.

Πρῶτον μὲν οὖν, ὡς οὐκ ἐνόμιζεν οὓς ἡ πόλις νομίζει θεούς, ποίῳ ποτ' ἐχρήσαντο τεκμηρίῳ· θύων τε γὰρ φανερὸς ἦν, πολλάκις μὲν οἴκοι, πολλάκις δὲ ἐπὶ τῶν κοινῶν τῆς πόλεως βωμῶν, καὶ μαντικῇ χρώμενος οὐκ ἀφανὴς ἦν. — MEMOR., I. i. 1, 2.

Give the date of the trial of Socrates, and the name and occupation of each of his accusers. Give the construction of the Participle θύων (line 7).

GREEK PROSE. 373

3. TRANSLATE: —

ἐνθαῦτα ἀναγκαίη ἐξέργομαι γνώμην ἀποδέξασθαι ἐπίφθονον μὲν πρὸς τῶν πλεόνων ἀνθρώπων, ὅμως δὲ τῇ γ' ἐμοὶ φαίνεται εἶναι ἀληθές, οὐκ ἐπισχήσω. εἰ Ἀθηναῖοι καταρρωδήσαντες τὸν ἐπιόντα κίνδυνον ἐξέλιπον τὴν σφετέρην, ἢ καὶ μὴ ἐκλιπόντες ἀλλὰ μείναντες ἔδοσαν σφέας αὐτοὺς Ξέρξῃ, κατὰ τὴν θάλασσαν οὐδαμοὶ ἂν ἐπειρῶντο ἀντιεύμενοι βασιλέϊ. εἰ τοίνυν κατὰ τὴν θάλασσαν μηδεὶς ἠντιοῦτο Ξέρξῃ, κατά γε ἂν τὴν ἤπειρον τοιάδε ἐγίγνετο· εἰ καὶ πολλοὶ τειχέων κιθῶνες ἦσαν ἐληλαμένοι διά τοῦ Ἰσθμοῦ Πελοποννησίοισι, προδοθέντες ἂν Λακεδαιμόνιοι ὑπὸ τῶν συμμάχων οὐχ ἑκόντων, ἀλλ' ὑπ' ἀναγκαίης, κατὰ πόλις ἁλισκομένων ὑπὸ τοῦ ναυτικοῦ στρατοῦ τοῦ βαρβάρου, ἐμουνώθησαν, μουνωθέντες δὲ ἂν αἱ ἀποδεξάμενοι ἔργα μεγάλα ἀπέθανον γενναίως. — HEROD., VII. 139.

Explain the use of ἂν (line 5) and the case of βασιλέϊ (line 6). Give the Attic equivalents of all the peculiarly Ionic forms in this passage.

4. TRANSLATE: —

Μένων δὲ ὁ Θετταλὸς δῆλος ἦν ἐπιθυμῶν μὲν πλουτεῖν ἰσχυρῶς, ἐπιθυμῶν δὲ ἄρχειν, ὅπως πλείω λαμβάνοι, ἐπιθυμῶν δὲ τιμᾶσθαι, ἵνα πλείω κερδαίνοι· φίλος τε ἐβούλετο εἶναι τοῖς μέγιστον δυναμένοις, ἵνα ἀδικῶν μὴ διδοίη δίκην. ἐπὶ δὲ τὸ κατεργάζεσθαι ὧν ἐπιθυμοίη συντομωτάτην ᾤετο ὁδὸν εἶναι διὰ τοῦ ἐπιορκεῖν τε καὶ ψεύδεσθαι καὶ ἐξαπατᾶν· τὸ δ' ἁπλοῦν καὶ ἀληθὲς τὸ αὐτὸ τῷ ἠλιθίῳ εἶναι. στέργων δὲ φανερὸς μὲν ἦν οὐδένα, ὅτῳ δὲ φαίη φίλος εἶναι, τούτῳ ἔνδηλος ἐγίγνετο ἐπιβουλεύων. — ANAB., II. vi. 21–23.

Give the construction of the Participle ἐπιθυμῶν (line 1). Explain the use of the Optative φαίη (line 7). Give a brief sketch of the life of *Cyrus the Younger*.

GREEK POETRY.

[*N. B.* — *You are advised to do the translation first, and answer the questions* (a–f) *afterward. Candidates in Course II. will do the translation in* 1 *and* 2, *and answer the first four questions* (a–d).

1. TRANSLATE:—

" τέτλαθι, μῆτερ ἐμὴ, καὶ ἀνάσχεο κηδομένη περ,
μή σε φίλην περ ἐοῦσαν ἐν ὀφθαλμοῖσιν ἴδωμαι
θεινομένην, τότε δ᾽ οὔ τι δυνήσομαι ἀχνύμενός περ
χραισμεῖν· ἀργαλέος γὰρ Ὀλύμπιος ἀντιφέρεσθαι.
590 ἤδη γάρ με καὶ ἄλλοτ᾽ ἀλεξέμεναι μεμαῶτα
ῥῖψε ποδὸς τεταγὼν ἀπὸ βηλοῦ θεσπεσίοιο.
πᾶν δ᾽ ἦμαρ φερόμην, ἅμα δ᾽ ἠελίῳ καταδύντι
κάππεσον ἐν Λήμνῳ. ὀλίγος δ᾽ ἔτι θυμὸς ἐνῆεν.
ἔνθα με Σίντιες ἄνδρες ἄφαρ κομίσαντο πεσόντα."

ILIAD, I. 586–594.

(*a*) Select from this passage six Homeric *forms*, and give their Attic equivalents.

(*b*) Give a brief outline (half a page) of the action of Iliad I.

2. TRANSLATE:—

τοὺς δ᾽ ὥς τ᾽ αἰπόλια πλατέ᾽ αἰγῶν αἰπόλοι ἄνδρες
475 ῥεῖα διακρίνωσιν, ἐπεί κε νομῷ μιγέωσιν,
ὣς τοὺς ἡγεμόνες διεκόσμεον ἔνθα καὶ ἔνθα
ὑσμίνηνδ᾽ ἰέναι, μετὰ δὲ κρείων Ἀγαμέμνων,
ὄμματα καὶ κεφαλὴν ἴκελος Διὶ τερπικεραύνῳ.

Ἄρεϊ δὲ ζώνην, στέρνον δὲ Ποσειδάωνι.
480 ἠΰτε βοῦς ἀγέληφι μέγ' ἔξοχος ἔπλετο πάντων
ταῦρος· ὃ γάρ τε βόεσσι μεταπρέπει ἀγρομένῃσι·
τοῖον ἄρ' Ἀτρείδην θῆκε Ζεὺς ἤματι κείνῳ,
ἐκπρεπέ' ἐν πολλοῖσι καὶ ἔξοχον ἡρώεσσιν.

ILIAD, II. 474-83.

(c) What is the force of the suffix of ὑσμίνηνδε (v. 477) and that of ἀγέληφι (v. 480)?

(d) Give the stems, with their meanings, from which the following words are *derived:* αἰγῶν (v. 474), τερπικεραύνῳ (v. 478), ἔξοχον (v. 483).

3. TRANSLATE: —

" ἤλυθες ἐκ πολέμου· ὡς ὤφελες αὐτόθ' ὀλέσθαι,
ἀνδρὶ δαμεὶς κρατερῷ, ὃς ἐμὸς πρότερος πόσις ἦεν·
430 ἦ μὲν δὴ πρίν γ' εὔχε' ἀρηϊφίλου Μενελάου
σῇ τε βίῃ καὶ χερσὶ καὶ ἔγχεϊ φέρτερος εἶναι·
ἀλλ' ἴθι νῦν προκάλεσσαι ἀρηΐφιλον Μενέλαον
ἐξαῦτις μαχέσασθαι ἐναντίον. ἀλλά σ' ἐγώ γε
παύεσθαι κέλομαι, μηδὲ ξανθῷ Μενελάῳ
435 ἀντίβιον πόλεμον πολεμίζειν ἠδὲ μάχεσθαι
ἀφραδέως, μή πως τάχ' ὑπ' αὐτοῦ δουρὶ δαμήῃς."

ILIAD, III. 428-436.

(e) Give the difference of meaning of πολεμίζειν and μάχεσθαι (v. 435).

(f) Explain the use of the Subjunctive δαμήῃς, and analyze its form (v. 436).

LATIN COMPOSITION.

TRANSLATE INTO ENGLISH:—

Publius Scipio Nasica consul Iugurthae bellum indixit, matrem, Idaeam e Phrygiis sedibus ad nostras aras focosque migrantem sanctissimis manibus excepit, multas et pestiferas seditiones auctoritatis suae robore oppressit.

When he was-a-candidate [1] for the edileship in his youth, and had grasped [2] the hand of a-man,[3] hardened [4] by field-work,[5] rather tight,[6] he asked [7] him in-sport,[8] whether he was-in-the-habit-of [9] walking [10] on his hands. This [11] joke,[12] picked-up [13] by the bystanders,[14] leaked-out [15] to [16] the public,[17] and caused [18] a defeat [19] for Scipio. For all the rustic tribes, thinking [20] their poverty [21] was thrown-in-their-teeth,[22] were-indignant-at [23] his insulting [24] raillery.[25]

[1] *be a candidate for*, petere. [2] adprehendere. [3] quidam. [4] durare. [5] rusticum opus. [6] tenaciter. [7] interrogare. [8] ioci gratia. [9] solere. [10] ambulare. [11] *relative*. [12] dictum. [13] excipere. [14] circumstantes. [15] manare. [16] ad. [17] populus. [18] *cause*, causam adferre. [19] repulsa. [20] iudicare. [21] paupertas. [22] exprobrare. [23] *be indignant at*, moleste ferre. [24] contumeliosus. [25] urbanitas.

LATIN GRAMMAR.

[*Write Latin words very distinctly, particularly the endings. Do not crowd your work.*]

1. How do you pronounce *per* and *cul* in *perculsus*; *vo* and *ces* in the substantive *voces*?

2. Give the meanings of the derivative suffixes of *tonsor, certamen, regulus*. Give the root and the stem of *stabulum*, of *dux*, of *natura;* give any other words you may remember from the same roots. Form from the stem of *pars* a noun meaning *a small portion;* one from the stem of *solus* meaning *loneli-ness;* one from a stem of *ulcisci* meaning *revenge*.

[*Mark the quantities of the penult and ultima of every Latin word you write in answering the 3d and 5th questions.*]

3. Decline *tota cohors; duplex acies* (in the singular only) *; uterque currus; quisquam*. What is the Abl. Sing. of *puppis?* of *mare?* the Gen. Pl. of *pons?* Compare *piger, beneficus, facile, feliciter*. Give rules for the gender of *labor, Aquilo, Delos*.

4. What is the root and what are the stems of *tangere?* of *arare?*

5. State where each of the following forms is made, and give the principal parts of the verb to which it belongs:. *coepisset, moreris, abscissum est, caverit, iaceretis, ordire*.

Form (α) 2d Pl. Pf. Ind. Act. of *sinere;*
(β) 3d Pl. Pf. Subj. of *solere;*
(γ) 3d S. Plupf. Ind. Act. of *pascere;*
(δ) 3d S. Impf. Subj. Act. of *edere;*
(ε) 3d S. Fut. Ind. Pass. of *miscere;*
(ζ) 3d Pl. Pres. Subj. Pass. of *fodere*.

Inflect (η) the Pres. Imperat. of *esse;* (θ) the Pres. Ind. of *potiri;* (ι) the Pres. Subj. Pass. of *domare*. Give (κ) all the participles of *sequi*.

6. What case or cases follow *credere, peritus, opus est, pudet, accusare, ante, coram.* Write in Latin: *At Rome; at Corinth; at Carthage; at Thebes; to Carthage; to the island.*

CAESAR, SALLUST, AND OVID.

[*Translate* two *pieces of prose and* two *of poetry; and write out the first two lines of III. or VII., marking feet, quantity, ictus, and caesura.*]

I. Ibi cum alii fossas complerent, alii multis telis conjectis defensores vallo munitionibusque depellerent, auxiliaresque, quibus ad pugnam non multum Crassus confidebat, lapidibus telisque subministrandis et ad aggerem cespitibus comportandis speciem atque opinionem pugnantium praeberent; cum item ab hostibus constanter ac non timide pugnaretur telaque ex loco superiore missa non frustra acciderent; equites circumitis hostium castris Crasso renunciaverunt, non eadem esse diligentia ab decumana porta castra munita facilemque aditum habere.—CAESAR, B. G. III. 25.

Who is the Crassus mentioned?

II. Licuit nobis cum summa turpitudine in exsilio aetatem agere; potuistis nonnulli Romae amissis bonis alienas opes exspectare; quia illa foeda atque intoleranda viris videbantur, haec sequi decrevistis. Si haec relinquere voltis, audacia opus est; nemo nisi victor pace bellum mutavit. Nam in fuga salutem sperare, quum arma quis corpus tegitur ab hostibus avorteris, ea vero dementia est. Semper in proelio iis maxumum est periculum, qui maxume timent; audacia pro muro habetur.—SALLUST, CAT. 58.

Who is the speaker?

III. At tu, funesti ne sim tibi muneris auctor,
Nate, cave, dum resque sinit, tua corrige vota.
Scilicet ut nostro genitum te sanguine credas,
Pignora certa petis. Do pignora certa timendo,
Et patrio pater esse metu probor. Adspice vultus
Ecce meos, utinamque oculus in pectora posses
Inserere, et patrias intus deprendere curas!
Denique, quidquid habet dives, circumspice, mundus,
Eque tot ac tantis coeli, terraeque, marisque
Posce bonis aliquid : nullam patiere repulsam.
OVID, MET. II. 88 - 97

IV. Cur, quem modò denique vidi,
Ne pereat, timeo? quae tanti causa timoris?
Excute virgineo conceptas pectore flammas,
Si potes, infelix. Si possem, sanior essem,
Sed trahit invitam nova vis, aliudque cupido,
Mens aliud suadet. Video meliora, proboque,
Deteriora sequor. Quid in hospite, regia virgo,
Ureris, et thalamos alieni concipis orbis?
OVID, MET. VII. 14 - 22.

V. Tristis abis; oculis abeuntem prosequor udis;
Et dixit tenui murmure lingua, 'Vale.'
Ut positum tetigi thalamo malè saucia lectum,
Acta est per lacrymas nox mihi, quanta fuit.
Ante oculos taurique truces, segetesque nefandae,
Ante meos oculos pervigil anguis erat.
Hinc amor, hinc timor est: ipsum timor auget amorem.
Mane erat, et thalamo cara recepta soror,
Disjectamque comas adversaque in ora jacentem
Invenit, et lacrymis omnia plena meis ;
Orat opem Minyis. Petit altera, et altera habebat.
OVID, HER. XII. 55 - 65.

[*Those who have not read Sallust and Ovid may substitute the following for II.–V.*]

VI. Caesar consilio ejus probato, etsi opinione trium legionum dejectus, ad duas redierat, tamen unum communis salutis auxilium in celeritate ponebat. Venit magnis itineribus in Nerviorum fines. Ibi ex captivis cognoscit, quae apud Ciceronem gerantur, quantoque in periculo res sit. Tum cuidam ex equitibus Gallis magnis praemiis persuadet, uti ad Ciceronem epistolam deferat. Hanc Graecis conscriptam litteris mittit, ne intercepta epistola nostra ab hostibus consilia cognoscantur.—CAESAR, B. G. V. 48.

Who is the Cicero mentioned?

VII. Tum sic Hyrtacides: Audite o mentibus aequis,
Aeneadae, neve haec nostris spectentur ab annis,
Quae ferimus. Rutuli somno vinoque soluti
Conticuere; locum insidiis conspeximus ipsi,
Qui patet in bivio portae, quae proxuma ponto;
Interrupti ignes, aterque ad sidera fumus
Erigitur; si fortuna permittitis uti,
Quaesitum Aenean et moenia Pallantea:
Mox hic cum spoliis, ingenti caede peracta,
Adfore cernetis. Nec nos via fallit euntis:
Vidimus obscuris primam sub vallibus urbem
Venatu adsiduo et totum cognovimus amnem.
VIRGIL, AEN. IX. 234–245.

VIII. Tum sic exspirans Accam, ex aequalibus unam,
Adloquitur; fida ante alias quae sola Camillae,
Quicum partiri curas; atque haec ita fatur:
Hactenus, Acca soror, potui; nunc volnus acerbum
Conficit, et tenebris nigrescunt omnia circum.
Effuge et haec Turno mandata novissima perfer:

Succedat pugnae Troianosque arceat urbe.
Iamque vale. Simul his dictis linquebat habenas,
Ad terram non sponte fluens.
VIRGIL, AEN. XI. 820-828.

IX. At regina, nova pugnae conterrita sorte,
Flebat, et ardentem generum moritura tenebat:
Turne, per has ego te lacrimas, per si quis Amatae
Tangit honos animum, — spes tu nunc una, senectae
Tu requies miserae; decus inperiumque Latini
Te penes; in te omnis domus inclinata recumbit —
Unum oro: desiste manum committere Teucris.
Qui te cumque manent isto certamine casus,
Et me, Turne, manent; simul haec invisa relinquam
Lumina, nec generum Aenean captiva videbo.
VIRGIL, AEN. XII. 54-63.

CICERO AND VIRGIL.

[*Translate two pieces from Cicero and two from Virgil, answering the questions on the pieces selected. Candidates for Course II. may take the piece from Caesar instead of one from Virgil.*

1. Quamobrem, Quirites, quoniam ad omnia pulvinaria supplicatio decreta est, celebratote illos dies cum conjugibus ac liberis vestris. Nam multi saepe honores dis immortalibus justi habiti sunt ac debiti, sed profecto justiores numquam: erepti enim estis ex crudelissimo ac miserrimo interitu, erepti sine caede, sine sanguine, sine exercitu, sine dimicatione; togati me uno togato duce et imperatore vicis-

tis. Etenim recordamini, Quirites, omnes civiles dissensiones, non solum eas, quas audistis, sed eas, quas vosmet ipsi meministis atque vidistis.

Celebratote; why not the shorter form, *celebrate? Erepti, togati;* explain allusions.

2. Vide quam non reformidem; vide quanta lux liberalitatis et sapientiae tuae mihi apud te dicenti oboriatur: quantum potero voce contendam, ut hoc populus Romanus exaudiat: suscepto bello, Caesar, gesto etiam ex parte magna, nulla vi coactus, judicio ac voluntate ad ea arma profectus sum, quae erant sumpta contra te. Apud quem igitur hoc dico? Nempe apud eum, qui cum hoc sciret, tamen me, ante quam vidit, rei publicae reddidit; qui ad me ex Aegypto litteras misit, ut essem idem qui fuissem.

Exaudiat, sciret, essem; explain mood and tense. *Bello, ea arma;* explain allusions.

3. An C. Falcidius, Q. Metellus, quos omnes honoris causa nomino, cum tribuni plebi fuissent, anno proximo legati esse potuerunt; in uno Gabinio sunt tam diligentes, qui in hoc bello, quod lege Gabinia geritur, in hoc imperatore atque exercitu, quem per vos ipse constituit, etiam praecipuo jure esse deberet? De quo legando consules spero ad senatum relaturos; qui si dubitabunt aut gravabuntur, ego me profiteor relaturum.

Honoris causa, relaturum; explain meaning. *Deberet;* explain mood and tense. *Lege Gabinia;* explain what is meant.

4. Candidus insuetum miratur limen Olympi
Sub pedibusque videt nubes et sidera Daphnis.
Ergo alacris silvas et cetera rura voluptas

Panaque pastoresque tenet Dryadasque puellas.
Nec lupus insidias pecori, nec retia cervis 60
Ulla dolum meditantur; amat bonus otia Daphnis.
Ipsi laetitia voces ad sidera iactant
Intonsi montes; ipsae iam carmina rupes,
Ipsa sonant arbusta: deus, deus ille, Menalca!
Sis bonus o felixque tuis! en quattuor aras:

Mark the metrical scheme of lines 59 and 60 with caesura, feet, and quantities.

5. Fama est Enceladi semiustum fulmine corpus
Urgueri mole hac, ingentemque insuper Aetnam
Inpositam ruptis flammam exspirare caminis; 580
Et fessum quotiens mutet latus, intremere omnem
Murmure Trinacriam, et caelum subtexere fumo.
Noctem illam tecti silvis inmania monstra
Perferimus, nec, quae sonitum det causa, videmus.
Nam neque erant astrorum ignes, nec lucidus aethra
Siderea polus, obscuro sed nubila caelo,
Et Lunam in nimbo nox intempesta tenebat.

Mark lines 585 and 586 as in 4.

6. Talibus Allecto dictis exarsit in iras.
At iuveni oranti subitus tremor occupat artus;
Deriguere oculi: tot Erinys sibilat hydris,
Tantaque se facies aperit; tum flammea torquens
Lumina cunctantem et quaerentem dicere plura
Reppulit, et geminos erexit crinibus anguis,
Verberaque insonuit, rabidoque haec addidit ore:
En ego victa situ, quam veri effeta senectus
Arma inter regum falsa formidine ludit;

7. Hac confirmata opinione timoris idoneum quendam hominem et callidum delegit Gallum ex his quos auxilii causa secum habebat. Huic magnis praemiis pollicitationibusque persuadet uti ad hostes transeat, et quid fieri velit edocet. Qui ubi pro perfuga ad eos venit, timorem Romanorum proponit, quibus angustiis ipse Caesar a Venetis prematur docet, neque longius abesse quin proxima nocte Sabinus clam ex castris exercitum educat et ad Caesarem auxili ferendi causa proficiscatur.

Educat; explain mood and tense. *Huic, auxili;* explain construction.

ARITHMETIC.

1. Find the simplest expression for $\dfrac{1}{3\frac{1}{8}} - \dfrac{2\frac{1}{4}}{9} + \dfrac{3\frac{5}{8}}{2} - \dfrac{\frac{4}{7}}{4\frac{4}{7}}$.

2. If 6 iron bars, 4 feet long, 3 inches broad, and 2 inches thick, weigh 288 pounds, find the weight of 15 bars, each $6\frac{1}{2}$ feet long, $2\frac{1}{2}$ inches broad, and $1\frac{1}{2}$ inches thick.

3. Find, by logarithms, the fourth power of $\dfrac{0.1397 \times 14}{4.379}$.

4. Find, by factoring, the greatest common divisor and least common multiple of 936 and 2925.

5. An alloy contains 325 parts of copper to 175 parts of zinc. How much of each metal is contained in 43 kilogrammes, 850 grammes of this alloy?

6. What sum of money, at 10 per cent compound interest, will amount to $8651.50 in three years?

ALGEBRA.

[*Write legibly and without crowding; give the whole work clearly; find all possible answers, and reduce them to their simplest forms. The shortest methods are preferred.*]

1. FIND the first four terms of $\left(\dfrac{a\sqrt{a}}{\sqrt[6]{b^5}} - \dfrac{\sqrt[4]{b}}{2a} \right)^{21}$, reducing each term of the result to its simplest form, and freeing it from negative and fractional exponents.

2. Reduce the following expression to its simplest form as a single fraction: —
$$\dfrac{x^2+y^2}{x^2-y^2} - \dfrac{1}{2}\left(\dfrac{x-y}{x+y} + \dfrac{x+y}{x-y} \right)\left(\dfrac{x-y}{x+y} - \dfrac{x+y}{x-y} \right);$$
and divide the result by $\dfrac{x^4 + 2x^2y^2 + y^4}{x^2 - 2xy + y^2}$, reducing the answer to its lowest terms and freeing it from parenthesis.

3. Two horsemen start at the same time, on the same road, from two places 15 miles apart. At the end of ten hours, the second horseman overtakes the first, and on comparing their rates, they find that there has been a difference of five minutes in the time of going every seven miles Find their rates, and the distances they have gone.

4. Solve the equation
$$\dfrac{2a - x - 19b}{ax - 2bx} = \dfrac{a - 2b - x}{a^2 - 4b^2} - \dfrac{5b - x}{ax + 2bx}.$$

5. Find the greatest common divisor and the least common multiple of
$$2x^3 + x^2 - 5x + 2 \text{ and } 4x^3 - 4x^2 - 5x + 3.$$

ADVANCED ALGEBRA.

[*Give the whole work neatly and clearly; find all possible answers, and reduce them to their simplest forms. The shortest methods are preferred.*]

1. The sum of eight numbers in arithmetical progression is 44; and the mean proportional between the 5th and 6th terms of the series is 2. Find the eight numbers.

2. Solve the equations
$$x^5 - y^5 = 275,\ x - y = 5.$$

3. Prove the general rule of arithmetic and algebra for finding the *greatest common measure* of two quantities (without considering the modifications which are necessary when the rule is applied to polynomials).

4. To find the sum of a geometric progression when the first term, the ratio, and the number of terms are given.

To find the sum of an infinite decreasing geometric progression.

5. Find the cube root of
$8x^6 - 36x^4y + 102x^2y^2 - 171y^3 + 204x^{-2}y^4 - 144x^{-4}y^5 + 64x^{-6}y^6.$

PLANE GEOMETRY.

[*Number your answers carefully, but do not restate any proposition which is stated in full on this paper.*]

1. A PERPENDICULAR is the shortest line which can be drawn from a point to a line. Prove.

2. If of two angles of a triangle the first is greater than the second, the side opposite the first is greater than the side opposite the second. State and prove the converse.

3. Define similar polygons. If two triangles have an angle in the one equal to an angle in the other and the sides about these angles proportional, the triangles are similar. Prove.

4. If in two similar triangles a side of one is three times as long as the homologous side of the other, what is the ratio of the areas? State and prove the proposition on which your answer depends.

5. If from a point without a circle a tangent and a secant be drawn, the tangent will be a mean proportional between the whole secant and its external segment. Prove.

6. If two circles intersect and from any point of their common chord extended a tangent be drawn to each circle, these tangents will be equal. Prove.

7. The areas of circles are to each other in what ratio? Why? The radius of one circle is two feet, of another is four feet; the area of the second is how many times the area of the first? Confirm your answer by computing and comparing the two areas.

SOLID GEOMETRY.

[*N. B. — Give as full proofs as you can for the theorems stated in Nos.* 1, 2, 3, *and* 5.]

1 The angle of two planes, which cut each other, is measured by the angle of two lines drawn perpendicular to

the common intersection of the two planes, at the same point, one in one of the planes, and one in the other.

2. If a solid angle is formed by three plane angles, the sum of either two of these angles is greater than the third.

3. A truncated triangular prism is equivalent to the sum of three pyramids, which have for their common base the base of the prism, and for their vertices the three vertices of the inclined section.

4. Two similar pyramids have altitudes of six and eight feet. Give the ratio of their surfaces, and also of their volumes.

5. The surface of a spherical triangle is measured by the excess of the sum of its three angles over two right angles.

ANALYTIC GEOMETRY.

1. Prove the formula for the tangent of the angle between two straight lines whose equations are given.

2. Find the equation of a straight line passing through the point, (4, 3) and perpendicular to the line whose equation is $3x - 2y = 12$.

3. Explain fully the method of finding the points of intersection of two curves given by their equations.

4. From the rectangular equation of an ellipse, deduce the polar equation, the centre being the pole.

5. Find the equation of a tangent to an hyperbola at any point of the curve. Give the reasoning in full.

TRIGONOMETRY

1. OBTAIN formulas that will enable you to compute the cosine and tangent of an angle when the sine is given.

Example.— Given sin 30° = .5. Compute cos 30° and tan 30° correct to two decimal places.

2. What are the sine, cosine, and tangent of 150°? of 210°?

3. Write the formulas for sin $(x + y)$ and sin $(x - y)$, and from them obtain a formula for sin A + sin B in terms of $\frac{1}{2}(A + B)$ and $\frac{1}{2}(A - B)$.

4. Having given two sides of a triangle and the included angle, write the formula by which the remaining angles are obtained, and state and prove the theorem from which it is derived.

5. At a point 103.8 feet from the foot of a flag-pole the angle of elevation of its top is found to be 62° 34'. Required the height of the pole.

6. Two sides of a triangle are 25 feet and 40 feet, respectively, and the angle opposite the second side is 120.° Solve the triangle.

ENGLISH COMPOSITION.

You are required to write a short English composition, correct in spelling, punctuation, division by paragraphs, and expression. You are recommended to arrange what you have to say before beginning to write, to pay more attention to quality than to quantity of work, and to make a fair copy from a rough draft.

One of the following subjects must be taken:—

    I. The Battle of Philippi.

    II. The Meeting of Saladin and Richard.

    III. The Moral of Rip Van Winkle.

FRENCH.

[*The translation should be in* good *English.* *French idioms should be rendered by corresponding English ones, whenever it is possible. If you fear you are going too far from the literal sense, the word-for-word meaning may be enclosed in parentheses. Leave blanks for the words you do not know.*]

TRANSLATE:—

IL était une fois un homme qui avait de belles maisons à la ville et à la campagne, de la vaisselle d'or et d'argent, des meubles en broderie, et des carrosses tout dorés. Mais, par malheur, cet homme avait la barbe bleue; cela le *rendait* si laid, qu'il n'était femme ni fille qui ne s'enfuît devant lui. Une de ses voisines avait deux filles parfaitement belles. Il lui en demanda une en mariage, en lui laissant le choix de celle qu'elle voulait lui donner. Elles n'en voulaient point toutes deux, et se le renvoyaient l'une à l'autre, ne pouvant se résoudre à prendre un homme qui eût la barbe bleue. Ce qui les dégoûta encore, c'est qu'il avait déjà épousé plusieurs femmes, et qu'on ne savait ce que ces femmes étaient devenues. La Barbe Bleue, pour faire connaissance, les mena, avec quelques jeunes gens du voisinage, à une de ses maisons de campagne, où on demeura huit jours entiers. Ce n'étaient que promenades, que parties de chasse et de pêche, que danses et festins : on ne dormait point, et on passait toute la nuit à se faire des malices les uns aux autres ; enfin, tout *alla* si bien, que la cadette commença à trouver que le maitre du logis n'avait plus la barbe si bleue, et que c'était un fort honnête homme. Dès qu'on fut de retour à la ville le mariage se *conclut*. Au bout d'un mois, la Barbe Bleue dit à sa femme qu'il était obligé de faire un voyage en province, de six semaines au moins, pour une affaire de conséquence ; qu'il la priait de se bien divertir pendant son absence ; qu'elle fît venir ses amies, qu'elles les menât à la campagne si elle voulait ; que partout elle *fît* bonne chère. — PERRAULT.

[*Any one may pass who does well the above translation, even should he not answer a single question in Grammar. The following questions are set to give a better chance to those whose translation may not be quite satisfactory, and who by answering them will show that they have a fair knowledge of elementary grammar. Even should the candidate feel that his translation is good enough, it will be well for him, if he can, to answer the last question (No. 5), and thereby make sure of a high mark for this examination.*]

2. Give in full the tense of the italicized verbs in the above.

3. Give the principal tenses of *marcher, savoir, lire,*

mentir (thus : INF., *être;* PRES. PART., *étant;* PAST PART., *été ;* IND. PRES., *je suis;* PRET., *je fus*).

4. Say what you know about the position of adjectives.

5. TRANSLATE INTO FRENCH : —

(*a*) Have you ever read Perrault's Fairy Tales? (*b*) They are very pretty and easy. (*c*) Everybody in France has read them. (*d*) Perrault was born in Paris, Jan. 12, 1628.

GERMAN.

[*State how long you have studied German, under whose instruction, and what German book or books you have read.*]

I.

TRANSLATE INTO GERMAN : —

1. On[1] this hill[2] were many[3] houses[4] which were beautiful[5] and large, but the merchant[6] did not wish to purchase[7] them because[8] they were exposed[9] to the North-wind.[10]

2. He has not-yet[11] been at home, but his brother[12] will be in this house to-morrow.[13]

3. At two o'clock[14] yesterday[15] the soldiers[16] of the enemy[17] were in their trenches,[18] but at three o'clock we had already[19] defeated[20] them and driven-them-back[21] into the city.[22]

4. Your father [23] is now in the garden, [24] but he will soon [25] go into his room. [26]

5. He became [27] ill [28] while [29] he was (still) here, and, since [30] he has no money, [31] I shall give [32] him some if he remains [33] much longer.

[1] auf. [2] der Hügel. [3] viele. [4] das Haus. [5] schön. [6] der Kaufmann. [7] kaufen. [8] weil. [9] aufsetzen (separate verb). [10] der Nordwind (use dative). [11] noch nicht. [12] Bruder. [13] morgen. [14] Uhr. [15] gestern. [16] der Soldat. [17] der Feind. [18] die Schanze. [19] schon. [20] schlagen. [21] zurücktreiben (separable verb, irregular or strong conjunction). [22] die Stadt. [23] Vater. [24] der Garten. [25] bald. [26] das Zimmer. [27] werden. [28] krank. [29] während. [30] da. [31] das Geld. [32] geben. [33] bleiben.

II.

1. You are an amiable [1] flatterer, [2] and your good opinion [3] of [4] us shows [5] that the Germans [6] and the British [7] are capable [8] of a true sympathy [9] (gen.).

2. Both-the [10] great nations [11] will derive [12] ample [13] profit [14] from [15] the friendly [16] fostering [17] of a sincere [18] understanding. [19]

3. Oh! that is the blackest [20] calumny! [21] said he; she depicts [22] me as the greatest epicure. [23]

4. Translate: Whose house is that? Which book have you? The man whose book I have is not at home.

[1] Liebenswürdig. [2] Schmeichler. [3] Meinung, f. [4] von. [5] zeigen. [6] der Deutsche. [7] der Britte. [8] fähig. [9] Sympathie. [10] both-the, beide. [11] Nation, f. [12] ziehen. [13] reichlich. [14] Gewinn, m. [15] aus. [16] freundschaftlich. [17] Pflege, f. [18] aufrichtig. [19] Verständniß, n. [20] schwarz. [21] Verleumbung. [22] schildern. [23] Epikuräer.

III.

3. Translate:—

Kosciusko, der edle Pole, wollte einst einem Geistlichen zu Solothurn einige Flaschen guten Weines übersenden. Er wählte dazu einen jungen Mann mit Namen Zeltner, und überließ ihm für die Reise sein eigenes Reitpferd. Als Zeltner zurückkam, sagte er: „Mein Feldherr! ich werde Ihr Pferd nicht wieder reiten, wenn Sie mir nicht zugleich Ihre Börse[1] mitgeben." „Wie meinst Du das?" fragte Kosciusko. Zeltner antwortete: „Sobald ein armer Mann auf der Straße den Hut abnahm und um ein Almosen bat, stand das Pferd augenblicklich still und ging nicht eher von der Stelle, als bis der Bettler etwas empfangen hatte, und als mir endlich das Geld ausging,[2] wußte ich das Pferd nur dadurch zufrieden zu stellen und vorwärts zu bringen, daß ich that, als ob ich den Bittenden etwas gäbe.

[1] purse. [2] was gone.

CHEMISTRY AND PHYSICS.

CHEMISTRY.

1. Describe experiments to show what happens when a candle burns.

2. Describe the preparation of carbonic acid gas.

3. Describe the element phosphorus and its uses.

4. Describe Davy's safety-lamp.

PHYSICS.

5. Describe the barometer.

6. Describe an experiment to show a metal expands on heating.

7. Define the terms specific heat and latent heat.

8. Describe the gold-leaf electroscope.

9. Describe the action of an electric current on water.

PHYSICS AND ASTRONOMY.

1. What is the principle of the Bramah press?

2. How is the rate of vibration corresponding to a given musical note determined?

3. What is specific heat?

4. What is the principle of the electric telegraph?

5. What is the action of points on electricity?

6. How is the velocity of light determined?

7. Explain the phases of the moon.

8. How is the moon's distance determined?

9. What is known of the sun's chemical constitution, and by what means?

BOTANY.

1. Describe the germination of a *cherry* and of a *horse-chesnut*.

2. How does a tuber differ from a fleshy root?

3. Describe, or give an outline sketch of, a palmately veined, once-compound leaf.

4. Describe the fruits of the *mulberry* and *apple*.

5. Is the flower given for examination regular and perfect?

6. What kind of a flower-cluster does the plant given have?

7. Is the calyx of the given flower free from, or adherent to, the ovary?

8. Describe the arrangement, venation, and shape of the leaves; and write answers to the following questions: —

(*a*) Is the plant *exogenous* or *endogenous?* (Give your reasons for the answer.)

(*b*) Upon what are the stamens inserted?

(*c*) Is the flower *monopetalous, polypetalous,* or *apetalous?*

APPENDIX.

HARVARD COLLEGE,
Cambridge, Mass.
1877–78.

REQUISITIONS FOR ADMISSION.

Candidates for admission to College in 1878, 1879, and 1880 will be examined in either of the two following methods at their option, the second method being that already in use:—

METHOD 1.

This method prescribes for the candidate a minimum requisition in every study, and a maximum requisition in two, selected from four principal studies at his option.

Every candidate will be required to pass a satisfactory examination in the following eleven subjects:—

1, 2. *Latin.* (1) Caesar, Gallic War I.-IV., with questions on the subject-matter, and on construction and grammatical forms; — Virgil, Eclogues and Aeneid I.-IV., with questions on the subject-matter and on prosody; — (2) the translation at sight of easy Latin prose; — and the translation into Latin of simple English sentences, to test the candidate's practical knowledge of grammar. (The passages set for translation at sight will be suited to the proficiency of those who have studied the prescribed books; and candidates will be supplied with a vocabulary of such words as they cannot reasonably be expected to know.)

3, 4. *Greek.* For five years, beginning with 1878, every candidate may offer himself for examination in either of the two following requirements: —

A. (3) The translation at sight of easy passages of Xenophon (suited to the proficiency of those who have studied the first 111 pages of Goodwin's Greek Reader or the first four books of the Anabasis), with a vocabulary of the less usual words; — and (4) the translation into Greek of simple sentences, such as those in the first 51 lessons of White's "First Lessons in Greek," to test the candidate's practical knowledge of grammar.

B. (3) The first 111 pages of Goodwin's Greek Reader (or Xenophon's Anabasis I.-IV.) and Iliad I. and II. vss. 1-493, with questions on the subject-matter, and on construction and grammatical forms; — and (4) the translation into Greek of simple sentences, such as those in the first 51 lessons of White's "First Lessons in Greek," to test the candidate's practical knowledge of grammar.

---

5. *Ancient History and Geography.* Greek History to the death of Alexander; Roman History to the death of Commodus. Smith's smaller histories of Greece and Rome will serve to indicate the amount of knowledge demanded.

6' 7, 8. *Mathematics.* (6) Arithmetic (Prime and Composite Numbers; Factors, Divisors, and Multiples; Proportion; Decimals, including Percentage, Simple and Compound Interest, and Discount, but not the technical parts of Commercial Arithmetic; Compound Numbers and the Metric System, the necessary tables and data being given on the papers; Square roots); — (7) Algebra, through quadratic equations; — (8) Plane Geometry, as much as is contained in the first thirteen chapters of Peirce's Geometry.

9. *Physics.* Rolfe and Gillett's Manual, or parts I. and II. of Arnott's Physics.

10. *English Composition.* The candidate will be required to write a short English composition, correct in spelling, punctuation, grammar, division by paragraphs, and expression. The subject for 1878 will be taken from one of the following works: Shakspeare's Macbeth, Coriolanus, or As You Like It; Irving's Sketch Book; Scott's Kenilworth, or Lady of the Lake.

11. *French* or *German.* The translation at sight of easy French prose, or of easy German prose, if the candidate prefer to offer German. Proficiency in elementary grammar, a good pronunciation, or facility in speaking, will be accepted as an offset for some deficiency in translation. There will be no required examination in pronunciation, but it is recommended that attention be given to pronunciation from the outset. Candidates who offer German in place of French will be required to study French in place of German during the Freshman year.

Every candidate will also be required to pass a satisfactory examination in at least two of the four following groups of subjects:—

I. LATIN.

1, 2. (1) Cicero's orations against Catiline, and Virgil's Aeneid V.–IX., with questions as in the minimum requirement stated in subjects 1 and 2 above;—(2) translation at sight of average passages from Cicero's orations;—and Latin Composition.

II. GREEK.

1, 2. (1) The translation at sight of average passages from Herodotus, with such help in notes as should be needed by those who have studied the Herodotus in Goodwin's Reader (pp. 112 – 191) or pp. 201 – 280 of Blakesley's text of Herodotus; — (2) simple Greek prose composition; — and *either* the translation at sight of average passages from the Iliad (the candidate being supplied with a vocabulary of the less usual words), *or* Iliad, I., II. vss. 1 – 493, and III., with questions on the passages set for translation.

[Candidates who are examined in Iliad I. and II., in the course marked *B* under 3, 4 (above), will be examined in Iliad III., IV., and VI., in place of Iliad I.-III.]

III. MATHEMATICS.

1, 2. (1) Logarithms and Plane Trigonometry; — **and** (2) Solid Geometry.

IV. PHYSICAL AND NATURAL SCIENCE.

1, 2. (1) Arnott's Physics, as far as Part IV., Section III., and (2) *either* Chemistry (Nichols' Abridgment of Eliot and Storer's Manual) *or* Botany (Gray's "How Plants Grow," with analysis of simple specimens).

In each of the four departments, Latin, Greek, Mathematics, and Physical and Natural Science, two courses will be carried on in the Freshman year; an ordinary course adapted to the state of preparation of those who are admitted with the minimum requirements, stated in subjects 1 – 9 above; and an advanced course for those who are admitted with the maximum requirement, last given above.

METHOD 2.

Candidates will be examined as heretofore in one of the two following courses of study, each embracing sixteen subjects:—

COURSE I.

1. *Latin Grammar* (including Prosody).
2. *Latin Composition* and *Latin at sight*. This will include the translation at sight of some passage in prose not included in the following requisitions.
3. *Caesar, Sallust* and *Ovid*. Caesar, Gallic War, Books I.-IV.; Sallust, Catiline; Ovid, four thousand lines.
4. *Cicero* and *Virgil*. Cicero, eight orations and the Cato Major; Virgil, Eclogues, and the Aeneid, Books I.-VI.

---

5. *Greek Grammar* (including metres).
6. *Greek Composition* (with the accents).
7. *Greek Prose.* Goodwin and Allen's Greek Reader; or Xenophon's Anabasis, Books I.-IV., and the seventh book of Herodotus.
8. *Greek Poetry.* Homer's Iliad, Books I.-III., omitting the catalogue of ships.
9. *Arithmetic* (including the metric system of weights and measures, and the use and rudiments of the theory of logarithms). The examples requiring the use of logarithms at the examination will be adapted to a four-place table.
10. *Algebra* (through quadratic equations).
11. *Plane Geometry* (as much as is contained in the first thirteen chapters of Peirce's Geometry).

---

12. *Ancient History and Geography.* Greek History to the death of Alexander; Roman History to the death of Commodus. Smith's smaller histories of Greece and Rome

will serve to indicate the amount of knowledge demanded in history.

13. *Modern and Physical Geography.* The following works will serve to indicate the amount of knowledge demanded in this subject: — in modern geography, Guyot's Common School Geography, or Miss Hall's Our World, No. 2; in physical geography, Guyot's Physical Geography, Parts II. III., or Warren's Physical Geography, the first forty-nine pages.

14. *English Composition.* Each candidate will be required to write a short piece of English, correct in spelling, punctuation, grammar, division by paragraphs, and expression. The subject for 1878 will be taken from one of the following works: Shakspere's Macbeth, Coriolanus, or As You Like It; Irving's Sketch Book; Scott's Kenilworth, or Lady of the Lake.

15. *French* or *German.* The translation at sight of easy French prose, or of easy German prose if the candidate prefer to offer German. Proficiency in elementary grammar, a good pronunciation, or facility in speaking, will be accepted as an offset for some deficiency in translation. There will be no required examination in pronunciation, but it is recommended that attention be given to pronunciation from the outset. Candidates who offer German in place of French will be required to study French in place of German during the Freshman year.

16. *Physical Science.* One of the three following subjects, the selection of the subject being left to the candidate: —

1. *Elementary Botany.*
2. *Rudiments of Physics and of Chemistry.*
3. *Rudiments of Physics and of Descriptive Astronomy.*

The following books will serve to indicate the nature and extent of this requisition: —

In Botany, Gray's How Plants Grow.
In Physics, Balfour Stewart's Primer of Physics.
In Chemistry, Roscoe's Primer of Chemistry.
In Astronomy, Rolfe and Gillet's Handbook of the Stars (first 124 pages).

Candidates who offer Botany will be required to give evidence that they can analyze simple specimens; and those who offer Physics or Chemistry, that they can perform simple experiments like those described in the Primers referred to above.

COURSE II.

1. *Latin Grammar* (including Prosody).
2. *Latin Authors.* Caesar, Gallic War, Books I. and II.; Cicero, six orations and the Cato Major; Virgil, Aeneid, Books I.–VI.
3. *Greek Grammar* (including metres).
4. *Greek Authors.* Goodwin and Allen's Greek Reader, first 111 pages, or Xenophon's Anabasis, Books I.–IV.; Homer's Iliad, Books I. and II., omitting the catalogue of ships.
5. *Arithmetic.* This requisition is the same as No. 9 of Course I.
6. *Elementary Algebra.* This requisition is the same as No. 10 of Course I.
7. *Advanced Algebra.* This subject, with the preceding, is regarded as embracing as much Algebra as is contained in the advanced text-books, such as the *larger* Algebras of Todhunter, Loomis, Greenleaf, etc.
8. *Plane Geometry.* This requisition is the same as No. 11 of Course I.
9. *Solid Geometry* (as much as is contained in Peirce's Geometry).
10. *Plane Trigonometry* (by the Analytic Method; as much as is contained in the first six chapters of Peirce's

Trigonometry, or in the large print of the first eight chapters of Chauvenet's Trigonometry).

11. *The Elements of Plane Analytic Geometry* (as much as is contained in Peck's Analytic Geometry, pages 1–151, omitting articles 40–43, 54, 57–61, 72, 74–76, and the more difficult problems).

12–16. These requisitions are the same as Nos. 12–16 Course I.

Candidates who enter College on Course II. substitute elective studies, amounting to four exercises a week, either in Mathematics or in some other subject, for the Mathematics of the Freshman year.

No particular text-book in Grammar is required; but either Allen's or Harkness's Latin Grammar, and either Goodwin's or Hadley's Elementary Greek Grammar, will serve to indicate the nature and amount of the grammatical knowledge demanded.

In Latin the following pronunciation is recommended: $\bar{a}$ as in *father*, $\breve{a}$ the same sound but shorter; $\bar{e}$ like $\hat{e}$ in *fête*, $\breve{e}$ as in *set* ; $\bar{\imath}$ as in *machine*, $\breve{\imath}$ as in *sit* ; $\bar{o}$ as in *hole*, $\breve{o}$ as in *nor*; $\bar{u}$ as in *rude*, $\breve{u}$ as in *put*; *j* like *y* in *year* ; *c* and *g* like Greek κ and γ.

Instructors are requested to teach their pupils in pronouncing Greek to use the *Greek accents*, and to give (for example) α the sound of *a* in *father*, η that of *a* in *fate*, ι that of *i* in *machine*, etc.

It is earnestly recommended that the requisitions in Latin and Greek Authors be accurately complied with; real equivalents, however, will be accepted, as, for example, under Method 2, Caesar's Gallic War, Books V. and VI., in place of Sallust's Catiline ; two additional orations of Cicero in place of the Cato Major; the seventh book of the Aeneid in place of the Eclogues; the last five books of the Aeneid in place of Ovid.

No partial substitutions or interchanges between Courses I. and II. of Method 2 will be allowed; but candidates can present themselves on *both* courses, or on one course with additional subjects belonging to the other.

No partial substitution or interchange between Method 1 and Method 2 will be allowed.

If a candidate passes with high credit in any one or more of the following groups of subjects (in Method I), namely,—

Prescribed Classics (subjects 1, 2, 3, 4),

Prescribed Mathematics (subjects 6, 7, 8),

Prescribed and Elective Latin (subjects 1, 2, and Group I.),

Prescribed and Elective Greek (subjects 3, 4, and Group II.),

Prescribed and Elective Mathematics (subjects 6, 7, 8, and Group III.),

Prescribed and Elective Physics (subject 9 and Group IV.),— such distinction will be noted on his certificate of admission.

A set of recent examination papers will be sent to any address, on application to JAMES W. HARRIS, Secretary.

PRELIMINARY EXAMINATIONS.

[*These examinations are held only in June.*]

Any candidate for admission to College may, at his option, pass the entire examination at one time; or he may pass a preliminary examination, on a part of the course, and be examined on the remaining subjects in some subsequent year (not the same year). But no candidate will be admitted to examination on a part of any subject; and no account will be made of, nor certificate be given for, the preliminary examination, unless the candidate has passed satisfactorily in at least five subjects.

Candidates will be received for the preliminary examination only in subjects in which their teachers give them certificates as prepared.

These certificates must be in the hands of Mr. J. W. HARRIS, Secretary, before June 21, 1878; *and no person will be received at the preliminary examination in* 1878 *who has not presented his certificate before that day.*

The preliminary examination (in Method 1) will be limited to the prescribed subjects, and will not be allowed to include any of the advanced or elective subjects; and certificates will be granted for any five prescribed subjects.

Teachers are advised and requested not to present their pupils for preliminary examination on the *whole* minimum in either Greek, Latin, or Mathematics, unless they mean to prepare them in the advanced subjects in the same study.

Candidates are encouraged to offer *more than two* of the four elective groups of Method 1 at the final examination, provided they have time to do the necessary work in a thorough manner.

OPTIONAL EXAMINATIONS.

A principal aim in providing these optional examinations is to encourage teachers to carry the studies of their brighter and more diligent pupils beyond the bare requisitions for admission, in whatever direction taste or opportunity may suggest. Full employment may thus be secured for the most capable student until he is thought mature enough to enter College, while his greater progress in school will make his College course more profitable, by enabling him to take up his studies at a more advanced stage, or to give more time to the studies of his choice.

THE CLASSICS.

Candidates who present themselves upon Course I. will be at liberty to offer themselves for additional examination upon one or both of the following classical courses:—

(1) *Latin.*

Livy, two books.
Horace, Odes and Epodes.
The translation at sight of a passage from the philosophical works of Cicero.
The retranslation of the English of a similar passage into Latin.

(2) *Greek.*

Plato, Apology and Crito.
Homer, Iliad, Books IV.-VIII., or Odyssey, Books IV., IX.-XII.
Euripides, Alcestis; or Homer, Odyssey, Books V.-VII.
The translation at sight of a passage from the works of Xenophon.
Translation from English into Greek.

No candidate will be *required* to present himself at these examinations; but those who pass them with high credit, in addition to the other classical examinations of Course I. above, will be admitted, immediately on entering College, to advanced sections in Latin and Greek, or to elective studies either in the Classics or in other departments, in place of the Freshman studies thus anticipated. All those who wish to attain distinction in classical studies, or to graduate with classical honors, are advised to pass these examinations on entering.

MATHEMATICS.

An advanced section in Mathematics is formed in the Freshman class, consisting of those who receive high marks in the Mathematical subjects required in Method 2, Course I., and also pass a creditable examination in Advanced Algebra, Plane Trigonometry, and Solid Geometry, as required in Method 2, Course II. Candidates who do not present Solid Geometry may, however, be admitted to the section, on condition of making up that subject.

The design of this section is not to anticipate the subjects taught in the elective courses, but to afford to students of good ability and preparation a more valuable training and a greater practical command of their Mathematics than they can gain in the ordinary course, and to facilitate the taking of Second-Year Honors in Mathematics at the end of the Sophomore year. It is recommended to those who desire to attain special distinction in Mathematics or Physical Science, and to all who would turn their Mathematical study to the best account.

☞ Freshmen will hereafter be allowed to take elective studies in place of their Mathematics, if they anticipate *all* the Mathematical subjects of the Freshman year, but not otherwise.

PHYSICS.

Candidates who pass a satisfactory examination at admission upon the course in Physics of the Freshman year may substitute for that course an elective study.

GERMAN.

Candidates for admission who present French may offer themselves for examination also in German Grammar and the translation of simple German prose; upon passing such examination with credit they will be allowed to substitute some elective course or courses in place of the Freshman course in German.

PRESCRIBED STUDIES OF THE SOPHOMORE AND JUNIOR YEARS.

Candidates for admission to the Freshman Class who are prepared to pass a creditable examination upon any of the prescribed studies of the Sophomore and Junior years may pass such examination at the beginning of the Freshman Year, instead of at the beginning of the year in which the study is pursued, and thereby relieve themselves from attendance at the exercises in that study in College.

ADVANCED STANDING.

A candidate may be admitted to the Sophomore, Junior, or Senior class, if he appear on examination to be well versed in the following studies: —

1. In the studies required for admission to the Freshman Class.

2. In all the prescribed studies already pursued by the class for which he is offered; and in as many elective studies as he would have pursued if he had entered at the beginning of the course, including, if he is offered in Course II., elective studies substituted for the Mathematics of the Freshman year.

All candidates for admission to advanced standing must be examined at the times of the regular examinations for admission, and in conformity with the following rules: —

1. All candidates for admission to advanced standing must be examined for admission to the Freshman Class; for this examination they may offer themselves at either the first or the second examination.

2. The examination on the studies of the Freshman, Sophomore, and Junior years is held *only in the autumn*, at the time of the regular examination for admission.

3. All candidates for admission to advanced standing will assemble on Tuesday, September 24, 1878, at 8 o'clock A. M., in Harvard Hall.

In the case of *graduates* of other colleges who seek admission to Harvard College, the examination will be directed to ascertaining whether their previous course of study has been sufficiently extensive, and their proficiency in it sufficiently great, to fit them to join the class for which they offer themselves; a minute acquaintance with all the ground they have previously gone over not being essential. Such candidates should bring evidence of their standing at the colleges where they received their degree.

TIMES AND PLACES OF EXAMINATION.

Two regular examinations for admission to the *Freshman* Class are held each year, one at the beginning of the summer vacation, and the other at the beginning of the academic year in the autumn.

In 1878 the first examination will be held in Cambridge, and in Cincinnati, Ohio, on Thursday, Friday, and Saturday, June 27, 28, and 29; and the second in Cambridge only, on Wednesday, Thursday, and Friday, September 25, 26, and 27. For each examination attendance on the three days is required. Candidates will assemble punctually at 8 o'clock A. M., — in Cambridge, in Harvard Hall; in Cincinnati, at some place to be announced in the daily papers of that city.

Candidates who propose to be examined in Cincinnati are requested to send their names to the Secretary of the University before June 15.

Persons who do not intend to enter College will be admitted, on payment of a fee of ten dollars, to the examination at Cincinnati; and if successful will receive certificates to that effect.

The *Optional Examinations* will be held at the time of the second examination for admission; those in Mathematics, at the first examination also.

The *Preliminary Examinations* will be held at the time of the first examination only.

No person will be examined for admission to College at any other time than those above specified.

ADMISSION WITHOUT MATRICULATION.

The *elective* courses of study are open to persons *not less than twenty-one years of age,* who satisfy the Faculty, without passing the usual examination for admission, that they are fitted for the courses they select. They will receive no *degree;* but at the end of each academic year they will receive a certificate of proficiency in those courses which they pursue during the year, and in which they attain not less than seventy-five per cent.

Persons who wish to avail themselves of this provision must present themselves at Harvard Hall, September 26, 1878, at 10 A. M., with the necessary testimonials as to age, character, and fitness to attend the courses they wish to pursue.

TESTIMONIALS AND BOND.

All candidates for admission are required, *at the time of the final examination for admission* to the Freshman Class, to produce certificates of good moral character; and students from other colleges are required to bring certificates from those colleges of honorable dismission.

Every candidate, if admitted, must furnish to the Bursar a bond for *four hundred dollars,* executed by two bondsmen, one of them a citizen of Massachusetts, as security for the payment of College dues. If the student prefer, however, he may in place of the bond make a deposit with the Bursar for the same purpose. A similar bond for *two hundred dollars*, or a deposit, will be required of unmatriculated students.

ANTICIPATION OF PRESCRIBED STUDIES.

As the prescriped studies of the Sophomore and Junior years are of an elementary character, students who wish to be relieved from attendance at College exercises in one or more of them will be so excused, if they pass a satisfactory examination in such study or studies at the time of their examination for admission, or at the beginning of the year in which the study is pursued in College. Preparation for such examination can often be made while the student is preparing for College, or in the long vacation, and time may be thus gained for higher courses of study.

For information concerning the College not contained in this circular the Catalogue should be consulted.

The College itself no longer issues au annual catalogue. "The Harvard University Catalogue" (price in cloth, 75 cents; in paper, 50 cents) is published by Mr. CHARLES W. SEVER, bookseller, Cambridge, Mass., to whom orders for it may be addressed.

Circulars giving information about the professional schools of the University, the Scientific School, and the Bussey Institution, may be obtained on application to J. W. HARRIS, Secretary, Cambridge, Mass.

ALLEN & GREENOUGH'S LATIN GRAMMAR.

The first edition was published in 1872, and was widely adopted, reaching a sale of *over 30,000 copies*. In 1877, the editors completed a revision, which has made it *virtually a new work* while *retaining all the important features of the old* Attention is invited to the following merits of the book:

1. *The Supplementary and Marginal Notes on Etymology, Comparative Philology, and the meaning of forms.* In this department it is believed to be more full and complete than any other school text-book, and to embody the most advance views of comparative philologists.

2. *Numerous Introductory Notes in the Syntax, giving a brief view of the theo of constructions.* These Notes are original contributions to the discussion of th topics of which they treat; they illustrate and greatly simplify syntactical construction, and are not based upon *abstract theory*, or "metaphysics of the subjunctive,' but upon *linguistic science*, or upon the actual historical development of language from its simplest forms.

3. *Treatment of Special Topics of Syntax.* On these points we invite comparison with other school grammars on the score of simplicity and clearness.

4. *The extended, and often complete, lists of forms and constructions.*

5. *Tabulated examples of peculiar or idiomatic use.*

6. *The full and clear treatment of Rhythm and Versification*, correspondin with the latest and best authorities on the subject.

7. *The unusual brevity attained without sacrifice of completeness or clearness.*

This Grammar expresses *the results of independent study of the best origina sources*. It has been *strictly subordinated to the uses of the class-room* through th advice and aid of several of our most experienced teachers. The rapid adoptio of this Grammar in *over three-fourths of the leading colleges and preparatory school of the country* is believed to be a full guaranty for its adaptation to the purposes o instruction.

ALLEN & GREENOUGH'S LATIN COURSE.

Leighton's Latin Lessons (designed to accompany the Grammar).

Six Weeks' Preparation for Reading Cæsar (designed to accompan the Grammar, and also to prepare pupils for reading at sight).

Allen & Greenough's Cæsar,* Cicero,* Virgil,* Ovid,* Sallust, Cato Major, Latin Composition, Preparatory Latin Course, No. II (with Vocabulary), containing four books of Cæsar's Gallic War, and eight Orations of Cicero.

Keep's Parallel Rules of Greek and Latin Syntax.

Allen's Latin Reader. Selections from Cæsar, Curtius, Nepos, Sallust, Ovid, Virgil, Plautus, Terence, Cicero, Pliny, and Tacitus. With Vocabulary.

Crowell & Richardson's Brief History of Roman Literature.

Crowell's Selections from the Less Known Latin Poets.

Stickney's De Natura Deorum.

Allen's (F. D.) Remnants of Early Latin.

Leighton's Critical History of Cicero's Letters.

Leighton's Elementary Treatise on Latin Orthography.

White's Junior Student's Latin-English Lexicons.

* With or without Vocabulary.

A Full Descriptive Catalogue mailed on Application.

GINN & HEATH, Publishers, Boston, New York, and Chicago.

GREEK TEXT-BOOKS.

Goodwin's Greek Grammar. Revised and Enlarged Edition for 1879.
It states general principles clearly and distinctly, with special regard to those who an preparing for college.
It excludes all detail which belongs to a book of reference, and admits whatever wil aid a pupil in mastering the great principles of Greek Grammar.
The sections on the Syntax of the Verb are generally condensed from the author' larger work on the Greek Moods and Tenses. (See below.)
It contains a brief statement of the author's new classification of conditional sentences with its application to relative and temporal sentences, which appears now for th first time in an elementary form.
It contains a catalogue of irregular verbs, constructed entirely with reference to th wants of beginners.
All forms are excluded (with a few exceptions) which are not found in the strictl classic Greek before Aristotle.

White's First Lessons in Greek. Prepared to accompany Goodwin' Greek Grammar.
A series of Greek-English and English-Greek Exercises, *taken mainly from th first four books of Xenophon's Anabasis*, with Additional Exercises on Forms, an complete Vocabularies. The Lessons are carefully graded, and do not follow th order of arrangement of the Grammar, but begin the study of the verb with th second Lesson, and then pursue it alternately with that of the remaining parts o speech. *It contains enough Greek Prose Composition for entrance into any college.*

Leighton's Greek Lessons. Prepared to accompany Goodwin's Gree Grammar.
A progressive series of exercises (both Greek and English), mainly selecte from the first book of Xenophon's Anabasis. The exercises on the Moods ar sufficient, it is believed, to develop the general principles as stated in the Gramma

Goodwin & White's First Four Books of the Anabasis.

Goodwin's Greek Reader contains the first and second books of the An basis. Also, selections from Plato, Herodotus, and Thucydides; being th full amount of Greek Prose required for admission at Harvard University.

Goodwin's Selections from Xenophon and Herodotus contains th first four books of the Anabasis, the greater part of the second book of the He lenica of Xenophon, and extracts from the sixth, seventh, and eighth books o Herodotus.

Anderson's First Three Books of Homer's Iliad.

Goodwin's Greek Moods and Tenses. Gives a plain statement of th principles which govern the construction of the Greek Moods and Tenses, the most important and the most difficult part of Greek Syntax.

F. D. Allen's Prometheus of Æschylus.
Tarbell's Orations of Demosthenes.
Flagg's Public Harangues of Demosthenes.
Tyler's Selections from the Greek Lyric Poets.
Seymour's Selections from Pindar and the Bucolic Poets.
Whiton's Select Orations of Lysias.
White's Œdipus Tyrannus of Sophocles.
F. D. Allen's Medea of Euripides.
Sidgwick's Introduction to Greek Prose Composition.
White's Schmidt's Rhythmic and Metric of the Classical Lan guages.
Liddell & Scott's Greek-English Lexicons. Abridged and Unabridge

A Full Descriptive Catalogue mailed on application.

GINN & HEATH   ew York and Chica o.

ENGLISH GRAMMAR.

Elementary Lessons in the English Language, *for Home an School use,* by W. D. WHITNEY *of Yale College,* and Mrs. N. L. KNOX, *Graduate of the Oswego Normal School, late teacher of Methods in the Brockport Normal School, and a very successful primary teacher.*

This book is in two parts. Part I. contains *no technical Grammar.* It is designed to give children such a knowledge of the English Language as will enable them to *speak, write,* and *use* it with accuracy and force. It is made up of exercises to increase and improve the vocabulary, lessons in enunciation, pronunciation, spelling, sentence-making, punctuation, the use of capitals, abbreviations, drill in writing number and gender forms, and the possessive form, letter-writing, and such other matters pertaining to the art of the language as may be taught simply, clearly, and profitably. Many and varied oral and written exercises supplement every lesson. Part II. is an introduction to "The Essentials of English Grammar."

The Teacher's Edition, *prepared by* Mrs. N. L. KNOX, contains, beside the text, plans for developing the lessons in the book, matter for oral lessons and methods of giving them, impromptu test-exercises, dictation lessons, plans for conducting reviews, and other valuable aids to the easy, attractive, and successful teaching of Language.

The Essentials of English Grammar, *for the use of Schools,* by Prof. W. D. WHITNEY *of Yale College.*

This is an *English* Grammar of the English Language, prepared by the best philologist in this country, and has already been re-published in England. It is clear, practical, and complete. It proceeds from facts to principles, and from these to classifications and definitions. Mechanical forms, unnecessary classifications, and abstract definitions are avoided.

The exercises, selected from the best English writers, leave none of the usual and regular forms of English structure untouched.

The plan of analysis is simple. The ordinary method of Gender in Nouns is displaced by one truer and far simpler. The sharp distinction of verb-phrases or compound forms from the real verb-forms is original and scholarly.

The facts of English Grammar are presented in such a way as to lay the best foundation for the further and higher study of Language in all its departments. This book is accompanied by

A Manual for the use of Teachers, *prepared by* Mrs. N. L. KNOX. The Manual is designed to supply methods of developing and emphasizing the lessons of the text, to furnish material for every-day application and practice, questions and tests for review, and tabular views and outlines (for the blackboard) to guide the pupils in study, recitation, and review.

Gilmore's Outlines of the Art of Expression: *A Treatise on English Composition and Rhetoric, designed especially for Academies, High Schools, and the Freshman Class in Colleges.*

This book is admirably fitted to help students in English who are preparing for admission to college; or to give them, after they have entered college, such preliminary training as will enable them to profit by higher and more systematic instruction in Rhetoric and Linguistic Science.

A good book for reviewing English Grammar in the High School, and for studying Grammar from a historical standpoint.

A Full Descriptive Catalogue mailed on application.

GINN & HEATH, Publishers, Boston, New York, and Chicago.

MATHEMATICS AND SCIENCE.

We have in course of preparation a complete series of text-books in both of the above departments.

A Series of Arithmetics by Dr. THOMAS HILL, Ex-President of Harvard University, and GEORGE A. WENTWORTH, Professor of Mathematics in Phillips (Exeter) Academy. The Primary will probably be ready in June, 1879.

A Geometry for Beginners. *Adapted to lower and Grammar-School work.* By G. A. HILL of Harvard University. (Ready in June, 1879.)

Wentworth's Elementary Algebra. (In preparation.)

Wentworth's Plane and Solid Geometry is based upon the assumption that *Geometry is a branch of practical logic*, the object of which is to detect, and state clearly and precisely, the successive steps from premise to conclusion.

In each proposition *a concise statement of what is given is printed in one kind of type, of what is required in another, and the demonstration in still another.* The reason for each step is indicated in small type, between that step and the one following, thus preventing the necessity of interrupting the process of demonstration by referring to a previous proposition.

A limited use has been made of symbols, wherein symbols stand for words and not for operations. The propositions have been so arranged that *in no case is it necessary to turn the page in reading a demonstration.*

A large experience in the class-room convinces the author that, if the teacher will rigidly insist upon the *logical form* adopted in this work, *the pupil will avoid the discouraging difficulties which usually beset the beginner in Geometry.*

Wheeler's Elements of Plane and Spherical Trigonometry.

Byerly's Differential Calculus. (Ready in June, 1879.)
Used two years at Harvard in manuscript form.

Peirce's Tables of Logarithmic and Trigonometric Functions to three and four places of Decimals.

Peirce's Elements of Logarithms; with an explanation of the Author's *Three and Four Place Tables.*

Searle's Outlines of Astronomy.

The Annals of the Astronomical Observatory of Harvard College.

Elements of Natural Philosophy. A Text-Book for Common and High Schools. By Prof. A. E. DOLBEAR, A.M., of Tufts College, and A. P. GAGE, Instructor in Physics in English High School, Boston.

This treatise differs from most text-books on Natural Philosophy in being *based upon the doctrine of the conservation of energy;* this is made prominent in every department of it. Whenever it is practicable, the *experimental part precedes the statements* of the laws; that is to say, the laws are deduced from the experiments performed. (Ready in July, 1879.)

Stewart's Elementary Physics. American edition. With *questions and exercises* by Prof. G. A. HILL of Harvard University.

A Course in Scientific German. Prepared by H. B. HODGES, Instructor in Chemistry and German in Harvard University. With Vocabulary.

Prepared to supply a want long felt by English and American students of science, of some aid in the acquirement of a knowledge of the German language of a sufficiently practical nature to enable them to read with ease the scientific literature of Germany.

A Full Descriptive Catalogue mailed on application.

GINN & HEATH, Publishers, Boston, New York, and Chicago.

GOODWIN'S GREEK GRAMMAR.

By WILLIAM W. GOODWIN, Ph. D., *Eliot Professor of Greek Literature in Harvard University.*

The object of this Grammar is to state *general principles* clearly and distinctly, with special regard to those who are preparing for college. The plan has been to exclude all detail which belongs to a book of reference, and to admit whatever will aid a pupil in mastering the great principles of Greek Grammar. The Syntax has been allowed more space, proportionally, than the statement of the forms: this has been done from a conviction of the author that the chief principles of Syntax are a more profitable study for a pupil in the earlier years of his classical course than the details of vowel-changes and exceptional forms, which are often thought to be more seasonable. The sections on the Syntax of the Verb are generally condensed from the author's larger work on the Greek Moods and Tenses, to which advanced students, and especially teachers, are referred for a fuller exposition of many matters which are merely hinted at in the elementary grammar. The latter contains a brief statement of the author's new classification of conditional sentences, with its application to relative and temporal sentences, which is contained in full in the larger work, and which appears now for the first time in an elementary form. A catalogue of irregular verbs is added, which has been constructed entirely with reference to the wants of beginners. All forms are excluded (with a few exceptions) which are not found in the strictly *classic* Greek before Aristotle; and all forms which are not used by Attic writers are enclosed in brackets.

From Prof. J. T. Dunklin, Agricultural and Mechanical College of Alabama.

For class instruction it has, in my opinion, no equal. All the necessary facts and principles of the Greek language are stated and illustrated so *plainly* and *clearly*, yet in so *brief* a compass, that students will find everything easy of comprehension and application, and be relieved from the unnecessary detail found in many school grammars.

GUIDES FOR SCIENCE-TEACHING.

Designed to supplement Lectures given to Teachers of the Public Schools of Boston.

By the Boston Society of Natural History.

They are intended for the use of Teachers who desire to practically instruct classes in Natural History. Besides simple illustrations and instructions as to the modes of presentation and study, there are, in each pamphlet, hints which will be found useful in preserving, preparing, collecting, and purchasing specimens.

No. I. *About Pebbles.* By Alpheus Hyatt, Custodian of the Boston Society of Natural History, and Professor of Zoölogy and Paleontology in the Massachusetts Institute of Technology. This pamphlet is an illustration of the way in which a common object may be used profitably in teaching. This was the opening lecture of the course, and the one which gave rise to these little books.

No. II. *Concerning a Few Common Plants.* By George L. Goodale, Professor of Botany in Harvard University. This is complete in two parts (which are bound together), and gives an account of the organs or "helpful parts" of plants, and how these can be cultivated and used in the schoolroom for the mental training of children.

No. III. *Commercial and other Sponges.* By Prof. Alpheus Hyatt. This gives an account of the sponges in common use, and of their structure, &c. Illustrated by 7 plates.

No. IV. *A First Lesson in Natural History.* By Mrs. Elizabeth Agassiz. Illustrated by 40 woodcuts and 4 plates. With admirable clearness and brevity, it gives in narrative form for young children a general history of Hydroids, Corals, and Echinoderms.

No. V. *Corals and Echinoderms.* By Prof. Alpheus Hyatt. Illustrated by 12 plates. Intended to supply such information as teachers cannot get from other sources.

No. VI. *Mollusca.* Oyster, Clam, and Snail.

No. VII. *Worms and Crustacea.* Earthworm, Lobster, Common Crab.

No. VIII. *Insects.* Grasshopper.

No. IX. *Fishes.* Yellow Perch.

No. X. *Frogs.* Common Frog and Toad.

No. XI. *Reptiles.* Alligators and Tortoises.

No. XII. *Birds.*

Lightning Source UK Ltd.
Milton Keynes UK
UKHW032020221118
332795UK00006B/1045/P